Microwave Approach to Highly Irregular Fiber Optics

WILEY SERIES IN MICROWAVE AND OPTICAL ENGINEERING

KAI CHANG, Editor
Texas A&M University

A complete list of the titles in this series appears at the end of this volume.

Microwave Approach to Highly Irregular Fiber Optics

HUANG HUNG-CHIA

A WILEY-INTERSCIENCE PUBLICATION
JOHN WILEY & SONS, INC.
NEW YORK / CHICHESTER / WEINHEIM / BRISBANE / SINGAPORE / TORONTO

Library of Congress Cataloging in Publication Data:

Huang, Hung-chia
  Microwave approach to highly irregular fiber optics / Huang Hung
-Chia
    p. cm.— (Wiley series in microwave and optical engineering)
  Includes bibliographical references and index.
  ISBN 0-471-31023-9 (cloth : alk. paper)
  1. Fiber optics. 2. Microwave optics. 3. Coupled mode theory.
I. Title. II. Series.
TA1800.H8 1998
621.36'92—dc21   97-15393

Printed in the United States of America

10 9 8 7 6 5 4 3 2 1

*To the memory of
my father Huang You-chang,
professor of law in
Peking University, 1912–1930,
praetor of the ex-Government,
eminent scholar,
and patriotic poet of renown*

Contents

Preface

The past decades have witnessed miraculous achievements in the exploitation of different ranges of the electromagnetic spectrum, of which microwaves and optical waves are the direct concern of this book. For stable transmission at any wavelength, it is often a prerequisite that a wave maintains its field structure, mode, or state of polarization (SOP) throughout the entire course of transmission.

STRUCTURE OF THIS BOOK

In considering conventional and nonconventional single-mode optical fibers, current research, as displayed clearly in the literature of fiber optics, is directed at the stable transmission of *linear* light. The fact that linear light is being investigated may not always be explicitly stated, but it is nevertheless tacitly assumed. The reason for this is all too obvious in view of the technical background of the art. Existing polarization-maintaining fibers actually are used for the polarization maintenance of linear light. In contrast, the assiduous efforts made in the search for a birefringent fiber for *circular* light have been virtually fruitless. The only currently known method of maintaining a circular light in fiber is still the inefficient means of "twisting," which was developed over one and a half decades ago.

Only recently, I had the good fortune to discover a novel fiber structure, called a "Screw fiber," which is capable of maintaining the stable transmission of circular light. The United States patent for my new invention has caused me to structure this book in a nontraditional way by giving adequate prominence to the hitherto not sufficiently emphasized circular light excitation and transmission, as can be seen in the Table of Contents.

I should mention, however, that the objective of this book is not merely a narrow exposition of my own invention. As the Table of Contents makes equally clear, this book is comparatively comprehensive and self-contained. The topic areas discussed in the book extensively cover low birefringent (lo-bi) fiber, existing circular-bi fiber by twisting, elliptical-bi fiber, circular highly birefringent (hi-bi) fiber, followed by the very interesting topic of polarization transformation and control by variable coupling of modes. In fact, it would be correct to add to the title of this book a subtitle: "Treatise on Special Fiber Optics."

Indeed, different problems will not be addressed in equal detail. For those topics that can be conveniently referenced in professional journals and books, I shall give a brief description only, supplemented wherever necessary by a list of readily available references. On the other hand, for those topics not known or not having been sufficiently dealt with in the scientific literature, I shall give a more detailed account, oftentimes with the aid of a rigorous analytic treatment. My inclination in the course of this endeavor has been to write a book that is essentially new and fresh.

THE METHOD OF APPROACH

I have adopted a Microwave Approach to the development of this book, which, as a matter of fact, deals exclusively with fiber optics. We recall that in the research and development of fiber optics, both microwaves and optics play important roles. We often speak of the union of microwaves and optics in describing how these two sciences complement each other in solving problems that otherwise would be extremely difficult, if not impossible, to solve.

In this microwave approach to fiber optics, I freely borrow from microwaves the coupled-mode theory to derive analytic and useful solutions of the many highly irregular fiber-optic problems concerned. The coupled-mode formalism has already been widely used in the literature of fiber optics. But many published papers merely consider the coupled-mode equations for specific fibers and their eigenmodes, without paying attention to the initial-value, or excitation condition, of the very process. As a matter of fact, stable transmission of light is secured only under the condition of *single-*eigenmode excitation. In this book, I shall put sufficient emphasis on this fundamentally important but sometimes neglected point.

While many physical concepts and mathematical methods employed in the text can be traced back to the early microwave literature, the specific problems of interest in microwaves and fiber optics are dramatically different. Microwaves and lightwaves differ in wavelength by orders of magnitude. It is thus only natural that they employ completely different technologies. For example, in the art of fiber optics, it is now an established technical practice to spin a fiber during the linear draw at a rate as high as 2000 rpm, or higher, yielding different spun fibers that find no equivalents in microwaves. This book attempts to adequately discuss how the early established coupled-mode theory was enriched through its various applications in the new field of highly irregular fiber optics.

APPLICATION-ORIENTED SCHEME

The book is intended to be application-oriented. It is my firm belief that the real worth of a technical science is largely measured by its practical usefulness. Mathematics as a tool, however refined and sophisticated, will lose much of its charm if it solves problems merely of academic interest, unlikely to be applicable in practice at the moment or in the forseeable future. In writing this book, I have taken care to use a minimum amount of mathematics, while at the same time giving as much intuitive and physical description as is required for elucidation of the relevant underlying principles. In an application-oriented text, it is probable that using too much mathematics would incur the risk of obscuring the physical meaning of a problem that may actually not be so difficult to understand. The relevant mathematical background is included in this book in the first two appendices.

CLOSING WORDS AND ACKNOWLEDGMENTS

The book is based on my research over the past several years on a key project sponsored by the Natural Science Foundation of China (NSFC). It is hoped that it will prove not only useful as a reference text for microwave-optical scientists familiar with the coupled-mode theory, but also interesting and readable for the general reader who is not familiar with this theory.

I owe special thanks to Prof. Kai Chang for inviting me to contribute a book to the Wiley Series on Microwave and Optical Engineering, and for his useful advice. I also make grateful acknowledgment to my colleagues and friends, abroad and at home, for their unfailing help and encouragement. Miss He Ying-chun and Mr. Young Chi-bang helped in checking the derivations, as well as in computer computation for preparing the numerical figures in Chapter 7. I also appreciate the interesting and helpful discussions that I had via e-mail with Dr. Chen Yijiang at Canberra while writing the analytic part of Chapter 3. Dr. Li Qingning and Dr. Ji Minning read the entire manuscript and made extensive suggestions. Ms. Wang Guoping proofread the text with admirable patience. Last but by no means least, thanks are expressed to Mr. George J. Telecki, Executive Editor at Wiley, for effectual cooperation between the publisher and the writer during the course of preparing the manuscript.

HUANG HUNG-CHIA, MS, HON. DSC.

Academician, Academia Sinica
Professor and Honorary President, Shanghai University

CHAPTER ONE

Introduction to the Modal Theory of Optical Fiber

This book attempts to address systematically a variety of highly irregular fiber-optic problems by the microwave approach. The text is intended to be relatively comprehensive and self-contained for the topics concerned, so that a wide range of readers will not find it difficult to grasp the theme and subject matter of the book, nor will they need to rely heavily on the abundance of published literature and patent documents to understand most of the detailed particulars in the text. Indeed, to say "self-contained" is not to mean "all-inclusive," to which the author makes no pretension within the scope of a single volume.

Along the general line of microwave approach, the mathematical technique mostly employed throughout the text is the theory of coupled modes. This theory has been well developed to form a unified analytic framework that reflects the common features of many seemingly unrelated phenomena. It is only natural that this theory is almost universally applicable to the multitude of problems found in modern high technology, in view of the fact that practically every branch of science involves the study of mutually coupled oscillating or wave-propagating systems.

Coupled-mode theory with constant coefficients has been frequently applied to fiber-optic problems, as can be seen in the existing publications. However, coupled-mode theory with variable coefficients, which was early established in the microwave art, does not seem to have received sufficient attention. In this book, it will be shown that, in dealing with certain highly irregular fiber-optic problems, the variably coupled-mode theoretical approach really proves powerful and unique.

1.1   THEORETICAL BACKGROUND OF MODAL ANALYSIS

Analytic study of a macroscopic field problem requires judicious use of Maxwell's equations. The form of Maxwell's equations adopted in fiber optics does not differ in any way from that adopted in microwaves, except that the relative dielectric permittivity, ε_r, is now expressed in terms of the refractive index n in fiber-optic problems. The equation that relates these two parameters is implied in Maxwell's initial work, and is part of Maxwell's electromagnetic theory of light:

$$n = \sqrt{\varepsilon_r} \tag{1.1}$$

This equation, simple as it appears, is of basic importance in that it connects the fundamental parameter, n, which describes the optical property of the medium, to the fundamental parameter, ε_r, which describes the electrical property of the medium. In a source-free region, Maxwell's equations are

$$\nabla \times \mathbf{H} = n^2 \varepsilon_0 \frac{\partial \mathbf{E}}{\partial t} \tag{1.2a}$$

$$\nabla \times \mathbf{E} = -\mu_r \mu_0 \frac{\partial \mathbf{H}}{\partial t} \tag{1.2b}$$

The mathematical formulations of Maxwell's equations in different coordinate systems can be readily found in any standard text on electromagnetic theory, and so are not written out here. For our immediate purpose, we shall only use some simplified forms of Maxwell's equations suitable for the idealized fiber model chosen in our analysis. More description of Maxwell's equations in relation to the formulation of the coupled-mode equations will be given in Appendix A.

On the basis of Fourier analysis, we can consider a simple harmonic function of time, so that in complex notation the mathematical operation for partial differentiation with respect to time is simply equivalent to a complex number, that is, $\partial / \partial t \rightarrow j\omega$. As is usual in waveguide theory, the ideal fiber model (Fig. 1.1a) is straight and infinitely long. If we further assume that the waveguiding medium is uniform in the transmission direction, or the property of the medium is z-invariant, then the transverse and longitudinal parts of the field components are separable. The solution of the z-dependent part of the field components is then an exponential function, $e^{-j\beta z}$, with β denoting the propagation constant. Therefore, partial differentiation of the field components with respect to z is again equivalent to a complex number:

$$\frac{\partial}{\partial z} \rightarrow -j\beta \tag{1.3}$$

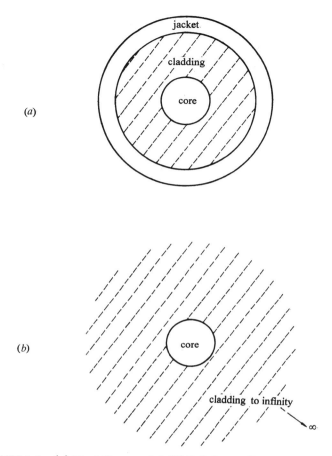

FIGURE 1.1   (*a*) Ideal fiber model. (*b*) Infinite medium approximation.

The (t, z) part of the field components is then $e^{j(\omega t - \beta z)}$, called the wave factor. The exponential wave factor that takes a negative sign in the exponent (i.e., $-\beta$) describes a forward wave, while for a backward wave, $+\beta$ is used instead.

We note, in passing, that some authors use $e^{j\beta z}$ for forward wave, and $e^{-j\beta z}$ for backward wave. Correspondingly, the complex notation for a harmonic function of time becomes $e^{-j\omega t}$. From the mathematical viewpoint for complex variables, this system of notation is equivalent to the system of notation that we have chosen in our book. Nevertheless, we prefer our choice, simply because a negative sign associated with the angular frequency of time might appear a little more puzzling.

Thus, with the (t, z) part separated out, and with partial differentiations with respect to t and z replaced by the respective complex numbers, what is left to deal with is a set of partial differential equations of the two transverse variables only. These transverse variables are (x, y) in Cartesian coordinates, or (r, φ) in cylindrical coordinates. The solutions of the set of two-dimensional partial differential equations describe the transverse field patterns whose motion in the longitudinal direction is described by the wave factor. Each transverse solution associated with the corresponding wave factor represents a complete picture of the wave field in the waveguiding medium concerned.

We then proceed to the two-dimensional transverse field equations. This seemingly not so difficult mathematical problem for an idealized fiber model is actually very difficult to solve, because what we are now looking for are solutions that not only satisfy the reduced two-dimensional Maxwell's equations, but also satisfy the specific boundary conditions of the fiber model. In differential calculus, functions that satisfy the differential equations are the general solutions, and functions that satisfy the differential equations plus the prescribed boundary conditions are the particular solutions. Differential equations plus boundary conditions constitute the so-called boundary-value problem. To be exact, what we are looking for are independent solutions of the boundary-value problem for the fiber model concerned. Such solutions are called eigensolutions, or eigenmodes. In actual mathematical derivation, the difficulty is that the solutions for field components are required to satisfy the boundary conditions.

We recall that, in the circular, metallic, hollow waveguide employed in microwave transmission, there exist a set of E or TM modes, and a set of H or TE modes, whose eigenvalues are simply given by the roots of Bessel functions and of the derivatives of Bessel functions, respectively. An optical fiber is an open waveguiding structure whose boundary condition is far more complicated. The eigenmodes in optical fiber are HE or EH hybrid modes that generally comprise all six field components. The result is that the mathematical derivation would be prohibitively difficult to carry out if we did not have recourse to judicious simplifying steps and early approximations.

One major simplifying step is the "infinite-medium approximation" (Fig. 1.1b) with which the outer diameter of the fiber cladding is assumed to extend to infinity. Consequently, we are able to get rid of the mathematical complication that otherwise would have been imposed at the outer boundary of the fiber cladding of finite diameter. The mathematical model is thus idealized to such an extent that the fiber under study is simply a dielectric cylinder of infinite length situated in an infinite medium of smaller refractive index. The early work by Hondros and Debye on dielectric waveguide [1] is not specific to the frequency used, and is therefore applicable to microwaves, as well as optical fiber. Despite the mathematical rigor, however, the solutions obtained in this early paper are far too complicated for practical use.

A. W. Snyder first derived the approximate eigenvalues and eigenmodes of dielectric optical waveguide [2]. The derivation of simple asymptotic solutions is made possible by taking advantage of the circumstance that in practical optical fibers the refractive index of the core and the refractive index of the clad are actually very nearly identical. A crucial step of approximation is thus made possible by the "weakly guiding" property of fiber, in the sense that the tiny index-difference between core and clad leads to weak guidance of guided modes. This approximation is justifiable, inasmuch as modern optical fiber actually requires a very small concentration of doping for the fiber core, so that the relative difference in refractive index between core and clad is only of the order of 10^{-3}.

1.2 THE LP MODES

The theoretical contribution toward mathematical simplification of optical fiber problems is also due to D. Gloge, who made early approximations to the extent that the modes or field structures in fiber are simply approximated by light of "linear polarization" called the LP modes [3]. Apparently, the LP modes cannot represent exact solutions of the boundary-value problem of fiber, because the longitudinal field components are actually nonzero in the general case of an open dielectric waveguide, and the electric lines over a cross section of the waveguide are necessarily more or less curved. As a matter of fact, the field components that describe an LP mode are not self-consistent within the theoretical frame of Maxwell's equations. Nevertheless, the LP modes are mathematically so much easier to deal with that they are widely used in fiber-optic design and applications for multimode fiber, and for single-mode fiber as well. We therefore follow Gloge in giving an outline of the crucial steps in the mathematical derivation that yields the eigenvalues and eigenfunctions of the LP modes. A step-by-step derivation leading to these modes is fairly lengthy and somewhat roundabout. Yet in the founding and in the research and development (R & D) of fiber optics, the establishment of the LP modes is so important from the analytic and application viewpoints that we regard it worthwhile to give some space to the following derivation. Also note that, even for highly irregular nonconventional fibers, the use of the LP modes is still of help in structuring a kind of canonical problem preliminary to the study of the more complicated actual fiber-optic problem.

LP Modes Based on Early Approximations

Consider the idealized step-index optical-fiber model shown in Figure 1.1b. As was said earlier, with the aid of the "infinite medium approximation," the only boundary condition that is left for consideration is the one at the core and clad interface. The boundary condition requires continuity of tangential electric and magnetic fields. Thus, we need to know the field expressions for E_z and H_z, as well as for E_φ and H_φ, and have them matched at the core–clad interface. Mathematical derivation of such field components by the regular way of solving Maxwell's equations is lengthy and tedious. The LP-mode approach adopts early approximation by introducing the weakly guiding condition at the very beginning:

$$\Delta = (n_1 - n_2)/n_2 \ll 1 \qquad (1.4)$$

where Δ is the relative index difference, and n_1, n_2 are the refractive indices of the core and the clad, respectively. Under the weakly guiding condition, it is surmised by intuition that the transverse field of modes is likely to be predominantly polarized in one direction, say, y. In the sense of a *trial* function, E_y in the core is put in the form:

$$E_y = CJ_l(ur/a)\cos l\varphi \qquad (1.5)$$

where C is a proportionality constant, and is equal to $E_l/J_l(u)$ if we use the symbol E_l to denote the electric field at the core−clad interface $r = a$, $\varphi = 0$. Bessel functions J_l are suitable to represent the standing-wave field in the core. For guided modes, the field outside the core should be evanescent. Modified Hankel functions K_l are suitable to represent this field, as such functions behave at large r like exponential functions with negative real arguments. Even supposing that the transverse electric field is represented by a single component E_y, as given by Eq. (1.5), it is still not that simple to express the transverse magnetic field in a consistent form. The next approximation is then to assume that the transverse field structure in a weakly guiding fiber is similar to the simple structure of fields of a plane wave. Then, if the electric field is in the y direction, the magnetic field will be x-directed such that $H_x = -(n/Z_0)E_y$, where $Z_0 = \sqrt{\mu_0/\varepsilon_0}$ is the plane-wave impedance in vacuum, and n is the refractive index of the medium concerned. With all the preceding postulations, *trial* functions for the transverse fields are written as

$$E_y = -H_x \begin{Bmatrix} Z_0/n_1 \\ Z_0/n_2 \end{Bmatrix} = E_l \begin{Bmatrix} J_l(ur/a)/J_l(u) \\ K_l(wr/a)/K_l(w) \end{Bmatrix} \cos l\varphi \qquad (1.6)$$

where u and w are related to the propagation constant β of any mode by

$$u = a(k^2 n_1^2 - \beta^2)^{1/2} \qquad (1.7a)$$

$$w = a(\beta^2 - k^2 n_2^2)^{1/2} \qquad (1.7b)$$

where $k = \omega\sqrt{\mu_0\varepsilon_0} = 2\pi/\lambda$ is the wave number in free space. In Eqs. (1.6), as in the following, the upper line refers to the core and the lower line refers to the clad. The parameters u/a and w/a are separation constants introduced in separating the transverse and the longitudinal parts of the fields inside and outside the core, respectively. The choice of cosine functions is suitable for describing the circumferential variation of the field that is necessarily periodic.

For every nonzero l, the relevant trial function in Eqs. (1.6) has a fourfold degeneracy, in view of the fact that the angular part of the function can be $\sin l\varphi$ as well as $\cos l\varphi$, and the linear polarization state can be E_x as well as E_y. For $l = 0$, the transverse field is only twofold degenerate, implying the existence of two orthogonal polarized states of linear light. The modes derived from the postulated transverse fields are called the LP modes.

For guided modes, u and w in Eqs. (1.7a) and (1.7b) must be real, requiring that the propagation constant lie within the limits

$$n_1 k \geq \beta \geq n_2 k \tag{1.8}$$

This equation refers to the forward modes of our present concern.

Accuracy of the LP Modes

Before going further, let us see whether the initial postulation of the trial function (1.6) is actually justifiable. In the postulated case wherein $E_x = H_y = 0$, source-free Maxwell's equations are simplified to the extent that the two curl equations are

$$\frac{\partial H_z}{\partial x} - \frac{\partial H_x}{\partial z} = -j\omega\varepsilon_0 n^2 E_y \tag{1.9a}$$

$$\frac{\partial H_x}{\partial y} = -j\omega\varepsilon_0 n^2 E_z \tag{1.9b}$$

$$\frac{\partial E_z}{\partial y} - \frac{\partial E_y}{\partial z} = -j\omega\mu_0 H_x \tag{1.9c}$$

$$\frac{\partial E_y}{\partial x} = -j\omega\mu_0 H_z \tag{1.9d}$$

where, for Eqs. (1.9a) and (1.9b), $n = n_1$ in the core, and $n = n_2$ in the clad. From Eq. (1.9c), it can be seen that

$$H_x \approx \frac{1}{j\omega\mu_0} \frac{\partial E_y}{\partial z} = -\frac{\beta}{\omega\mu_0} E_y \tag{1.10}$$

provided $\partial E_z/\partial y \ll \partial E_y/\partial z$. Further, the equivalence expression $\partial/\partial z \to -j\beta$ is used in Eq. (1.10). To derive H_x from E_y we resort to the weakly guiding condition (1.4) in order to introduce the following approximation:

$$\beta \approx nk, \qquad n \approx n_1 \approx n_2 \tag{1.11}$$

which appears justifiable for guided modes whose propagating constants are confined to the very small range given by Eq. (1.8). Using this rather bold approximation step, and with the aid of Eq. (1.10), we have $H_x \approx -(n/Z_0)E_y$ as having been used in the trial function in Eq. (1.6). The above lines of derivation involve the several early approximations on which the postulation of the trial field function (1.6) is based.

For convenience in further derivation, we note that the quadratic summation of u and w yields a third parameter v, in which the propagation constant β is eliminated:

$$v^2 = u^2 + w^2 \tag{1.12}$$

$$v = ak\left(n_1^2 - n_2^2\right)^{1/2} \approx \left(2\pi a n\sqrt{2\Delta}/c\right)f \tag{1.12a}$$

in which $k = 2\pi/\lambda = (2\pi/c)f$, where λ is the operating wavelength as before, c is the velocity of light in free space, and f is the operating frequency. The last approximate identity of Eq. (1.12a) holds under the weakly guiding approximation, $n_1 \approx n_2$, for which n in the bracket can be either n_1 or n_2.

The adimensional parameter v is termed the "normalized frequency" for the obvious reason that, for a fiber whose characteristic figures (a, n, Δ) are given, v is inversely proportional to the operating wavelength λ, or directly proportional to the operating frequency, f. Since the physical meaning of v is the same as f, except for a constant proportionality, we often refer to v simply as frequency, taking it for granted that v is a normalized value.

In the LP-mode approach, the derivation is substantially simplified by the mathematical artifice, wherein the field functions are expressed in cylindrical coordinates suitable for the fiber geometry, but in deriving the mathematical expression of a new field component from some known field components, the Cartesian coordinates are used to carry out the differential operation. Thus, the longitudinal fields can be derived from the transverse fields by Eqs. (1.9b) and (1.9d):

$$E_z = \frac{jZ_0}{k}\begin{Bmatrix} 1/n_1^2 \\ 1/n_2^2 \end{Bmatrix}\frac{\partial H_x}{\partial y}$$

$$H_z = \frac{j}{kZ_0}\frac{\partial E_y}{\partial x} \tag{1.13}$$

Because the trial field components are expressed as functions of the cylindrical coordinate variables (r, φ), differentiation with respect to the Cartesian coordinate variables (x, y) needs the following formulas from elementary differential calculus:

$$\frac{\partial}{\partial x} = \frac{\partial r}{\partial x}\frac{\partial}{\partial r} + \frac{\partial \varphi}{\partial x}\frac{\partial}{\partial \varphi}$$

$$\frac{\partial}{\partial y} = \frac{\partial r}{\partial y}\frac{\partial}{\partial r} + \frac{\partial \varphi}{\partial y}\frac{\partial}{\partial \varphi}$$

(1.14)

where, by the simple relations of geometry:

$$r = \left(x^2 + y^2\right)^{1/2}$$

$$\varphi = \arctan\left(\frac{y}{x}\right)$$

(1.15)

Using the above relevant equations and through simple and straightforward but slightly tedious derivations, we obtain the longitudinal field components E_z and H_z in the following form:

$$E_z = \frac{jE_l}{2ka}\left\{ \begin{array}{l} (u/n_1)J_+\sin(l+1)\,\varphi + (u/n_1)J_-\sin(l-1)\,\varphi \\ (w/n_2)K_+\sin(l+1)\,\varphi - (w/n_2)K_-\sin(l-1)\,\varphi \end{array} \right\}$$

$$H_z = \frac{-jE_l}{2kZ_0a}\left\{ \begin{array}{l} uJ_+\cos(l+1)\,\varphi - uJ_-\cos(l-1)\,\varphi \\ wK_+\cos(l+1)\,\varphi + wK_-\cos(l-1)\,\varphi \end{array} \right\}$$

(1.16)

where

$$J_\pm = \frac{J_{l\pm1}(ur/a)}{J_l(u)}$$

(1.17a)

$$K_\pm = \frac{K_{l\pm1}(wr/a)}{K_l(w)}$$

(1.17b)

Under the weakly guiding condition $\Delta \ll 1$, the longitudinal components given by Eqs. (1.16) are small compared with the transverse components. The factors involved are $u/(ak)$ and $w/(ak)$, which are both of the order $\Delta^{1/2}$ according to Eqs. (1.7a) and (1.7b). By Eqs. (1.9a) and (1.9c), in which $\partial/\partial z \to -j\beta \approx -jkn$, we can eliminate H_x such that E_y is derivable from derivatives of the longitudinal fields E_z and H_z. Similarly, we can also derive H_x from derivatives of the longitudinal fields. As expected, the transverse fields thus derived do not reproduce the initial postulated functions. This is only natural, in view of the various approximations adopted in the course of derivation. Thus, field expressions for the LP modes are not self-consistent in the theoretical frame of Maxwell's equations. Nevertheless, the error is small (being only of the order Δ, that is about 10^{-3}).

Characteristic Equation for LP Modes

To match the tangential field components at the core–clad interface, we still need to derive E_φ and H_φ in the cylindrical coordinate system. Simple geometrical relations that relate field expressions in Cartesian coordinates and those in cylindrical coordinates are given by

$$F_r = F_x \cos\varphi + F_y \sin\varphi$$
$$F_\varphi = -F_x \sin\varphi + F_y \cos\varphi$$

(1.18)

where the symbol F stands for either E or H. By Eqs. (1.6) and (1.18), we have

$$E_\varphi = \frac{1}{2}E_l \left\{ \begin{matrix} J \\ K \end{matrix} \right\} [\cos(l+1)\varphi + \cos(l-1)\varphi]$$

$$H_\varphi = \frac{1}{2}\frac{E_l}{Z_0} \left\{ \begin{matrix} n_1 J \\ n_2 K \end{matrix} \right\} [\sin(l+1)\varphi - \sin(l-1)\varphi]$$

(1.19)

where $J = J_l(ur/a)/J_l(u)$, and $K = K_l(wr/a)/K_l(w)$.

Using the weakly guiding approximation $n_1 \approx n_2$ in the above field expressions again, and using the recurrence relations for J_l and K_l, it is found that matching all of the tangential field components at the core–clad interface can be achieved by one equation of the form [3]:

$$u\frac{J_{l-1}(u)}{J_l(u)} = -w\frac{K_{l-1}(w)}{K_l(w)}$$

(1.20)

This is the characteristic equation for the LP modes. Using this equation it is simple to obtain the limiting solutions for u corresponding to $w = 0$ and $w \to \infty$. For $w = 0$, the characteristic equation (1.20) is reduced to $J_{l-1}(u) = 0$, which determines the cutoff wavelengths or cutoff frequencies of the LP modes. This can be seen from Eq. (1.12). At $w = 0$, v assumes a minimum value ($v = u$), so that f is minimum according to Eq. (1.12a). Therefore, the reduced characteristic equation $J_{l-1}(u) = 0$, resulting from Eq. (1.20) by setting $w = 0$, yields the cutoff solutions for u. In the limit $w \to \infty$, the characteristic equation (1.20) is reduced to $J_l(u) = 0$. Thus, the solutions for u are between the zeros of $J_{l-1}(u)$ and $J_l(u)$.

For $l = 0$, the cutoff values of u at $w = 0$ are the roots of $J_{l-1}(u) = J_{-1}(u) = -J_1(u)$. The root $u = 0$ of $J_1(u)$ is counted as the first root that yields the principal mode, to be designated by LP_{01}. For any mode LP_{lm}, the subscript number l minus 1 denotes the order of the relevant Bessel function, and the subscript number m denotes the number of the roots of this Bessel function.

Analytic Determination of Eigenvalues of LP Modes

The amazingly simple mathematics for determining the eigenvalues of the LP modes in optical fiber outlined in the preceding subsections is reminiscent of the early-established simple mathematics for determining the TE and TM modes in a microwave circular guide, whose eigenvalues are the roots of Bessel functions or of the derivatives of Bessel functions. To aid numerical computation in the design work of fiber optics, we find it useful to use the early paper "Table of First 700 Zeros of Bessel Functions" [4], a publication of Bell Labs that served the need of circular electric wave H_{01} transmission decades ago.

In using the same mathematical table as a tool to treat both microwave guide and optical fiber, we are always aware of the radically distinct transmission features of these two kinds of waveguiding medium, as described below.

In an idealized circular metallic-tube model, the eigenmodes are H (or TE) modes and E (or TM) modes. The principal or lowest order mode is the H_{11} mode whose cutoff wavelength is determined by the smallest nonzero root (1.841184) of the derivative of J_1. The wave field whose wavelength is longer than the cutoff wavelength of the H_{11} mode cannot propagate in the microwave circular waveguide. The zero root of J_1 is precluded from consideration, as it represents a trivial solution of the null field inside the metallic tube. The first higher mode is the E_{01} mode, whose cutoff wavelength is determined by the first zero of J_0.

In distinct contrast, the eigenmodes in optical fiber are hybrid modes HE and EH, plus TE_{0n} and TM_{0n}. The LP modes defined with the aid of the weakly guiding condition $n_1 \approx n_2$ are approximate eigenmodes of optical fiber, wherein the "small field" descriptive of the slightly curved electric lines over the fiber cross section, as well as the small longitudinal fields, are all neglected in the modal analysis. The principal or lowest order mode LP_{01} in optical fiber has no cutoff, that is, in the mathematical sense this mode can propagate in fiber without a bound for its wavelength.

In using the said Table of Zeros [4] for eigenvalues of the LP modes, we chose only the roots of Bessel functions, not the roots of the derivatives of Bessel functions. Also, we need to include the zero root of J_1 as the first root of Bessel functions in our table for the LP mode set, as we said earlier.

Mode Sets in Optical Fiber

Under this fairly theoretical topic, we wish to give a sketch, in the form of a short summary, of the possible sets of modes that may exist in optical fiber as an open waveguide. The entirety of the mode sets consists of the guided modes and the radiation modes. Our description of radiation modes can only be very brief, inasmuch as only the guided modes will be of concern in this book. For a more detailed mathematical treatment of the theory of radiation modes, see the original publication [5].

The modes associated with a discrete β-spectrum are the guided modes of optical fiber. For a step-index fiber model, Eq. (1.8) gives the range of the discrete β values for forward waves. For backward guided waves, the range is $-n_2 k > -\beta > -n_1 k$. Optical fiber is an open waveguiding structure that also has a continuous β-spectrum associated with the radiation field [5]. Fiber-optic problems mostly concern the guided modes. But, if we wish to deal with a problem involving the process of radiation, we will need a relatively complete knowledge of the entirety of the mode sets existing in optical fiber.

Because of the multiple reflections of waves at the core–clad boundary, fields inside the fiber core are necessarily standing waves described in cylindrical coordinates by Bessel functions. It is only the difference in the external field patterns outside the core that makes the difference in the types of modes in optical fiber.

External fields with undiminishing but not growing standing waves in the transverse directions are described by Hankel functions of the first kind and second kind in cylindrical coordinates. These transverse external fields with a factor of exponential function $e^{\mp j\beta z}(\beta$, real), which is indicative of the traveling wave behavior in the longitudinal direction, are possible solutions of Maxwell's equations. The corresponding β-spectrum is continuous over the range $-n_2 k \leq \beta \leq n_2 k$. Such fields are associated with the so-called "radiation modes" of the first kind, called the "propagating" radiation modes, and are responsible for the radiation losses caused by optical fiber imperfections.

Possibilities exist for β to take imaginary values over the range $-j\infty < \beta < j\infty$. Such modes are called radiation modes of the second kind, or the "evanescent" radiation modes. Like the propagating radiation modes, the evanescent radiation modes also have standing-wave patterns in the transverse directions in the core and in the clad. The evanescent behavior of the field is only in the longitudinal direction. This can be seen from the exponential wave factor, wherein β now takes an imaginary value. The set of evanescent radiation modes is useful in describing the fine field structure in the vicinity of fiber imperfections.

The undiminishing behavior of the external fields associated with the radiation modes may cause questions regarding the qualification of such field solutions to be a mode, inasmuch as the radiation modes do not satisfy the radiation condition. Individually, therefore, the radiation modes are not independent field structures as required by the strict definition of an eigenmode. This conceptual difficulty is circumvented by considering the radiation modes as an entirety. Marcuse [5] shows that radiation modes when taken as a whole do satisfy the radiation condition.

In summary, the sum of discrete guided modes plus the continuum of radiation modes form a complete set of field patterns by which an arbitrary field in optical fiber can be represented. Such mode constituents relating to an open fiber-optic structure differ radically from the sets of modes in closed guides employed in microwaves. We recall that in microwave guide transmission, there are two types of modes—the guided modes with real β values and the evanescent modes (or evanescent fields) with imaginary β values. Both types of modes in closed waveguides have discrete β-spectrum.

1.3 MULTIMODES IN CONVENTIONAL FIBER

One way of classifying optical fibers is according to the number of modes in transmission, namely, the multimode fiber whose diameter is large compared to the operating wavelength, and the single-mode fiber whose diameter is comparable to the operating wavelength. Until the late 1970s, attention was focused on the multimode fiber version because it was easier to connect or splice the fibers of a larger core, and because of the then-available more rugged and cheaper light-emitting diodes (LED) suitable for multimode transmission of light in fiber. Multimode fibers are commonly designed with a nonhomogeneous transverse refractive-index distribution approximating a parabolic law, in order to minimize the wave path differences of the multitude of modes in transmission, or to minimize the intermodal dispersion.

In this section, we are not prepared to discuss the multimode fiber version whose transverse index distribution is nonhomogeneous. We limit our modal analysis on the basis of the LP modes in the step-index model. Our present interest is to investigate the possible occurrence of the higher-order modes as the operating wavelength decreases to become small in comparison to the core diameter. An analysis of this sort will be relevant to understanding the single-mode transmission in optical fiber, a topic area to be discussed in the next section. For convenience in our present study, we put the normalized frequency v given by Eq. (1.12) in the following form:

$$v \approx \pi n \sqrt{2\Delta} \, \frac{D}{\lambda} = \pi N \frac{D}{\lambda} \qquad (1.21)$$

where $D = 2a$ is the core diameter, and $N \approx n\sqrt{2\Delta}$ is another useful parameter termed the numerical aperture of fiber. It is thus seen that, for given n and Δ, the normalized frequency v is proportional to (D/λ), the ratio of the core diameter to the operating wavelength.

Consider the case of a fairly large value of this ratio, say $D/\lambda \approx 20$. Let $\Delta \approx 0.3\%$ and $n \approx 1.46$ (for silica). Equation (1.21) then gives $v \approx 7.106$. The largest cutoff frequency below this v value is 7.016 for the LP_{41} mode, and there are over 20 guided modes (degenerate modes counted) existing in this fiber, according to the Table of Zeros of Bessel Functions in Ref. [4]. If the ratio (D/λ) is enlarged to 50, the number of guided modes in the same fiber will be well over 100 according to the Table of Zeros.

TABLE 1.1   Cutoff Frequencies of LP Modes

$0 -(LP_{01})-2.405-(LP_{11})-3.832-(LP_{21}L-5.135-(LP_{02})$	
$5.520-(LP_{31})-6.380-(LP_{12})-7.016-(LP_{41})-7.588-(LP_{22})$	
$8.417-(LP_{03})-8.654-(LP_{51})-8.722-(LP_{32})-9.751-(LP_{61})$	

The principal mode LP_{01} relating to the first root $u = 0$ of $J_1(u)$ has no cutoff frequency in the mathematical sense ($v = u = 0$, at $w = 0$). As the operating frequency rises higher and higher, or the operating wavelength becomes shorter and shorter, more and more higher-order modes are permitted in the fiber. As regards which higher-order modes will actually propagate in the fiber, it depends on the excitation condition.

To show the sequence of modes that can occur in fiber as the frequency increases, we need to know the sequence of the relevant roots of Bessel functions. With the help of Ref. [4], the first 12 cutoff frequencies ($v = u$ at $w = 0$) are listed in Table 1.1. For each LP_{lm} mode in the table, the figure on its left is the cutoff value of the normalized frequency v of this mode. Thus, the principal mode LP_{01} has a zero cutoff frequency (or no cutoff). The cutoff frequencies of the higher modes are $v = 2.405$ for the LP_{11} mode; $v = 3.832$ for the LP_{21} mode; $v = 5.135$ for the LP_{02} mode, and so on.

In Figure 1.2 of Gloge's initial paper [3], the regions of the parameter u for modes of order $l = 0, 1$ are shown. This figure is based on the roots of two Bessel functions (J_0 and $J_1 = -J_{-1}$) only, and therefore is not complete if we wish to show the entire sequence of the LP modes. For completeness, Bessel functions of higher order are needed. I therefore prepared Figure 1.2

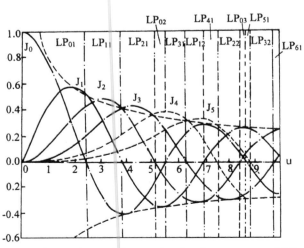

FIGURE 1.2   The regions of the u-value for the sequence of LP modes.

with a view of showing the complete division of the regions of the parameter u for the sequence of the LP modes.

The division of the regions of u for the LP modes (shown in Fig. 1.2) is not to be confused with the frequency or wavelength ranges of the modes. By intuition, there is no lower bound of the working wavelength (or upper bound of the working frequency) for any mode. To show the wavelength or frequency range of each mode, we need to find the corresponding regions of the parameter v from the regions of the parameter u. Note that the lower bound of v is the same as the lower bound of u, inasmuch as this lower bound is defined by the cutoff values of u, where $v = u$ for $w = 0$. But the equality $v = u$ is true only at this point where $w = 0$. Except for this point, v differs from u, but is defined by Eq. (1.12). It is thus apparent that the upper bounds of v and u for any LP mode are not the same. The upper bound of u for any LP mode is finite, and is to be determined by the root of the Bessel function that defines the cutoff u value of the next higher LP mode. On the other hand, the upper bound of v tends to infinity as w tends to infinity, that is, according to Eq. (1.12), $v \to \infty$, for $w \to \infty$. Thus, for any mode there is only a lower bound referring to the lowest allowable frequency (or cutoff frequency), but not a higher bound of frequency. This means that any mode will not be cut off as the frequency increases indefinitely.

The mathematical construct for the LP modes is just as simple as outlined through Eqs. (1.4)–(1.20), using only elementary differential calculus and the Table of Zeros of Bessel Functions. Looking back to the mathematical steps, one is naturally impressed by the unusually large number of approximations and intuitive assumptions implied in deriving the LP modes. From the angle of an electromagnetics theoretician, it is perhaps almost unacceptable that the wave-field components of the LP modes are inconsistent in the framework of Maxwell's equations. One is therefore likely to suspect if such approximate LP modes are sufficiently accurate for use in actual fiber practice. Fortunately, it turns out that, in most practical cases, the LP modes prove simple and very useful.

In support of the adequacy of the series of LP modes in representing the wave fields in optical fiber, it is interesting to note that such LP modes are actually *observable* physical entities. Their existence can be *seen* directly by the naked eye in a laboratory, if a visible light source is used to excite the fiber. The intensity patterns of the LP modes of quite high order have been observed under accurately and delicately arranged excitation conditions. The microphotos prepared by W. J. Steward of the Plessey Company [6] fascinatingly display intensity patterns of high-order LP modes like $LP_{47,7}$ $LP_{61,5}$ $LP_{80,3}$, among others.

The use of early approximations is the tactic by which the LP modes are successfully formulated. We close this section by noting that early approximations are possible in many cases only when the understanding of the problem has achieved such maturity that we feel free to adopt a "short-cut" to address the problem in a simple way. Each early step of mathematical approximation implies some loss of the content in the problem. The precedent work helps us to judge whether a simplifying step can or cannot be taken. In this sense, we are indebted to the earlier work that uses tedious and sometimes roundabout mathematical manipulations on field equations.

1.4   SINGLE-MODE FIBER

It is always of major interest to know the principal mode, or the mode of lowest cutoff frequency, for any given waveguiding medium. As indicated in the foregoing section, the principal mode LP_{01} of a step-index fiber model is associated with the root $u = 0$ of $J_1(u) = -J_{-1}(u)$. By Eq. (1.12), $v = u$ at $w = 0$, such that the cutoff value of v for the principal mode is given by the root zero of $J_1(u)$. The principal mode LP_{01} therefore has no cutoff wavelength or cutoff frequency. The existence of the noncutoff principal mode in optical fiber is by nature distinct from the case of a closed metallic tube employed in microwaves.

Meanwhile, it should be noted that the noncutoff property of the principal mode in fiber is actually a mere mathematical idealization. From the viewpoint of physics, as the wavelength increases indefinitely, the portion of the power within the core will become vanishingly small, such that the field structure is virtually not a real guided mode in the practical sense.

By "single-mode transmission" it is naturally meant that transmission of light is only in the principal mode LP_{01}. However, because of the possible existence of two orthogonal polarized principal modes in fiber, the so-called "single-mode" fiber actually supports two propagating modes, even if the operating wavelength is long enough such that all higher modes are cutoff. In the research and development of fiber optics, it has been a long-wished for objective to have a real single-mode fiber in order to ensure a stable transmission of light. Despite assiduous efforts in this direction, however, this seemingly simple objective has not yet been achieved. For almost two decades, this objective has been a major impetus of the search for special fibers of diversified versions.

As the operating frequency increases, it becomes possible for the higher modes to propagate in fiber. The first higher mode above the principal mode is the LP_{11}, whose cutoff value is given by the first zero (2.405) of J_0. When the frequency increases such that $v \geq 2.405$, the LP_{11} mode begins to be a propagating mode. To be exact, this cutoff frequency ($v = 2.405$) refers to the first higher mode LP_{11}, not to the principal mode LP_{01}. Therefore, a fiber cannot be classified as single-mode or multimode if without a specification of the operating wavelength. A nominal single-mode fiber will become a multimode fiber if the operating frequency increases to a certain higher value. On the other hand, a supposed multimode fiber would become single mode if a lower operating frequency is used. The operating frequency range in which only the principal mode propagates in fiber is given by

$$0 < v < 2.405 \tag{1.22}$$

As the frequency increases to $v > 2.405$, both the principal mode LP_{01} and the first higher mode LP_{11} coexist in the fiber until the cutoff frequency of the second higher mode is reached, beyond which three modes begin to coexist, and so forth. By Eq. (1.21), we have

$$\frac{\lambda_c(LP_{11})}{D} \approx \frac{\pi n \sqrt{2\Delta}}{2.405} \tag{1.23}$$

The above equation is put in a form useful in designing and fabricating single-mode fibers. The cutoff wavelength λ_c referring to the first higher mode LP_{11} is the least allowable value of the operating wavelength that keeps single-mode transmission of light in the fiber. According to Eq. (1.23), if the refractive index n and the core-clad relative index difference Δ are fixed up, then the ratio λ_c/D is a constant.

Suppose the measured value of $\lambda_c(LP_{11})$ of a fabricated fiber specimen is 10% too large for the operating wavelength of the source. The simplest way to achieve a reduction of the value of $\lambda_c(LP_{11})$ by 10% is to decrease the core diameter by 10%. The index difference Δ can also be changed in order to meet the requirement for a suitable value of $\lambda_c(LP_{11})$, but this method of adaptation needs more deliberation in the design framework. In practice, a specified change of the refractive index n ($\approx n_1 \approx n_2$ for slight doping) is generally more delicate.

The foregoing discussions are based on the idealized step-index fiber model. The refractive-index profile of an actual fiber deviates more or less from a step function. Nevertheless, the simplified equations thus far derived remain useful, sometimes approximately, in actual fiber design and fabrication. Because of the utmost simplicity of the step-index model, it is a usual practice to approximate the actual (analytically almost intractable) index profile of a practical fiber by the so-called "equivalent" step-index profile.

1.5   COUPLED-MODE THEORY FOR SINGLE-MODE FIBER

The preceding sections concern the determination of independent wave fields, or eigenmodes, that are supported by a straight and uniform fiber of infinite length. An eigenmode in an idealized uniform fiber model is z-invariant in its transverse field structure, undergoing only a phase change of $-\beta z$ (for forward mode) in the longitudinal or transmission direction. Such transmission behavior apparently implies that each eigenmode is simply an exponential function of the form $Ce^{-j\beta z}$ (where C is a constant) satisfying an independent or "uncoupled" ordinary differential equation of the first order:

$da_i/dz = -j\beta_i a_i$ ($i = 1, 2$ if the fiber works in single mode in the nominal sense). Any deviation of the actual fiber from the idealized fiber model, usually called a "perturbation," will cause interchanges of power between the modes such that each of them no longer satisfies an independent differential equation. What results is that both modes are interrelated by a pair of simultaneous differential equations called the "coupled-mode equations."

Classical Method and Phenomenological Method

The concept of mode is a frequent topic of discussion in microwaves as well as in fiber optics. During the late 1950s when I was working on modal analysis of electromagnetic fields, I was very much inspired by a report on the Round Table Discussion on Mode and Field Problems in Non-Conventional Waveguides, which was prepared by L. B. Felsen, with an introduction by N. Marcuvitz [7]. This report is probably the earliest recorded recognition of the importance of the concepts relating to "characteristic" modes (or "normal" modes, or "eigenmodes") versus the coupled modes. While N. Marcuvitz was in favor of using eigenmodes for nonconventional waveguiding structures, S. A. Schekunoff emphasized the importance of the concept of coupled modes. There was difficulty at the Round Table in agreeing on the definition of a mode—a difficulty that recurs even now—but there seemed to be agreement that the eigenmode is preferable whenever a set of eigenmodes can be found, because such a representation involves no mode coupling. However, for those configurations in which the eigenvalue problem cannot be solved, the coupled-mode approach is more appropriate.

Examples discussed at the said Round Table included nonconventional waveguides characterized by the invariance of their cross-sectional shape, and of other parameters, with respect to displacement along the guide axis. Also discussed were periodic structures wherein there is invariance to axial displacement only over a finite distance—the period of the particular structure.

Practical waveguide systems do not often fall into the above-mentioned category. Many waveguides or waveguide devices are employed whose parameters vary not only in the transverse direction but also in the longitudinal or transmission direction. In such general cases, it becomes extremely difficult, if not impossible, to find the set of eigenmodes. The perturbation approach, which is suitable for nonconventional but otherwise regular waveguides, is not applicable here because, even though the variation of parameters is small over a short section of the waveguide, the overall change of the waveguide properties may not be small.

The effectiveness of solving a waveguiding problem using the concept and technique of coupled modes is quite apparent when the problem involves nonconventional and nonuniform waveguide regions. The founding of the coupled-mode theory may be traced back to Schekunoff's "generalized telegraphist's equations" [8], which are coupled-mode equations put in standing-wave form. This set of equations can be readily transformed into coupled-mode equations in terms of traveling waves, in the form that is commonly adopted today. The major procedure used in deriving these equations is shown clearly in Ref. [8]. Such a procedure achieves at one blow the derivation in a rigorous way of the coupled-mode equations, while at the same time establishing the analytic expressions for all the coefficients involved in the equations.

The mathematical technique of deriving the coupled-mode equations by conversion of Maxwell's equations is referred to as the *classical* method. This method has been used effectively in solving a multitude of microwave problems, particularly in the millimeter wavelength range, in which the waveguiding structures are usually in the form of a shielded metallic tube. In fiber optics, A. W. Snyder and then D. Marcuse derived in a fairly general way the coupled-mode equations for anisotropic as well as isotropic fibers [9, 10]. With the help of vector notation, the formulations appear succinct and neat, as compared with the initial formulations in microwaves whose myriad mathematical lines for the partial differential equations are rather tedious looking. The spirit of the conversion of Maxwell's equations into coupled-mode equations is indeed retained in the coupled-mode theory for modern fiber optics.

While the classical method is theoretically rigorous and general, the complexity in mathematical manipulation by this method often prohibits acquisition of useful analytic solutions for certain coupled-mode problems encountered in practice. We are indebted to S. E. Miller for his coupled-wave theory [11], in which the coupled-mode equations are not derived by the classical method of converting Maxwell's equations, but are directly established at the initial stage of analysis, with or without apparent connection with Maxwell's equations. Such an approach is referred to as the *phenomenological* method of the coupled-mode theory. In this approach, the coefficients in the directly formulated coupled-mode equations are assumed to be known a priori.

In the phenomenological formulation of coupled-mode equations, the relevant coefficients can be determined separately any way that happens to be feasible, such as by physical intuition or postulation, by analyzing a simpler model taken as a kind of canonical problem, or by adopting empirical expression, and eventually by making use of experimental data.

Sometimes, but not always, the phenomenological method of coupled-mode theory yields end results equivalent or closely equivalent to the results by the classical method. The phenomenological method is favored for its mathematical simplicity, and is therefore more widely used in practical applications, both in microwaves and in fiber optics. For example, for a microwave or fiber-optic transmission medium involving longitudinally nonuniform waveguiding regions, establishing the coupled-mode equations by converting Maxwell's equations is hardly achievable. For such highly irregular wave-guiding problems, we are forced to establish the coupled-mode equations in the phenomenological way. Thus, we consider a set of coupled-mode equations with constant coefficients that represents the coupled-mode formulation in a simplified fiber model that is longitudinally uniform, and then, for the actual longitudinally varying fiber, we simply allow all or some coefficients in the equations to become functions of the longitudinal coordinate. Such a phenomenological method of approach is by no means rigorous from the mathematical standpoint. However, it does work in dealing with many practical problems. In its course of development, the phenomenological theory has been highly refined, and stands now as a useful scientific discipline in its own right. In the present book, this theory will be extensively used for solving quite many problems in practical fiber optics.

In this introductory chapter, we shall outline the formulation of coupled-mode equations by the phenomenological approach only. The classical approach to coupled-mode equations, by converting Maxwell's equations, is tedious-looking in mathematics, and is relegated to Appendix A.

Phenomenological Approach to Coupled-Mode Formulation

Our theory will be applied primarily to single-mode fiber, particularly the single-mode fiber of nonconventional versions. Why use coupled modes to deal with a "single"mode fiber? Recall that in Section 1.4 it was noted that a single-mode fiber actually supports two orthogonal linearly polarized modes, called the linear polarization modes, or linear modes, for short. These two linear modes are eigenmodes in the idealized fiber model, but are *coupled* to each other in the actual fiber, which deviates from the idealized fiber model either in geometry or in some or all fiber parameters. Coupled-mode equations for a single-mode fiber describe this coupling process.

It is easy to derive the coupled-mode equations in a phenomenological or intuitive way. In a one-dimensional transmission-line problem, a general all-inclusive coupled-mode formulation for single-mode fiber involves not only two forward modes, but also two backward modes. However, for problems of current interest, we are interested in those two modes that propagate in the same direction, say, the forward direction. The two backward modes are therefore disregarded, inasmuch as the exchange of power between two modes in opposite directions are negligibly small. This is a crucial step that simplifies the coupled-mode formulation from involving four modes to involving two modes. Such an early approximation greatly simplifies the mathematical derivations without sacrificing the physical essence that is required in the problems we are concerned with here.

Let $a_1 \equiv a_1(z)$ and $a_2 \equiv a_2(z)$ be the two modes propagating in the forward direction. Over any infinitesimally short section Δz of fiber, either of the modes, say a_1, will undergo a small change, Δa_1, partly as a result of wave propagation in this mode along the short fiber section, and partly due to power exchange with the other mode a_2 coupled to this mode, such that $\Delta a_1 = k_{11} a_1 \Delta z + k_{12} a_2 \Delta z$. Similarly, the other mode undergoes a small change given by $\Delta a_2 = k_{22}a_2\Delta z + k_{21}a_1\Delta z$. At the moment, k_{11}, k_{22}, k_{12}, and k_{21} are taken simply as proportionality coefficients. Dividing these two incremental equations by Δz, taking the limit $\Delta z \rightarrow 0$, and rearranging, we have

$$\frac{da_1}{dz} = k_{11}a_1 + k_{12}a_2$$

$$\frac{da_2}{dz} = k_{21}a_1 + k_{22}a_2$$

(1.24)

This is in fact a standard form of the coupled-mode equations for single-mode fiber established almost effortlessly by intuition.

Description of the Coupled-Mode Equations

The coupled-mode equations given by Eq. (1.24) are usually taken as the starting equations in dealing with the transmission problems of microwaves, and of lightwaves as well. In the equations, z denotes the transmission direction. With x and y denoting the transverse coordinates, a_1 and a_2 are customarily taken as linear modes referring to the coordinate system (x, y, z).

The coefficients k_{11} and k_{22} are propagation constants of the modes, and are generally complex such that

$$k_{11} = -\alpha_1 - j\beta_1$$

$$k_{22} = -\alpha_2 - j\beta_2$$

(1.25)

where the real part of either complex expression denotes the attenuation constant, and the imaginary part, the phase constant. Here, the choice of negative sign for phase constants conforms to the previous definition of wave factor for forward mode. Refer to Eq. (1.3). In passing, we note that $+j\beta_1$ and $+j\beta_2$ for backward modes are not considered. In many cases of practical interest, it is legitimate to put aside the attenuation constants in the analytic treatment, such that the propagation constants are simply given by the phase constants, that is, $k_{11} \rightarrow -j\beta_1$ and $k_{22} \rightarrow -j\beta_2$. These constants are modifications of the phase constants of the eigenmodes of the system in the absence of the coupling mechanism, with the modification terms large or small depending on the kind of coupling mechanism concerned.

For the majority of problems, the so-called phase constants or propagation constants β_1 and β_2 are really constants independent of the transmission coordinate z. However, if Eq. (1.24) is applied to a nonuniform fiber by allowing β_1 and β_2 to become functions of z, then β_1 and β_2 should be called "coefficients" rather than "constants."

The coefficients k_{12} and k_{21} are coupling coefficients between the two modes. In the early study of microwaves, a fundamental relation was derived with the help of the principle of the conservation of power. This relation shows that these two coefficients are not independent, but are correlated in the way that they are *negative conjugate* to each other (see, for example, Refs. [11 and 12]):

$$k_{12} = -k_{21}^* \qquad (1.26)$$

where the asterisk (*) denotes the complex conjugate of a complex number. Two cases are thus possible. In one case, both coefficients are real, but opposite in sign. In another case, both coefficients are imaginary, and the same in sign. We therefore have either of the following two relations for the coupling coefficients:

$$k_{12} = -k_{21} = c$$
$$k_{12} = k_{21} = jk \qquad (1.26a)$$

where either c or k is real.

On the basis of the preceding description, we are ready to put the coupled-mode equations for the two forward modes in the following form:

$$\frac{da_1}{dz} = -j\beta_1 a_1 + ca_2 \qquad (1.27a)$$

$$\frac{da_2}{dz} = -ca_1 - j\beta_2 a_2 \qquad (1.27b)$$

or

$$\frac{da_1}{dz} = -j\beta_1 a_1 + jka_2 \qquad (1.28a)$$

$$\frac{da_2}{dz} = jka_1 - j\beta_2 a_2 \qquad (1.28b)$$

where both c and k are real. Both of these two formulations are employed in this book. The former formulation occurs if linear modes are used as the base modes, while the latter formulation occurs if circular modes are used as the base modes (see Sections 2.3 and 4.6).

From the standpoint of practical application, there are two different kinds of coupled-mode problem. One kind concerns the *intentional* coupling of modes. Such is the case for a fiber-optic device, notably a directional coupler. The other kind concerns the *unintentional* coupling of modes, such as the transfer of power from the desired mode to the undesired mode in a polarization-maintaining fiber. The latter kind of problem is to evaluate how well the fiber in question is capable of maintaining the desired SOP of light in the fiber.

An initial work on coupled-mode theory for microwaves, due to S. E. Miller of Bell Labs [11] applies equally well to the study of modern coupled-mode theory for fiber optics. As Miller put it in the abstract of his well-known paper: "Either $(\alpha_2 - \alpha_1)/c$ or $(\beta_2 - \beta_1)/c$ being large compared to unity is sufficient to prevent appreciable energy exchange between the coupled waves." (Here the original $\alpha_1 - \alpha_2$ in Ref. [11] is written as $\alpha_2 - \alpha_1$ for consistency with the context of this book.)

Miller's one simple sentence contains a wealth of basic concepts that can scarcely be elucidated without adequate description. The first thing we observe is that the capability of the power exchange between the modes depends not only on the magnitude of the coupling coefficient c, but also on the attenuation difference $(\alpha_2 - \alpha_1)$ or the phase-velocity difference $(\beta_2 - \beta_1)$. In fact, it is the ratio of either $(\alpha_2 - \alpha_1)$ or $(\beta_2 - \beta_1)$ to c that is the determinant.

There are therefore two mechanisms that can be utilized to prevent a power exchange from the desired mode to the undesired mode. The first is to design the transmission line, waveguide, or fiber so that the difference in attenuation $(\alpha_2 - \alpha_1)$ of the two modes is deliberately enhanced for a specified value of c. In this book we do not elaborate on the attenuation aspect of fiber, with the recognition that the principle implied in this aspect has been used in the design of special fibers called "polarizers." The second mechanism that deliberately enhances the phase-constant difference $(\beta_2 - \beta_1)$ in order to hold the power in the desired mode is in fact the underlying principle on which most, if not all, currently existing polarization-maintaining fibers work (see Chapter 3). The counterparts of such fibers in microwaves are called the "dephasing" waveguides, such as circular guides of the dielectric-coated and helix versions.

In coupled-mode analysis of either microwaves or fiber optics, the ratio $(\beta_2 - \beta_1)/c$ or its inverse $c/(\beta_2 - \beta_1)$ appears so often that it is convenient to introduce a specific parameter for this ratio:

$$Q = \frac{c}{\delta\beta} \qquad (1.29)$$

$$\delta\beta = \beta_2 - \beta_1 \qquad (1.29a)$$

where Q is called the "coupling capacity," to distinguish it from the coupling coefficient c (see Section 2.2).

In Eq. (1.29), which defines the parameter Q, it is taken for granted that $\delta\beta$ and c always assume the same sign so that Q is positive. Since the subscripts 1 and 2 for β are arbitrary as regards the modes that they denote, we can always let $\delta\beta$ be positive (implying the choice of $\beta_2 > \beta_1$), with c understood to be positive. In fact, the parameter Q can be defined as the absolute value of the quotient. For neatness in mathematical formulation, the notation indicating an absolute value is not included in Eqs. (1.29) and (1.29a).

Here it is important to note that the amount of power exchange taking place between modes in a fiber system cannot be judged by the strength of the coupling coefficient c alone. It is the coupling capacity Q ($= c/\delta\beta$) that determines. The underlying physical principle is apparent. The smaller $\delta\beta$ is, the more closely syncronized the modes will be, so that more power exchange takes place over a longer distance even if the coupling coefficient c is small. Alternatively, if $\delta\beta$ is large, the modes will be substantially out of step, such that power exchange cannot be large even if the coupling coefficient c is large.

The concept under discussion easily explains why only two forward modes are considered in our coupled-mode description of lightwave transmission in a nominal single-mode fiber, without regard to the backward modes. The reason is that two counterpropagating modes are entirely out of synchronism (with the phase-velocity difference $\delta\beta = \beta_2 + \beta_1$) such that, for the same value of c, power exchange between them is negligibly small as compared with power exchange between two modes propagating in the same direction. Coupling between counterpropagating modes becomes important in certain special waveguiding structures, (such as a periodic structure, which is a topic that does not concern us in this book).

Returning to the discussion of two forward modes, for the same strength of c, the power exchange from the desired mode to the undesired mode can therefore be diminished if $\delta\beta$ is sufficiently large. This simple rule established early in microwaves is actually the very principle that governs the researches on the modern polarization-maintaining optical fibers.

The parameter called the phase-velocity difference $\delta\beta$ in microwaves is exactly what is called "birefringence" in fiber optics. In special fiber optics, it is customary to classify the variety of fiber versions as *highly birefringent* (*hi-bi*) or low *birefringent* (*lo-bi*), depending on whether the birefringence $\delta\beta$ of the fiber is large or small. According to the previous discussion, a fiber being hi-bi is a desirable qualification for stronger polarization maintenance. In fiber-optic terminology, large $\delta\beta$ is described as small beat length L_b. The latter is defined as the length of fiber over which $\delta\beta$ changes by 2π, such that we have the simple relation $L_b = 2\pi/\delta\beta$. It is thus seen that shorter beat length is a qualification of stronger polarization maintenance (see Sections 2.1 and 3.1).

Solution of Coupled-Mode Equations with Constant Coefficients

In mathematical language, the coupled-mode equations concerned are two simultaneous first-order linear ordinary differential equations with constant coefficients. These equations can be solved by the standard method of integral calculus. To be specific, consider the coupled-mode equations (1.27).

By differentiation and substitution, the two simultaneous first-order differential equations become two independent second-order differential equations of the same form for a_1 and a_2:

$$\frac{d^2a_1}{dz^2} + j(\beta_1 + \beta_2)\frac{da_1}{dz} + (c^2 - \beta_1\beta_2)a_1 = 0$$

$$\frac{d^2a_2}{dz^2} + j(\beta_1 + \beta_2)\frac{da_2}{dz} + (c^2 - \beta_1\beta_2)a_2 = 0$$

(1.30)

A simple way of solving linear differential equations with constant coefficients is to assume a "trial" solution in the form of an exponential function $e^{\lambda z}$ (exponent λ is yet unknown). Putting this trial solution into either of the preceding equations yields:

$$\lambda_1 = -j\beta + j\sqrt{\left(\frac{\delta\beta}{2}\right)^2 + c^2}$$

$$\lambda_2 = -j\beta - j\sqrt{\left(\frac{\delta\beta}{2}\right)^2 + c^2}$$

(1.31)

where $\beta = (\beta_1 + \beta_2)/2$ is the average of the propagation constants, and $\delta\beta = (\beta_2 - \beta_1)$ is as defined earlier. The initially assumed exponential trial functions are therefore valid, if their exponents are given by Eq. (1.31). The *general* solution of the coupled mode equations is therefore a linear combination of the two exponential functions:

$$a_1(z) = C_1 e^{\lambda_1 z} + C_2 e^{\lambda_2 z}$$

(1.32)

where C_1 and C_2 are arbitrary constants of integration (generally complex). By Eq. (1.27a):

$$a_2(z) = \frac{1}{c}(\lambda_1 + j\beta_1)C_1 e^{\lambda_1 z} + \frac{1}{c}(\lambda_2 + j\beta_1)C_2 e^{\lambda_2 z}$$

(1.33)

The general solutions given by Eqs. (1.32) and (1.33), which involve undetermined integration constants, are descriptive of the transmission property of the fiber structure itself, regardless of the kind of input light that excites the fiber.

The integration constants C_1 and C_2 in Eqs. (1.32) and (1.33) can thus be determined by the given initial values:

$$a_1(0) = C_1 + C_2$$

$$a_2(0) = \frac{1}{c}(\lambda_1 + j\beta_1)C_1 + \frac{1}{c}(\lambda_2 + j\beta_1)C_2$$

(1.34)

By Eqs. (1.32) and (1.33), where the integration constants C_1 and C_2 are written using Eq. (1.34) in terms of the initial values, we have

$$
\begin{aligned}
a_1(z) = {} & \left[\frac{g + \delta\beta/2}{2g} a_1(0) - j\frac{c}{2g} a_2(0) \right] e^{\lambda_1 z} \\
& + \left[\frac{g - \delta\beta/2}{2g} a_1(0) + j\frac{c}{2g} a_2(0) \right] e^{\lambda_2 z} \\
a_2(z) = {} & \left[j\frac{c}{2g} a_1(0) + \frac{g - \delta\beta/2}{2g} a_2(0) \right] e^{\lambda_1 z} \\
& + \left[-j\frac{c}{2g} a_1(0) + \frac{g + \delta\beta/2}{2g} a_2(0) \right] e^{\lambda_2 z}
\end{aligned}
\tag{1.35}
$$

where $g = \sqrt{c^2 + (\delta\beta/2)^2}$, $\delta\beta = \beta_2 - \beta_1$. These solutions, which satisfy the coupled mode equations, are the *particular* solutions, with the initial condition $a_1(0)$ and $a_2(0)$ taken as known. By definition in integral calculus, a particular solution satisfies both the differential equation and the initial condition.

For the coupled-mode equations (1.28), the general solutions and the particular solutions can be derived in a similar way.

Mathematical Artifice by the Method of Diagonalization

From the foregoing derivation, we see that the coupled-mode equations (simultaneous ordinary differential equations with constant coefficients of first order) can be solved simply by a trial solution in the form of an exponential function. No advanced mathematics beyond the scope of elementary integral calculus is required in this derivation. For convenient reference, we shall call this simple and obvious method the "direct method."

In the direct method, while Eq. (1.35) is written in a form that provides mathematical expressions for the output light with given input light, numerical calculations for different specifications of the input light may be fairly lengthy and tedious. Moreover, much of the physical meaning underlying the lightwave transmission process in fiber is obscured by the bare mathematics. For this reason the "direct method," despite its methodological simplicity, is not used in the following chapters to attempt to solve the various special fiber-optics problems.

It is therefore of primary interest to frame the analytic theory of the book on a more refined method of approach, in which the physical meaning implied in the overall transmission process is clear in every mathematical step. This is the mathematical technique called the "method of diagonalization," which is known to be useful in almost every branch of theoretical physics. In the study of various kinds of fiber transmission problems, we have found this mathematical method to be equally efficient. As will be seen in the chapters that follow, it is exactly by the use of this mathematical artifice that much that is obscure in highly irregular special fiber optics is made analytic.

In mathematical physics, the "method of diagonalization" sometimes also refers to the "orthogonal expansion theory." We shall use the former term in this book, inasmuch as it is explicitly indicative of the principle on which the method operates.

At this point more should be said about the general solution and the particular solution of a set of differential equations. The constructs of the general solution can take limitlessly many different forms, inasmuch as this solution involves undetermined (generally complex) integration constants. In order to determine what specific lightwave field actually propagates in the fiber concerned, knowledge of the general solution is not sufficient. What is needed is the *particular* solution that satisfies not only the differential equations but also the boundary condition. As said, a set of differential equations plus the boundary condition constitute a "boundary-value problem." The solution of a given boundary-value problem fits the differential equations and the boundary condition simultaneously. In mathematical physics the effort required to fit the boundary condition is often not easier than the effort of finding the general solution of the given differential equations.

The major topics to be covered in this book can be generally categorized as boundary-value problems. If we consider a uniform fiber and disregard the backward waves, the differential equations concerned are linear first-order ordinary differential equations with constant coefficients. When two forward traveling waves are involved in single-mode operation (backward waves disregarded), only the boundary condition at the input end needs to be specified. The fiber can be of any length, without requiring a specification of the boundary condition at the output end. Such a boundary condition is called an "open" boundary condition, or "one-sided" boundary condition.

The boundary condition in a boundary-value problem can be expressed in different forms which are descriptive of the source, or of the wave field at some boundary. In fiber transmission optics, it is simple and practical to let the open boundary condition refer to the incident light that excites the fiber at the input. This kind of boundary condition is also variously called the "excitation condition," the "terminal condition," or sometimes the "initial condition" (while recognizing that the word "initial" is often used in specifying a time domain).

Normally, the coupled-mode equations and the incident light are given, and the task is to find the particular solution of the given one-sided boundary-value problem. Alternatively, in the opposite sense, the desired kind of lightwave to propagate in fiber is specified in the beginning, and the task is to determine what kind of incident light (the initial condition) should be applied at the input that will be just right to excite this desired lightwave.

An efficient technique to treat the variety of boundary-value (or "initial-value") problems occurring in special fiber optics is provided by the said "method of diagonalization." This mathematical method will be consistently employed in this book for all special fibers that are longitudinally uniform. For nonuniform fibers whose longitudinally structural characteristics vary, it is the spirit of this same method that helps us find a way by which very difficult fiber-optics problems become solvable through admirably simple and easy mathematical manipulation.

1.6   NOTES

For the sake of simplicity and neatness in mathematical representation, throughout the text we shall use the matrix formulation almost exclusively in treating the coupled-mode problems for a diversity of special fiber structures. In each problem the subject of *polarization* and *birefringence* will always be the key point of primary concern.

Hi-bi and lo-bi fibers with respect to light of linear polarization and circular polarization will be of particular interest, and will be treated separately in different chapters. A list of references for this introductory chapter includes several review papers [13–17] on birefringent fibers and a number of books [18–30] that address dielectric optical waveguides, coupled modes, and related topics.

REFERENCES

[1]   D. Hondros and P. Debye, "Elektromagnetische Wellen an dielektrischen Drahten," *Ann. der Phys.*, vol. 32, pp. 465–476 (1910).

[2]   A. W. Snyder, "Asymptotic expressions for eigenfunctions and eigenvalues of a dielectric optical waveguide," *IEEE Trans. Microwave Theory Tech.*, vol. MTT-17, pp. 1130–1138 (1969).

[3]   D. Gloge, "Weakly guiding fibers," *Appl. Opt.*, vol. 10, pp. 2252–2258 (1971).

[4]   C. L. Beattie, "Table of first 700 zeros of Bessel functions," *Bell Syst. Tech. J.*, vol. 37, pp. 689–697 (1958).

[5]   D. Marcuse, *Theory of Dielectric Optical Waveguides*, Academic Press, 257 pages (1974).

[6]   W. Stewart, "LP modes in optical fibres" (microscopic photos), The Plessey Company Ltd., Allen Clark Research Centre, Caswell, Towcester, Northants, England.

[7]   L. B. Felsen, Ed., "Mode and field problems in non-conventional waveguides (Round Table discussion)," *Proc. Symp. on Modern Advances in Microwave Techniques*, Polytechnic Institute of Brooklyn, 492 pages (1954).

[8]   S. A. Schelkunoff, "Conversion of Maxwell's equations into generalized telegraphist's equations," *Bell Syst. Tech. J.*, vol. 34, pp. 995–1043 (1955).

[9]   A. W. Snyder, "Coupled mode theory for optical fibers," *J. Opt. Soc. Am.*, vol. 62, pp. 1267–1277 (1972).

[10] D. Marcuse, "Coupled mode theory of round optical fibers," *Bell Syst. Tech. J.*, vol. 52, pp. 817–842 (1973); also, vol. 54, pp. 985–995 (1975).

[11] S. E. Miller, "Coupled wave theory and waveguide applications," *Bell Syst. Tech. J.*, vol. 33, pp. 661–719 (1954); also, "Waveguide as a communication medium," pp. 1209–1265 (1954).

[12] W. H. Louisell, *Coupled Mode and Parametric Electronics*, John Wiley & Sons, 264 pages (1960).

[13] I. P. Kaminow, "Polarization in optical fibers," *IEEE J. Quantum Electron.*, vol. QE-17, pp. 15–22 (1981).

[14] T. Okoshi, "Single-polarization single-mode optical fibers," *IEEE J. Quantum Electron.*, vol. QE-17, pp. 879–884 (1981).

[15] D. N. Payne, A. J. Barlow, and J. J. R. Hansen, "Development of low- and high-birefringence optical fibers," *IEEE J. Quantum Electron.*, vol. QE-18, pp. 477–488 (1982).

[16] S. C. Rashleigh, "Origins and control of polarization effects in single-mode fibers," *IEEE J. Lightwave Technol.*, vol. LT-1, pp. 312–331 (1983).

[17] J. Noda, K. Okamoto, and Y. Sasaki, "Polarization-maintaining fibers and their applications," *IEEE J. Lightwave Technol.*, vol. LT-4, pp. 1071–1089 (1986).

[18] A. Yariv, *Optical Electronics*, 3rd ed., HRW Book Co., 552 pages (1985).

[19] A. B. Buckman, *Guided Wave Photonics*, HBJ, Saundered College Publishing, 364 pages (1992).

[20] H. A. Haus, *Waves and Fields in Optoelectronics*, Prentice Hall, 402 pages (1984).

[21] H. G. Unger, *Planar Optical Waveguides & Fibers*, Oxford: Clarendon Press, 751 pages (1977).

[22] J. E. Midwinter, *Optical Fibers for Transmission*, John Wiley & Sons, 410 pages (1979).

[23] H. C. Huang, *Coupled Mode and Nonideal Waveguides*, Microwave Research Institute (MRI), Polytechnic of New York, 191 pages (1981).

[24] A. W. Snyder and J. Love, *Optical Waveguide Theory*, Chapman & Hall, 734 pages (1983).

[25] L. B. Jeunhomme, *Single-Mode Fiber Optics*, Marcel Dekker, 275 pages (1983).

[26] T. Okoshi, *Optical Fibers for Communication* (original in Japanese, 1983), Chinese translation by Post & Telecom Press, Beijing, 413 pages (1989).

[27] Huang Hung-chia and Allan W. Snyder (editors), *Optical Waveguide Sciences*, Martinus Nijhoff Publishers, 360 pages (1983).

[28] Huang Hung-chia, *Coupled Mode Theory as Applied to Microwave and Optical Transmission*, Science Press, 364 pages (1984).

[29] C. Vasallo, *Optical Waveguide Concepts*, Book 1 of Optical Wave Sciences and Technology (series editor: Huang Hung-Chia), Elsevier, 322 pages (1991).

[30] C. G. Someda and G. Stegeman (editors), *Anisotropic and Nonlinear Optical Waveguides*, Book 2 of Optical Wave Sciences and Technology (series editor: Huang Hung-chia), Elsevier, 236 pages (1992).

BIBLIOGRAPHY

C. S. Brown and F. Muhammad, "The unified formalism for polarization optics: Further developments," *SPIE*, vol. 2265, pp. 327–336 (1994).

Huang Hung-chia, "From microwaves to optics," invited presentation at the 2nd Annual Meeting of CIE, published in *Acta Electron. Sin.*, September Issue, pp. 1–22 (1979).

Huang Hung-chia and Wang Zi-hua, "Analytic approach to prediction of dispersion properties of step-index single mode optical fibres," *Electron. Lett.*, vol. 17, pp. 202–204 (1981).

K. C. Kao and G.A. Hockman, "Dielectric fibre surface waveguide for optical frequencies," *Proc. IEE*, vol. 113, pp. 1151–1158 (1966).

H. Kogelnik, "Coupled wave devices," *Fiber and Integrated Optics* (D. B. Ostrowsky, editor), Plenium Publishing Corporation, pp. 281–299 (1979).

M. Monerie and L. Jeunhomme, "Polarization mode coupling in long single-mode fibres," *Opt. Quantum Electron.*, vol. 12, pp. 449–461 (1980).

Qian Jing-reng, "On coefficients of coupled-wave equations," *Acta Electron. Sin.*, No. 2, p. 46 (1982).

H. E. Rowe and W. D. Warters, "Transmission in multimode waveguide with random imperfections," *Bell Syst. Tech. J.*, vol. 41, pp. 1031–1170 (1962).

E. Snitzer, "Cylindrical dielectric waveguide modes," *J. Opt. Soc. Am.*, vol. 51, pp. 491–498 (1961).

V. P. Tzolov and M. Fontaine, "Theoretical analysis of birefringence and form-induced polarization mode dispersion in birefringent optical fibers: A full-vectorial approach," *J. Appl. Phys.*, vol. 77, pp. 1–6 (1995).

A. Yariv, "Coupled-mode theory for guided-wave optics," *IEEE J. Quantum Electron.*, vol. QE-9, pp. 919–933 (1973).

A. Yariv, "On the coupling coefficients in the coupled-mode theory," *PIRE*, vol. 46, pp. 1956–1957 (1958).

CHAPTER TWO

Low Birefringent Spun Fiber

This chapter is devoted to the study of the general effect of spinning a fiber in the "hot" state during the linear draw in fiber fabrication. Such a spinning process is distinctively different from the postdraw twisting of a fiber in its "cool" state. The latter will be the subject matter in Chapter 4.

The main point to be shown is that, as far as the geometrical effect is concerned, the spinning process in the course of fiber drawing will always reduce the unspun linear birefringence of a fiber of any kind. In the case of a conventional fiber of almost circular geometry, spinning a fiber at a readily achievable spin rate will cause the fiber to become fairly low birefringent (lo-bi). In principle, should the spin rate be elevated sufficiently high, even a highly birefringent (hi-bi) fiber will become lo-bi through the spinning process.

As to practical application, it was thought that a lo-bi spun fiber should be an ideal medium for measuring the Faraday rotation in fiber wherein a magnetic field is applied. This has been proven true if the Faraday rotation effect alone is concerned. The actual circumstance is complicated, however, because a lo-bi spun fiber is too weak in its polarization-holding capacity to resist the various unavoidable outside perturbations, either intentional or unintentional. To our knowledge, this kind of fiber has not found application on a substantial scale in fiber systems. Nevertheless, the lo-bi spun fiber remains a special kind of fiber of fundamental importance. An understanding of the properties of lo-bi spun fiber is basic to the study of a variety of topics in special fiber optics that involve a rotation of the fiber structure.

2.1 CONVENTIONAL APPROACH TO Lo-Bi FIBER

While the theme of the present chapter almost exclusively concerns the lo-bi characteristics of *spun* fiber, it is useful to give a brief account of the earlier

31

technology when preparation of lo-bi fiber was attempted by the conventional approach. First we give some basic definitions.

Fundamental Definitions

In modern fiber optics, we borrow the term "birefringence" from classical optics to describe the phase-velocity difference between two propagating modes (see Section 1.5). In Section 1.2 we observed that, in an *idealized* fiber model of perfect circular symmetry, the principal mode LP_{01} has a twofold degeneracy with respect to the two orthogonal polarization directions (x, y). The exact equality of the phase velocity of two polarization modes would mean "zero birefringence." For *practical* fibers, however, zero birefringence can never be a reality, but can only be approximated more or less closely.

The birefringence of fiber can be described directly by the phase-velocity difference $\delta\beta$ or by the normalized birefringence B defined as the ratio of $\delta\beta$ and the free-space propagation constant $k = 2\pi/\lambda$, where λ is the operating wavelength. In passing, it is noted that there are several wavelengths that are commonly used in fiber practice. The so-called "short wavelengths" generally refer to 0.6328 μm and 0.85 μm: the former because of its visible nature is often a convenient wavelength for laboratory measurement, and the latter is of interest for the availability of small solid-state light source of appreciable power at this wavelength. The "long wavelengths" generally refer to 1.3 μm and 1.5 μm; the former is advantageous for zero-dispersion, while the latter is favorable for lowest attenuation. At any operating wavelength, short or long, it is always more convenient to characterize the birefringence property of fiber by a parameter that is directly measurable. This is the parameter L_b termed the "beat length." The said parameters $\delta\beta$, B, and L_b for description of birefringence are related by

$$B = \frac{\lambda}{2\pi}\delta\beta \tag{2.1}$$

$$L_b = \frac{2\pi}{\delta\beta} = \frac{\lambda}{B} \tag{2.2}$$

Beating is descriptive of the phenomenon of superposition of two waves that have different phase velocities. The superposed field pattern of the two waves is enveloped by curves in the form of rises and falls along the transmission direction. The beat length is the least length at which a field configuration reproduces itself. As a measure of birefringence, the beat length is inversely proportional to the phase-velocity difference, and so the shorter the beat length, the higher the birefringence will be. In practice, the beat length of a hi-bi fiber is a few millimeters or shorter than one millimeter. On the other hand, the beat length of a lo-bi fiber is in terms of meters at least. Very lo-bi fibers may have beat lengths up to some tens or hundreds of meters.

Early Lo-Bi Fibers by Conventional Approach

Lo-bi fiber can be achieved in a conventional way by a deliberate design and fabrication process. Conventional fiber is essentially characterized by its nominal circular symmetry, or azymuthal invariance, of the refractive-index distribution and other parameters over the fiber cross section. In the radial direction of conventional fiber, the refractive-index distribution is specified according to the required transmission characteristics.

Practically speaking, there are two kinds of index profile along a radius of conventional fiber cross section that are in common use. One is the "depressed-cladding" fiber, wherein the index of the cladding is lower than the indices of the core and of the substrate tube. Such fiber is more resistive to microbending within the operating range, and hence suitable for use in long fiber line. But it is not suitable for making fiber devices like the directional coupler, because the depression of the refractive-index will hamper the power flow from one fiber element to the other fiber element. The other conventional version is the "matched-index" fiber, wherein the indices of cladding and of substrate tube are matched or almost matched. Such fiber closely resembles the step-index fiber, and is suitable for making fiber devices. There are other conventional fibers, notably the multiple-cladding fibers, whose radial index distributions are specifically designed to meet loss and dispersion requirements at specified frequencies or within specified frequency ranges. Despite the varieties in radial index distribution, a conventional fiber is always lo-bi or nearly lo-bi as long as the circular symmetry is essentially maintained.

Prior to the advent of spun fiber, the conventional approach to lo-bi fiber was attempted to improve the circular symmetry of fiber under carefully controlled conditions [1, 2]. Two factors were found to be responsible for the occurrence of considerable residual birefringence in conventional fiber, namely, the core ellipticity and the residual thermal expansion mismatch between the core and clad regions. To make the fiber lo-bi, it is obvious that the initial preform should be highly circular-symmetrical over the transverse cross section, and longitudinally uniform. The setup of the fiber-drawing tower and accessories should be highly precise, so as to minimize nonuniformities produced in the course of fiber drawing.

A fairly large amount of analytic and computational work has appeared in the literature concerning calculation of the linear birefringence due to core ellipticity. However, in this chapter on lo-bi fiber, our interest is not to study the birefringence behaviors of elliptical-core fibers with intentionally enhanced ellipticity. The topic of our immediate interest is the case where the core ellipticity is so slight that it can be regarded as a perturbation of the conventional circular-core fiber. With this in view, a formula given in Ref. [3] is quoted here, which gives the phase-velocity difference caused by core ellipticity in the form:

$$\delta\beta = \varepsilon n k \Delta^2 G(v) \qquad (2.3)$$

where ε is the ellipticity ($\varepsilon = 1 - b/a$), with b and a denoting the half-minor and half-major axes, respectively, of the core ellipse; n is the refractive index; k is the free-space wave number; Δ is the relative index difference; and $G(v)$ is a function of the normalized frequency v. In the relevant publications, different methods have been devised for determining the function $G(v)$ for an arbitrary core ellipticity. An analytic formula of this function suitable for small core ellipticity is given in Ref. [4, table 1]:

$$G(v) = (w^2/v^4)\{u^2 + (u^2 - w^2)[J_0(u)/J_1(u)]^2 + uw^2[J_0(u)/J_1(u)]^3\}$$

$$(2.4)$$

where u, w, and v are given by Eqs. (1.7a), (1.7b), and (1.12).

In the design work for lo-bi fiber, it is preferable to choose a smaller value for the relative refractive-index difference (Δ, 10^{-4} by order), as well as a larger core diameter (D, 8–9 μm at $\lambda = 0.6328$ μm), in view of a relaxation of the otherwise too-severe tolerance on core circularity. Fairly low birefringence ($< 3°$ per meter) was reported in Ref. [2] for lo-bi fiber made by such a conventional approach.

2.2 THE SPUN FIBER

The advent of spun fiber marked a real success in making ultra-lo-bi fiber in a simple and practical way. The technique of spinning the preform during the linear drawing of fiber was invented independently at Bell Labs [5] and at Southampton [6–8]. Both of these places achieved making lo-bi fibers by the spinning technique, though aiming at rather different objectives toward application. The Bell scientists used the spinning technique for measuring the outer diameter of fiber with high resolution. The Southampton scientists, on the other hand, envisaged the application areas of spun fiber in Faraday rotation devices, and in nominal single-mode fiber transmission with diminished polarization mode dispersion.

In lo-bi fiber fabrication, the spinning technique is attractive for its particular effectiveness and simplicity, without the need of deliberate design and fabrication process. Probably, the only comparatively stringent technical requirement in making spun fiber is the geometrical alignment of the parts of the fiber-drawing setup, which requires the incorporation of a speed-controlled motor or spinner on top of a traditional drawing tower. A spinning rate of 2000 revolutions per minute (rpm) or even higher poses no technical difficulty. The effective spinning rate can be further enhanced by lowering the linear fiber-drawing speed.

A useful parameter in spun-fiber specification is defined as the ratio of the spinning rate to the residual linear birefringence inherent in the fiber at its unspun state:

$$Q = \frac{L_b}{L_S} = \frac{\tau}{\delta\beta} \qquad (2.5)$$

where L_b and L_S are the beat length of fiber in the unspun state and spin pitch, respectively; τ is the spin rate; and $\delta\beta$ is the phase-velocity difference (or residual intrinsic linear birefringence) of the fiber. Equation (2.5) is a special form of the general definition of Q given by Eq. (1.29). The parameter Q is called variously the "coupling capacity," the "quality factor," or the *normalized* spin rate.

In Eq. (2.5), the term $\delta\beta$ assumes a value that is different in different conventional fiber versions. Ordinarily, the corresponding range of L_b in conventional fibers is from submeter to several meters. For the quality factor to be sufficiently high, say, on the order of 10^1–10^2, Eq. (2.5) shows that a spin pitch is required in the decimeter or centimeter range. This requirement is fairly easy to achieve by employing the usual linear-drawing speed and a workable spin rate τ.

Take the data reported in Ref. [5], for example. The spin rate is 270 rpm and the linear fiber-drawing speed is 8.4 m/min, yielding a spin pitch that is approximately equal to 3 cm. The measured birefringence of the spun-fiber specimen on a winding drum (with a diameter of 26.8 cm) is about 30 m in beat length. Since the drum-induced strain may cause a birefringence in fiber of this level, the basic birefringence of the unconstrained fiber will be much smaller than this value.

The data reported in Refs. [6–8] appear more attractive in showing the dramatically powerful effect that spinning has on achieving the lo-bi characteristics of spun fiber. A fiber specimen of moderately high intrinsic birefringence is intentionally chosen for illustration. The fiber has an elliptical cladding whose aspect ratio (major to minor axis) is 2.5, such that the intrinsic birefringence falls in the short centimeter range ($\approx$ 19 mm at $\lambda = 0.6328$ μm). The spin pitch $L_S = 1.2$ mm, such that $Q \approx 15.8$. The resulting spun fiber exhibited a negligibly small birefringence (close to the limits of measurements, $\approx 1°$/meter). Note that the rather small spin pitch of 1.2 mm can be realized by lowering the linear-drawing speed to 2.4 m/min provided the spinning rate is kept at 2000 rpm.

The mechanism that shows how the spinning process makes the fiber low birefringent has been seriously examined in Refs. [6–8]. At first thought, it is natural to speculate that the spinning process at high temperature is likely to produce a kind of rounding effect that improves the circular symmetry of the fiber, resulting in a reduction of the polarization birefringence. Experimental evidence, however, does not support such a speculation. In measurement, a small residual ellipticity can be observed which rotates helically along the fiber axis at the spinning rate. This reveals that the internal birefringence is virtually unaltered on a microscopic local scale. According to the cited publications, the lo-bi property of spun fiber is primarily attributed to an *averaging* effect, in the sense that the local principal axes, defined by the residual birefringence, become "scrambled," so to say, when the spinning rate is sufficiently high, with the result that the fiber appears almost azymuthally (or circumferentially) isotropic on a macroscopic scale. A basically important consequence of this understanding is that an originally hi-bi fiber of any kind will eventually become lo-bi, if the fiber undergoes a spinning process of sufficiently high spin rate.

2.3   COUPLED-MODE EQUATIONS FOR SPUN FIBER

Coupled-mode equations were formulated in Section 1.5 in a heuristic way. The equations established therein are "formal" mathematical expressions, in which the propagation constants and the coupling coefficients in the equations are all taken to be known a priori. In dealing with a specific coupled-mode problem, these coefficients will have to be determined in order to carry out the analysis.

Local Coordinates and Linear Base Modes

Before starting to solve an analytic problem of coupled modes, a preliminary step is to choose the coordinate system (fixed or local) and to choose the base modes (linear or circular). In the coupled-mode formulation in existing publications, these choices (the coordinate system and the base modes) are often not explicitly stated, but only implied.

On the choice of coordinates, it can be seen from the relevant literature that the fixed coordinates are seldom chosen to attempt a derivation of the coupled-mode equations for spun fiber. The mathematical manipulation would be prohibitively complicated if the fixed coordinates are used directly at the very beginning of the analysis. On the other hand, coupled-mode equations with reference to the local coordinates are easy to derive by a few different approaches. Over any transverse cross-section these local coordinates (Fig. 2.1) are taken to be aligned with the principal axes of the fiber, and so are spinning or rotating in the transmission direction at the spin rate τ. Once these local coupled-mode equations are solved, the solutions with respect to the fixed coordinates can be readily derived with the aid of the coordinate transformation.

The local coordinates are related to the fixed coordinates by a "rotation" matrix of the form:

$$\mathbf{R}(\vartheta) = \begin{bmatrix} \cos(\vartheta) & \sin(\vartheta) \\ -\sin(\vartheta) & \cos(\vartheta) \end{bmatrix} \tag{2.6}$$

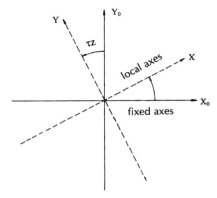

FIGURE 2.1 Local coordinates of a system involving a geometrical rotation: fixed coordinates, solid lines; local coordinates, dashed lines.

where ϑ is the rotation angle. Let the rotational rate (or spin rate) of the local axes be τ, then ϑ in Eq. (2.6) is simply equal to

$$\vartheta = \tau z \qquad\qquad (2.6a)$$

where z is the translation of the coordinate along the transmission direction.

It should be apparent that Eq. (2.6) applies not only for a transformation from the fixed coordinates to the local coordinates, but also for a transformation between two arbitrary sets of coordinates differently inclined (with the difference in inclination angles equal to ϑ).

On the choice of the base modes for the coupled-mode analysis of spun fiber, throughout this chapter we shall use the linear base modes only. As the following sections will show, a spun fiber whose spin rate is sufficiently high behaves like an isotropic medium, such that all kinds of polarizations are allowed to propagate along the fiber. The behaviors of spun fiber can therefore be fully understood through the use of linear base modes in the coupled-mode formulation. The use of circular base modes will prove advantageous when a fiber works predominantly in one circular mode, so that the simple weak-coupling theory applies (see Section 4.6).

Our task is therefore to derive the coupled-mode equations in the local coordinates, with the set of linear modes chosen as the base modes. Before starting this derivation, we examine more carefully the implication of the local coordinates in the case of a conventional fiber with a nominal circular core.

As said, the local coordinates are defined as the pair of rotating axes that coincide with the principal axes of the spun fiber for all z. In an idealized fiber model of perfect circular symmetry, no preferred directions exist that are definable as the principal axes. In practice, however, the nominal circular core of a conventional single-mode fiber is actually slightly-elliptical. Experimental evidence of the presence of a pair of principal axes in such a "nominal" circular guide can be found by performing a simple measurement on a length of conventional fiber with a linear input light. When the incident linear light is adjusted to a specific orientation, the emergent light from the output end of the fiber will be correspondingly also linear at a certain specific orientation. These input and output orientations of linear light thus define the local principal axes due to residual linear birefringence of the conventional fiber. Although this residual linear birefringence is too small to resist even a slight external perturbation, it serves the purpose of providing a pair of principal axes by which the local coordinates can be defined in a practical way.

Derivation of Coupled-Mode Equations

Several different ways have been used in the literature to derive the coupled-mode equations for spun fiber. In the initial publication by McIntyre and Snyder [9], a "twisted" (or rotating) medium of infinite transverse dimension is expressed as a stack of plates. The wave fields in any plate can be expressed in terms of those of an adjacent plate. Difference equations are established by matching the fields at the interface of two plates, which reduce to the desired coupled-mode (differential) equations by letting the thickness of every plate shrink to zero. In a later publication by Okoshi, the coupled mode equations are derived in a different way, that is, by matrix transformation of the local coordinates and the fixed coordinates [3, pp. 192–193, 407–408].

Here we follow the approach by McIntyre and Snyder [9]. For a "twisted" anisotropic medium, it is natural to choose the local coordinates as the reference coordinates. These coordinates are so defined that the transverse axes (x, y) coincide for any z with the birefringence axes (or principal axes) of the twisted anisotropic medium. In addition, linear modes are chosen as the base modes.

The medium is modeled as a stack of birefringent plates, each of thickness δz. As a result of twisting, the birefringence axes of each plate is at an angle $\delta \vartheta = \tau(\delta z)$ to those of the preceding plate at a small incremental distance δz. Consider two adjacent plates 1 and 2 in the "stack of plates" model. Let z^- and z^+ denote the coordinates of the infinitesimally thin plates 1 and 2, respectively. The linear modes in these two plates can be thus written as

$$\mathbf{a}_1(z^-) = \begin{bmatrix} a_x e^{-j\beta_x z^-} \\ a_y e^{-j\beta_y z^-} \end{bmatrix}$$

$$\mathbf{a}_2(z^+) = \begin{bmatrix} (a_x + \delta a_x)e^{-j\beta_x z^+} \\ (a_y + \delta a_y)e^{-j\beta_y z^+} \end{bmatrix} \tag{2.7}$$

where $a_x \equiv a_x(z^-)$, $a_y \equiv a_y(z^-)$, and δa_x, δa_y are incremental changes of the linear modes from z^- to z^+ in the local (twisting) coordinates.

Let (x_1, y_1) be the transverse local axes of plate 1, and (x_2, y_2) be the transverse local axes of plate 2. These two sets of local axes are related by a rotation matrix $\mathbf{R}(\delta\vartheta)$ according to Eq. (2.6), wherein ϑ is replaced by $\delta\vartheta = \tau(\delta z)$. At the interface of the two plates, the transverse fields on the two sides are required to match in the *same local axes*, either (x_1, y_1) or (x_2, y_2). Thus, $\mathbf{a}_1(z^-)$ which refers to (x_1, y_1), should be transformed into $\mathbf{R}(\delta\vartheta)\mathbf{a}_1(z^-)$ in order to match $\mathbf{a}_2(z^+)$, which refers to (x_2, y_2), or alternatively, $\mathbf{a}_2(z^+)$, which refers to (x_2, y_2), should be transformed into $\mathbf{R}(-\delta\vartheta)\mathbf{a}_2(z^+)$ in order to match $\mathbf{a}_1(z^-)$, which refers to (x_1, y_1). In the former way, we have the field-matching equation given by

$$\mathbf{a}_2(z^+) = \mathbf{R}(\delta\vartheta)\mathbf{a}_1(z^-) \qquad (2.8)$$

where $z^+ = z^- = z$ at the interface of the two adjacent plates 1 and 2. Putting Eq. (2.7) and Eq. (2.6) in Eq. (2.8), we have

$$(a_x + \delta a_x)e^{-j\beta_x z} = a_x e^{-j\beta_x z}\cos(\delta\vartheta) + a_y e^{-j\beta_y z}\sin(\delta\vartheta) \qquad (2.8a)$$

$$(a_y + \delta a_y)e^{-j\beta_y z} = -a_x e^{-j\beta_x z}\sin(\delta\vartheta) + a_y e^{-j\beta_y z}\cos(\delta\vartheta) \qquad (2.8b)$$

Here we note in passing that in Eq. (2.7), and hence Eqs. (2.8a) and (2.8b), the $(-)$ sign is chosen for the phase factor to be consistent with our earlier specification that the $(-)$ exponent refers to a forward wave, while the $(+)$ exponent refers to a backward wave [see Eq. (1.3) and the relevant description]. The matrix expressions in Eq. (2.7) are identifiable to Eqs. (1) and (2) in Ref. [9], except for the said choice of sign for the exponent of the exponential function. Also, the expression of $\mathbf{R}(\delta\vartheta)$ is identifiable to Eqs. (3) and (4) in Ref. [9].

Then the next step is to derive the differential equations for the coupled modes. For constant rotation rate of the stack of plates for the anisotropic medium, we have $\vartheta = \tau z$, $\delta\vartheta = \tau\delta z$, $\cos(\delta\vartheta) \approx 1$, and $\sin(\delta\vartheta) \approx \delta\vartheta$, such that Eqs. (2.8a) and (2.8b) can be written as

$$\delta a_x e^{-j\beta_x z} = a_y e^{-j\beta_y z}(\tau\delta z) \qquad (2.9a)$$

$$\delta a_y e^{-j\beta_y z} = -a_x e^{-j\beta_x z}(\tau\delta z) \qquad (2.9b)$$

We introduce the substitutions

$$\bar{a}_x = a_x e^{-j\beta_x z}$$

$$\bar{a}_y = a_y e^{-j\beta_y z}$$

from which we derive the following difference equations:

$$\delta \bar{a}_x = \delta a_x e^{-j\beta_x z} - j\beta_x (\delta z) e^{-j\beta_x z} a_x$$

$$\delta \bar{a}_y = \delta a_y e^{-j\beta_y z} - j\beta_y (\delta z) e^{-j\beta_y z} a_y$$

Putting Eqs. (2.9a) and (2.9b) in the above equations, dividing the resulting equations by δz, and taking the limit $\delta z \rightarrow 0$, we obtain

$$\frac{d\bar{a}_x}{dz} = -j\beta_x \bar{a}_x + \tau \bar{a}_y \tag{2.9c}$$

$$\frac{d\bar{a}_y}{dz} = -\tau \bar{a}_x - j\beta_y \bar{a}_y \tag{2.9d}$$

These are identical to Eqs. (7) and (8) in Ref. [9], except for a change of sign before j (a result quite predictable from our initial choice of the negative sign to denote the forward wave). It will be seen immediately below that this difference in the sign is trivial in the coupled-mode description of lightwave transmission in fiber.

Normalized Form of Coupled-Mode Equations

The coupled-mode equations, Eqs. (2.9c) and (2.9d), refer to the *local coordinates*. The modes that are coupled by the geometrical spinning of the coordinates are the *local modes*. The form of these equations is still slightly inconvenient for practical use. There are three coefficients or parameters (β_x, β_y, and τ) involved in these equations. In actual practice, τ (the spin rate) is a readily measurable quantity. However, it is often not convenient to measure, individually, the propagation coefficients β_x and β_y. What is readily available is their difference $\delta\beta$, which is directly related to the measurable quantity L_b (the beat length) by the simple relation $\delta\beta = 2\pi/L_b$. For more convenience in the practical application of the theory, it is obviously very desirable if the coupled-mode equations are reformulated such that they involve the two readily measurable quantities τ and $\delta\beta$ only. This can be easily done by introducing the following parametric definitions:

$$\delta\beta = \beta_y - \beta_x \tag{2.10a}$$

$$\beta = \frac{\beta_x + \beta_y}{2} \tag{2.10b}$$

where $\delta\beta$ is the difference of propagation constants (or phase-velocity difference), and β is the average of the propagation constants of the two modes. We also wish to extract the common phase factor $e^{-j\beta z}$ from $\bar{a}_x(z)$,

$\bar{a}_y(z)$ by letting

$$\bar{a}_x(z) = A_x(z)e^{-j\beta z}$$
$$\bar{a}_y(z) = A_y(z)e^{-j\beta z}$$

(2.11)

Substituting these equations into the coupled-mode equations, Eqs. (2.9c) and (2.9d), and using Eqs. (2.10a) and (2.10b), we have

$$\frac{dA_x}{dz} = j\frac{\delta\beta}{2}A_x + \tau A_y$$
$$\frac{dA_y}{dz} = -\tau A_x - j\frac{\delta\beta}{2}A_y$$

(2.12)

This is the standard form of the coupled-mode equations that is most commonly used in fiber-optics practice. Specifically, Eq. (2.12) is referred to as the normalized form of the coupled-mode equations. We observe that, while three parameters (β_x, β_y, τ) are involved in Eqs. (2.9c) and (2.9d), two parameters ($\delta\beta$, τ) are involved in Eq. (2.12) through extraction of the common phase factor by Eq. (2.11).

Suppose initially we did not make the change of sign of the exponent. That means that, following Ref. [9], we keep using $e^{j\beta z}$ to denote the forward wave. Then nothing will change in formulating the end result (the normalized coupled-mode equations), if we only define $\delta\beta$ as ($\beta_x - \beta_y$), instead of ($\beta_y - \beta_x$). This is surely a trivial matter of definition. We recall that in connection with Eq. (1.29a) in Section 1.5, we discussed the arbitrariness of the definition of $\delta\beta$ as regards its sign.

All the above equations are derived with respect to the *local coordinates*. The modes that are coupled by the geometrical spinning of the coordinates are the *local modes*. By intuition, we anticipate that the coupling coefficient for local modes should be proportional to the spin rate. But, before the derivation, we are not able to ascertain what this proportionality is. The above result shows that the intuitive anticipation is true. Moreover, we observe that the proportionality is simply equal to unity, such that the coupling coefficient is exactly identical to the spin rate. Meanwhile, Eqs. (2.9c) and (2.9d), and hence Eq. (2.12), also show that the coupling mechanism originated by spinning the principal axes of the medium does not modify the propagation constants of the local modes that are coupled by this mechanism.

Question may arise regarding the applicability of the derived coupled-mode equations to an actual fiber of finite transverse dimension. In the coupled-mode formulation, neglect of the waveguiding effect of the fiber structure is naturally a problem that needs to be carefully examined. Fortunately, it has been shown that the "infinite medium" approximation, essentially, does not sacrifice the analytic accuracy. It is estimated that the coupling coefficient of spun fiber that takes the waveguiding effect into account deviates little from τ derived on the basis of plane wave propagation in an infinite "twisted" medium, provided that the normalized frequency $v > 1.5$ and that the fiber ellipticity $\varepsilon \leq 5\%$.

We close this section by noting that the coupled-mode equations in the form of Eq. (2.12) apply to the case of pure *spinning* of fiber with a residual (large or small) linear birefringence. Prior to the advent of spun fiber, there was no need to differentiate the meanings of the two terms *spin* and *twist*. According to the present-day terminology, what was treated in Ref. [9] ("Light propagation in twisted anisotroptic media") is actually a problem of spinning, not twisting. In the case of twisting, the coupled-mode equations will have to be modified. This will be treated separately in Chapter 4.

2.4   THE METHOD OF DIAGONALIZATION

We recall that in Section 1.5 we attempted solving the coupled-mode equations simply by putting a trial exponential solution in the equations, thereby determining the two possible exponents λ_1 and λ_2 of the exponential function. The solution is then expressed as the linear superposition of the two exponential solutions $e^{\lambda_1 z}$ and $e^{\lambda_2 z}$, whose proportionality constants (generally complex) in the summation are determinable from the given initial values $A_x(0)$ and $A_y(0)$. Throughout the entire derivation only elementary differential calculus and simple algebra were used. While this method (referred to as the "direct method") is easy and familiar, a more elaborate and refined mathematical technique called the "method of diagonalization" will be used instead throughout this book for the following reasons.

First, the method of diagonalization will help us gain a physical insight into the coupled modes in the course of a "step-by-step" derivation. Second, this method is adaptable to mathematical simplifications. Third, the method allows an extension of the uniformly coupled-mode solution to the more general case of the variably coupled-mode solution.

The method of diagonalization is actually not new, but is a classical method commonly used in almost every branch of mathematical physics. Throughout this book, we shall use this method consistently to treat a variety of subject areas relating to special fiber optics. In this chapter, we begin by using this method in spun-fiber problems.

Matrix Formulation of Coupled-Mode Equations

For simplicity and neatness, as well as easier mathematical operation, we put the normalized coupled-mode equations (2.12) in matrix form:

$$\frac{d\mathbf{A}}{dz} = \mathbf{K}\mathbf{A} \qquad (2.13)$$

where $\mathbf{A}$ is a column matrix whose elements are the two linear modes being coupled by the geometrical spinning process, and $\mathbf{K}$ is a square matrix whose elements are descriptive of the fiber structure only, regardless of the applied excitation condition:

$$\mathbf{A} = \begin{bmatrix} A_x \\ A_y \end{bmatrix}$$

$$\mathbf{K} = \begin{bmatrix} j\dfrac{\delta\beta}{2} & \tau \\ -\tau & -j\dfrac{\delta\beta}{2} \end{bmatrix} \tag{2.14}$$

For conventional spun fiber, the phase-velocity difference $\delta\beta$ (or linear birefringence) is small, and is usually the slight residual ellipticity of the "nominal" circular core. Such residual ellipticity simply defines in practice the principal coordinate axes in an analytic study of the spun fiber.

Solution of Coupled-Mode Equations by Diagonalization

The idea implied in the method of diagonalization is fairly simple. In Eq. (2.14) $\mathbf{K}$ is a square matrix, with its diagonal elements representing the inherent residual linear birefringence of the fiber, and its off-diagonal elements representing the coupling coefficients between the modes in the local coordinates. If a mathematical technique can be constructed by which the square matrix $\mathbf{K}$ (coefficients all nonzero) in the coupled-mode equations can be transformed into a diagonal matrix (off-diagonal elements all equal to zero), the transformed matrix equation will represent independent differential equations. Then the solutions can be obtained by solving each independent differential equation separately, and the mathematical manipulation can be greatly simplified.

To do this, we introduce the following linear transformation:

$$\mathbf{A} = \mathbf{OW} \tag{2.15}$$

and conversely

$$\mathbf{W} = \mathbf{O}^{-1}\mathbf{A} \tag{2.15a}$$

where the square matrix $\mathbf{O}$ is the *diagonalizing matrix* that relates $\mathbf{A}$ and $\mathbf{W}$, the latter denoting a new column matrix whose elements are called the *normal modes*. The mathematical–physical implications of $\mathbf{O}$ and $\mathbf{W}$ will become clear in the course of the following derivations. For now, we derive the formal solution of the coupled-mode equations simply by substituting

Eq. (2.15) in Eq. (2.13), such that

$$\frac{d\mathbf{W}}{dz} = \Lambda\mathbf{W} \tag{2.16}$$

$$\Lambda = \mathbf{O}^{-1}\mathbf{K}\mathbf{O} \tag{2.17}$$

What is required now is that the resulting square matrix Λ in the form of Eq. (2.17) become diagonal with zero off-diagonal elements, such that Eq. (2.16) represents two independent ordinary differential equations. With this requirement, neither Λ nor $\mathbf{O}$ can be arbitrary. Only particular matrices will do. Without resorting to abstract theory, can these matrices be determined by a simple algebraic approach? To do this, Eq. (2.17) is written as

$$\mathbf{K}\mathbf{O} = \mathbf{O}\Lambda \tag{2.18}$$

That is,

$$\begin{bmatrix} j\dfrac{\delta\beta}{2} & \tau \\[2mm] -\tau & -j\dfrac{\delta\beta}{2} \end{bmatrix} \begin{bmatrix} o_{11} & o_{12} \\[2mm] o_{21} & o_{22} \end{bmatrix} = \begin{bmatrix} o_{11} & o_{12} \\[2mm] o_{21} & o_{22} \end{bmatrix} \begin{bmatrix} \lambda_1 & 0 \\[2mm] 0 & \lambda_2 \end{bmatrix} \tag{2.19}$$

where we have assumed a priori that Λ is diagonal with diagonal elements λ_1 and λ_2. Multiplication of matrices are performed on both sides of Eq. (2.19). Elements of the first and the second columns of the resulting matrix are then rearranged to yield:

$$\left(j\frac{\delta\beta}{2} - \lambda_1 \right) o_{11} + \tau o_{21} = 0$$

$$\tag{2.20}$$

$$-\tau o_{11} + \left(-j\frac{\delta\beta}{2} - \lambda_1 \right) o_{21} = 0$$

$$\left(j\frac{\delta\beta}{2} - \lambda_2 \right) o_{12} + \tau o_{22} = 0$$

$$\tag{2.21}$$

$$-\tau o_{12} + \left(-j\frac{\delta\beta}{2} - \lambda_2 \right) o_{22} = 0$$

Nonzero solutions of o_{11} and o_{21} require that the determinant of Eq. (2.20) vanish. Likewise, nonzero solutions of o_{12} and o_{22} require that the determinant of Eq. (2.21) vanish. These requirements yield

$$\lambda_1 = +j\left[\tau^2 + \left(\frac{\delta\beta}{2}\right)^2\right]^{1/2}$$

(2.22)

$$\lambda_2 = -j\left[\tau^2 + \left(\frac{\delta\beta}{2}\right)^2\right]^{1/2}$$

These are the *eigenvalues*. The diagonal matrix is then

$$\Lambda = \begin{bmatrix} jg & 0 \\ 0 & -jg \end{bmatrix}$$

(2.23)

$$g = \left[\tau^2 + \left(\frac{\delta\beta}{2}\right)^2\right]^{1/2}$$

(2.23a)

With λ_1 and λ_2 given by Eq. (2.22), the elements of the diagonalizing matrix $\mathbf{O}$ can be determined as follows. Equation (2.20) provides the ratio o_{21}/o_{11}, and Eq. (2.21) provides the ratio o_{22}/o_{12}. Thus, certain additional restrictive conditions still need to be applied to the elements in order to determine each element individually. This is simply a condition of normalization in the sense that the "module" of the elements in each column is defined as equal to unity, that is, $\sqrt{|o_{11}|^2 + |o_{21}|^2} = 1$, $\sqrt{|o_{12}|^2 + |o_{22}|^2} = 1$.

Thus, all the elements of $\mathbf{O}$ are determinable. By Eq. (2.20) and the condition of normalization:

$$o_{11} = \frac{\tau}{\sqrt{\tau^2 + \left(g - \frac{\delta\beta}{2}\right)^2}}$$

(2.24)

$$o_{21} = \frac{j\left(g - \frac{\delta\beta}{2}\right)}{\sqrt{\tau^2 + \left(g - \frac{\delta\beta}{2}\right)^2}}$$

Similarly, by Eq. (2.21) and the corresponding normalization condition, algebraic derivation leads to $o_{12} = o_{21}$ and $o_{22} = o_{11}$. The forms of the above expressions suggest that if τ and $(g - \delta\beta/2)$ in the numerators of o_{11} and o_{21} are taken as the base and the leg of a right triangle, then the denominator of either o_{11} or o_{21} will be equal to the hypotenuse of this triangle.

For succinctness, we can write $o_{11} = \cos \phi$, $o_{21} = j \sin \phi$, such that $\tan \phi = [g - (\delta\beta/2)]/\tau$, and $\tan(2\phi) = \tau/(\delta\beta/2)$. The diagonalizing matrix can thus be written as

$$\mathbf{O} = \begin{bmatrix} \cos \phi & j \sin \phi \\ j \sin \phi & \cos \phi \end{bmatrix} \qquad (2.25a)$$

$$\mathbf{O}^{-1} = \begin{bmatrix} \cos \phi & -j \sin \phi \\ -j \sin \phi & \cos \phi \end{bmatrix} \qquad (2.25b)$$

$$\phi = \frac{1}{2} \arctan\left[\frac{\tau}{\delta\beta/2}\right] = \frac{1}{2} \arctan(2Q) \qquad (2.26)$$

The structure of Eq. (2.25a) could have been predicted if in an earlier step the ratios of the elements had been put, according to Eqs. (2.20) and (2.21), in the form of $o_{21}/o_{11} = j \tan \phi$ and $o_{22}/o_{12} = -j \cot \phi$.

Returning to the differential equations (2.16), as the matrix Λ is diagonalized with diagonal elements λ_1 and λ_2, these equations become independent differential equations:

$$\frac{dW_1}{dz} = \lambda_1 W_1$$
$$\frac{dW_2}{dz} = \lambda_2 W_2 \qquad (2.27)$$

where $\lambda_1 = jg$, and $\lambda_2 = -jg$. The solutions are

$$W_1(z) = e^{jgz} W_1(0)$$
$$W_2(z) = e^{-jgz} W_2(0) \qquad (2.28)$$

which, previously called the normal modes in connection with the transformation, Eq. (2.15), are exactly the *eigenfunctions* associated with the eigenvalues given by Eq. (2.22). Equation (2.28) can be put in matrix form

$$\mathbf{W}(z) = \tilde{\Lambda}\mathbf{W}(0) \qquad (2.29)$$

$$\tilde{\Lambda} = \begin{bmatrix} e^{jgz} & 0 \\ 0 & e^{-jgz} \end{bmatrix} \qquad (2.29a)$$

Note that in Eq. (2.29) $\tilde{\Lambda}$ (with a tilde) is not to be confused with Λ (without a tilde) in Eq. (2.23). The latter Λ is a diagonal matrix whose diagonal elements are the eigenvalues given by Eq. (2.23a). The former $\tilde{\Lambda}$ is also diagonal, but is derived by integrating Eq. (2.23), and its diagonal elements are the corresponding eigenfunctions.

If the common phase factor $e^{-j\beta z}$ is written out, the resulting phase constants of the above-said two diagonal terms are, respectively, $(\beta - g)$ and $(\beta + g)$. Thus, $W_1(z)$ is called the "fast" normal mode, while $W_2(z)$ is called the "slow" normal mode.

We now summarize the method of diagonalization in simple terms. With the aid of the transformation Eq. (2.15), the original simultaneous differential equations for $\mathbf{A}$ (with elements (A_x, A_y) are transformed into independent (or diagonalized) differential equations for $\mathbf{W}$ (with elements W_1, W_2). The latter (independent) ordinary differential equations with constant coefficients are readily solvable. The solutions are exponential functions given by Eq. (2.28) or Eq. (2.29). Then, with the aid of the inverse transformation, Eq. (2.15a), the solutions of $\mathbf{A}$ are derived from the already obtained solutions of $\mathbf{W}$.

At this point, it is easier to elucidate the mathematical–physical implications of $\mathbf{O}$ and $\mathbf{W}$ introduced in the initial transformation, Eq. (2.15). We see from Eq. (2.17) that the diagonal matrix Λ is derived from $\mathbf{K}$ by the matrix operation $\Lambda = \mathbf{O}^{-1}\mathbf{KO}$. The matrix $\mathbf{O}$ thus plays the role of transforming the matrix $\mathbf{K}$ into the diagonal matrix Λ; hence, the name "diagonalizing" matrix. The matrix $\mathbf{W}$ has elements called the "normal modes." This is not difficult to understand when we see from the above derivation that the elements W_1, W_2 of $\mathbf{W}$ are exactly the eigenfunctions.

Nevertheless, some conceptual abstraction inevitably arises if we go one step further to ask: What are the coordinates to which the normal modes (or eigenfunctions) W_1, W_2 refer? We know that the local modes A_x, A_y refer to the local coordinates, which have a geometrical meaning. According to the theory of linear algebra, the normal modes refer to the "normal" coordinates. But a question still remains about the so-called normal coordinates. Such coordinates are certainly different from the local coordinates (x, y), though sometimes we also denote the normal modes (formerly denoted by W_1, W_2) by W_x, W_y, in an analogous sense. To be exact, the normal coordinates do not have a geometrical meaning. We cannot determine the geometrical location of the normal coordinates by actual measurement. In fact, the concept of normal coordinates is merely a mathematical abstraction. From a practical viewpoint, the normal modes (said to refer to the normal coordinates) can be simply taken to mean the matrix elements that are related to the local modes by the transformation, Eq. (2.15) (see Appendix B).

Transfer Matrix of Spun Fiber in Local Coordinates

Let the initial condition of the local modes be written as

$$\mathbf{A}(0) = \begin{bmatrix} A_x(0) \\ A_y(0) \end{bmatrix} \tag{2.30}$$

By Eqs. (2.15) and (2.29):

$$\mathbf{W}(0) = \mathbf{O}^{-1}\mathbf{A}(0)$$

$$\mathbf{W}(z) = \tilde{\Lambda}\mathbf{W}(0) \tag{2.31}$$

$$\mathbf{A}(z) = \mathbf{O}\mathbf{W}(z)$$

Thus, the end solution can be obtained simply by a sequence of matrix operation:

$$\mathbf{A}(z) = \mathbf{O}\tilde{\Lambda}\mathbf{O}^{-1}\mathbf{A}(0)$$

$$= \mathbf{T}_l \mathbf{A}(0) \tag{2.32}$$

$$\mathbf{T}_l = \mathbf{O}\tilde{\Lambda}\mathbf{O}^{-1} \tag{2.32a}$$

where $\mathbf{T}_l$ is the "transfer matrix" of the fiber in the local coordinates. This matrix connects the output light $\mathbf{A}(z)$ with the input light $\mathbf{A}(0)$ such that when the input light is given, the output light can be obtained by a simple matrix–algebraic calculation. Note that this matrix $\mathbf{T}_l$ describes the fiber structure only, regardless of what the input light is.

2.5 ASYMPTOTIC APPROXIMATION FOR FAST-SPUN FIBER IN LOCAL COORDINATES

A spun fiber is said to be "fast-spun" if the spin rate is high as compared to the inherent linear birefringence, that is, $\tau \gg \delta\beta$. In terms of the qualification factor, the fast-spun condition is expressed as

$$Q = \frac{\tau}{\delta\beta} = \frac{L_b}{L_S} \gg 1 \tag{2.33}$$

where it is taken for granted that either τ or $\delta\beta$ stands for the absolute value such that Q and L_b, L_S are always positive.

If a spun fiber in its unspun state is a conventional fiber of nearly circular symmetry, then the residual linear birefringence $\delta\beta$ is relatively small, so that a moderate spin rate τ will satisfy the fast-spun condition, $Q \gg 1$. On the other hand, if the spun fiber in its unspun state is linearly hi-bi, then it becomes necessary to resort to certain special technique in order to achieve the very high spin rate required by the fast-spun condition. For order of magnitude estimation, for example, the linear hi-bi fiber is assumed to have an unspun beat length in the short millimeter range, such that the fast-spun condition requires a spin pitch in the submillimeter range or, preferably, even shorter. If the linear drawing speed of fiber is some tens of meters/minute, then the required spin rate in terms of rpm would become impractically high (up to 10^5 rpm by order). Nevertheless, in principle at least, the fast-spun condition applies to a fiber of any kind, irrespective of its intrinsic linear birefringence, low or high.

In view of actual application, the customarily-called spun fibers refer specifically to the kind of "fast-spun" fibers that are conventional fibers of nominal circular geometry prior to the spinning process.

Under the fast-spun condition $\tau \gg \delta\beta$, the mathematical task required in solving the relevant coupled-mode equations by the method of diagonalization becomes greatly simplified. Referring to the diagonalizing matrix $\mathbf{O}$, the fast-spun condition yields $\varphi = \arctan(2\tau/\delta\beta) \to 45°$, such that $\cos\varphi = \sin\varphi \to 1/\sqrt{2}$, and

$$\mathbf{O} \to \frac{1}{\sqrt{2}}\begin{bmatrix} 1 & j \\ j & 1 \end{bmatrix} \tag{2.34}$$

The above matrix refers specifically to the case where the spinning is in the positive or clockwise sense. For negative or anticlockwise spinning, the same matrix applies if we put a negative sign before each off-diagonal term.

With Eq. (2.34) approximating the diagonalizing matrix $\mathbf{O}$, we also need the inverse of this matrix (i.e., $\mathbf{O}^{-1}$) in the course of mathematical analysis. Under the fast-spun condition, we have

$$\mathbf{O}^{-1} \to \frac{1}{\sqrt{2}}\begin{bmatrix} 1 & -j \\ -j & 1 \end{bmatrix} \tag{2.35}$$

Equations (2.34) and (2.35) are exact for $\varphi = 45°$. For a finite value of the spin rate, we put $\varphi = 45° - \bar{\varepsilon}$, where $\bar{\varepsilon}$ is a small deviation of φ from $45°$. Intuitively, the expected error of Eqs. (2.34) and (2.35) is in proportion to this small deviation $\bar{\varepsilon}$ (see Section 7.10).

In the limit of fast spinning, from Eq. (2.23a) we have:

$$g = \tau\left[1 + \left(\frac{\delta\beta}{2\tau}\right)^2\right]^{1/2}$$

$$\approx \tau\left[1 + \frac{1}{2}\left(\frac{\delta\beta}{2\tau}\right)^2 + \cdots\right] \to \tau \tag{2.36}$$

Thus, the asymptotic approximation $g \to \tau$ can be safely used with a negligible error of $(\delta\beta)^2/(8\tau)$, such that Eq. (2.29a) reduces to

$$\tilde{\Lambda}(z) \to \begin{bmatrix} e^{j\tau z} & 0 \\ 0 & e^{-j\tau z} \end{bmatrix} \tag{2.37}$$

With the aid of Eqs. (2.34)–(2.37), the mathematical analysis of a fast-spun fiber becomes simplified to such an extent that solving an initial-value problem is almost effortless.

2.6   FAST-SPUN Lo-Bi FIBER BEHAVIORS IN LOCAL COORDINATES

A spun fiber is intended to be lo-bi, and hence necessarily satisfies the fast-spun condition. With the aid of Eqs. (2.34)–(2.37), it is simple and straightforward to carry out the derivation of the transfer matrix of *fast-spun* lo-bi fiber in the local coordinates with the result:

$$\mathbf{T}_l = \mathbf{O}\tilde{\mathbf{\Lambda}}\mathbf{O}^{-1}$$

$$= \begin{bmatrix} \cos(\tau z) & \sin(\tau z) \\ -\sin(\tau z) & \cos(\tau z) \end{bmatrix} \tag{2.38}$$

This is a "rotation matrix" that represents a rotation of the coordinates by an angle (τz). For an opposite rotation of this angle, the signs before the off-diagonal elements of Eq. (2.38) should be interchanged.

Circular Light in Fast-Spun Fiber in Local Coordinates

Let a right-circular light be launched at the input of a fast-spun lo-bi fiber. The input light (initial condition) is given by

$$\mathbf{A}(0) = \frac{1}{\sqrt{2}} \begin{bmatrix} 1 \\ j \end{bmatrix} \tag{2.39}$$

Using Eqs. (2.39) and (2.38), the result is

$$\mathbf{A}(z) = \mathbf{T}_l \mathbf{A}(0)$$

$$= \frac{1}{\sqrt{2}} \begin{bmatrix} 1 \\ j \end{bmatrix} e^{j\tau z} \tag{2.40}$$

The equation shows that in a fast-spun lo-bi fiber a circular light (here right circular) in the local coordinates preserves its circular state of polarization (SOP). What changes is an exponential phase shift $e^{j\tau z}$. This implies that light transmission involves only the fast mode.

Similarly, if the incident light is left circular, the result is

$$\mathbf{A}(0) = \frac{1}{\sqrt{2}} \begin{bmatrix} 1 \\ -j \end{bmatrix}$$

$$\mathbf{A}(z) = \frac{1}{\sqrt{2}} \begin{bmatrix} 1 \\ -j \end{bmatrix} e^{-j\tau z} \tag{2.41}$$

The left-circular SOP is likewise preserved. What changes is the phase change factor $e^{-j\tau z}$, which indicates the involvement of the slow mode only.

Linear Light in Fast-Spun Fiber in Local Coordinates

Let a linear light excite the spun fiber at $z = 0$ in the local coordinates:

$$\mathbf{A}(0) = \begin{bmatrix} \cos \theta \\ \sin \theta \end{bmatrix} \tag{2.42}$$

where θ is the orientation of the incident linear light. The output light is derived by Eq. (2.38):

$$\mathbf{A}(z) = \begin{bmatrix} \cos(\tau z) & \sin(\tau z) \\ -\sin(\tau z) & \cos(\tau z) \end{bmatrix} \begin{bmatrix} \cos \theta \\ \sin \theta \end{bmatrix}$$

$$= \begin{bmatrix} \cos(\theta - \tau z) \\ \sin(\theta - \tau z) \end{bmatrix} \tag{2.43}$$

The result shows that in a fast-spun lo-bi fiber, the orientation of a linear light in the local coordinates rotates (or precesses) at the spin rate, but in opposite sense (handedness) to the spin of the fiber structure. An observer in the local coordinates therefore "sees" the linear light continuously counterrotating at an angular rate τ.

(θ, δ)-Representation of Elliptical Light

An arbitrary elliptical light can be expressed in different matrix forms. A simple matrix expression of elliptical light modifies the matrix for linear light, Eq. (2.42), by introducing a phase term $e^{j\delta}$ such that

$$\mathbf{A}(0) = \begin{bmatrix} \cos \theta \\ \sin \theta e^{j\delta} \end{bmatrix} \tag{2.44}$$

where δ denotes the difference of phase between the two linear polarization modes A_x and A_y. Note that, except for the special case of linear light, θ in the general case given by Eq. (2.44) is not the orientation angle of the polarization ellipse. Referring to a standard text on optics, it is found that this orientation angle is given by $\psi = 0.5 \arctan[\tan 2\,\theta \cos \delta]$.

Equation (2.44) implies linear light and circular light as limiting cases. If $\delta = 0$, Eq. (2.44) represents a linear light, and it is in this and only this limiting case that θ denotes the orientation angle. If $\theta = \pi/4$, $\delta = \pm\pi/2$, Eq. (2.44) represents a right- or left-circular light, depending on the $\pm$ sign of δ.

The (θ, δ)-representation of the SOP of light, as given by Eq. (2.44), is general in that it can be used to represent all kinds of the SOPs (elliptical, linear, and circular). It is also concise in mathematical form. Nevertheless, in this (θ, δ)-representation the two basic parameters (ellipticity ε and orientation ψ) that characterize an elliptical light are not shown in the matrix elements explicitly.

(ε, ψ)-Representation of Elliptical Light

Naturally, a matrix representation of an arbitrary elliptical light appears more illustrative if it displays the two characteristic features (ellipticity ε and orientation ψ) explicitly. Such a matrix, called the (ε, ψ)-representation, can be written as

$$\mathbf{A}(0) = \frac{1}{\sqrt{1 + \varepsilon^2}} \begin{bmatrix} \cos\psi - j\varepsilon\sin\psi \\ \sin\psi + j\varepsilon\cos\psi \end{bmatrix} \tag{2.45}$$

where, as before, ε and ψ denote the ellipticity and orientation of the elliptical light, respectively.

The (ε, ψ)-representation of elliptical light, as given by Eq. (2.45), is likewise general, which reduces to circular light and linear light as two limiting cases. Thus, putting $\varepsilon = 1$ in Eq. (2.45) yields a right-circular light with a phase factor $e^{-j\psi}$. On the other hand, putting $\varepsilon = 0$ in Eq. (2.45) yields a linear light with an orientation angle ψ.

We now have two different matrix forms to represent an elliptical light, that is, Eqs. (2.44) and (2.45). These matrices are equivalent, and can be used alternatingly in the mathematical treatment of different problems occurring in special fiber optics. By equating the corresponding elements of these two matrices, it is easy to find that the two parameters ε and ψ in Eq. (2.45) are related with θ and δ in Eq. (2.44) by the following equations:

$$\frac{1 - \varepsilon^2}{1 + \varepsilon^2} = \sqrt{1 - \sin^2(2\theta)\sin^2\delta}$$

$$\tan(2\psi) = \tan(2\theta)\cos\delta \tag{2.46}$$

From now on in this book, both Eq. (2.44) and Eq. (2.45) are used interchangeably. As said, Eq. (2.44) is short and neat, but not in a form that shows the two characteristic features (ε and ψ) of an arbitrary elliptical light explicitly. Equation (2.45) shows these two features (ε, ψ) explicitly, but the expression is somewhat longer. Because of these circumstances, our tactic is therefore to use Eq. (2.45) only when an explicit exposition of (ε, ψ) is required. In the other case, using Eq. (2.44) is preferred for formal simplicity.

Propagation of Arbitrary Elliptical Light in Fast-Spun Fiber

Let an incident arbitrary elliptical light be expressed by the (ε, ψ)-representation of Eq. (2.45). In a fast-spun lo-bi fiber, by Eqs. (2.38) and (2.45), the output light at the coordinate z is derived straightforwardly:

$$A(z) = O\tilde{\Lambda}O^{-1}A(0)$$

$$= \frac{1}{\sqrt{1 + \varepsilon^2}} \begin{bmatrix} \cos(\psi - \tau z) - j\varepsilon \sin(\psi - \tau z) \\ \sin(\psi - \tau z) + j\varepsilon \cos(\psi - \tau z) \end{bmatrix} \qquad (2.47)$$

This result explicitly shows that the ellipticity of the incident light is preserved in the local coordinates. What changes is only that the orientation angle of the polarization ellipse counterrotates about the z-axis by an angle $-\tau z$.

Summarizing the main results of this section, we observe a basic difference in circular light transmission and linear or elliptical light transmission in a fast-spun lo-bi fiber in the local coordinates. For the former, circular light transmission undergoes a phase shift, while for the latter, linear or elliptical light transmission undergoes a change of orientation. Intuitively, this is natural enough. Linear light or elliptical light is featured by an orientation. But a circular light has no preferred orientation. Thus circular light transmission in fast-spun lo-bi fiber can only manifest itself as a phase shift.

2.7   FAST-SPUN Lo-Bi FIBER BEHAVIORS IN FIXED COORDINATES

For a spun fiber, the local coordinates are so defined that they conform for all z with the local principal axes of fiber. The local coordinates rotate (or precess) about the z axis at the spin rate τ. The polarization mode solution in the fixed coordinates can therefore be obtained by turning back the local coordinates by an angle $-\tau z$ via the following "rotation matrix:"

$$R(-\tau z) = \begin{bmatrix} \cos(\tau z) & -\sin(\tau z) \\ \sin(\tau z) & \cos(\tau z) \end{bmatrix} \qquad (2.48)$$

Apparently, Eq. (2.48) is exactly the same as Eq. (2.6), except for a change in the sign of the rotation angle from τz to $-\tau z$.

At the origin $z = 0$, Eq. (2.48) reduces to a unit matrix:

$$R(-\tau z)|_{z=0} = \begin{bmatrix} 1 & 0 \\ 0 & 1 \end{bmatrix} \qquad (2.48a)$$

which implies that the local fields are the same as the fields in the fixed axes at this starting point $z = 0$:

$$\hat{A}(0) = R(-\tau z)|_{z=0}A(0) = A(0) \qquad (2.49)$$

where the circumflex ($\wedge$) denotes a quantity in the fixed coordinates.

Circular Light in Fast-Spun Fiber in Fixed Coordinates

By Eq. (2.48), we transform the wave field solution in the local coordinates given by Eq. (2.40) into the solution in the fixed coordinates:

$$\hat{\mathbf{A}}(z) = \mathbf{R}(-\tau z)\mathbf{A}(z)$$

$$= \frac{1}{\sqrt{2}}\begin{bmatrix} 1 \\ j \end{bmatrix} = \mathbf{A}(0) = \hat{\mathbf{A}}(0) \qquad (2.50)$$

Note that the phase factor $e^{j\tau z}$ in Eq. (2.40), which refers to the local coordinates, now disappears in the fixed coordinates.

Linear Light in Fast-Spun Fiber in Fixed Coordinates

By Eqs. (2.43) and (2.48), the corresponding wave field solution in the fixed coordinates is given by

$$\hat{\mathbf{A}}(z) = \mathbf{R}(-\tau z)\mathbf{A}(z)$$

$$= \begin{bmatrix} \cos \theta \\ \sin \theta \end{bmatrix} = \mathbf{A}(0) = \hat{\mathbf{A}}(0) \qquad (2.51)$$

Note that the rotation of the orientation angle $-\tau z$ in Eq. (2.43) referring to the local coordinates now disappears in the fixed coordinates.

Arbitrary Elliptical Light in Fixed Coordinates

By Eqs. (2.47) and (2.48), the polarization ellipse, which describes the output light in the fixed coordinates, are:

$$\hat{\mathbf{A}}(z) = \mathbf{R}(-\tau z)\mathbf{A}(z)$$

$$= \frac{1}{\sqrt{1 + \varepsilon^2}}\begin{bmatrix} \cos \psi - j\varepsilon \sin \psi \\ \sin \psi + j\varepsilon \cos \psi \end{bmatrix} = \mathbf{A}(0) = \hat{\mathbf{A}}(0) \qquad (2.52)$$

Note that the change of orientation $-\tau z$ in the local coordinates, shown by Eq. (2.47), now disappears in the fixed coordinates.

In summary, Eqs. (2.50)–(2.52) show that, in the fixed coordinates, an incident light of any polarization (circular, linear, or elliptical) will just "go straight" along the fast-spun fiber without any change whatsoever. It is understood that the nature of the propagating light of any SOP is a traveling wave described by the common wave factor $e^{j(\omega t - \beta z)}$, which is not written out for simplicity. So, in the fixed coordinates, the nature of the wave motion is implied, but the waveguiding effect, in the form of an extra phase term or extra angular change (appearing in the local coordinates), no longer exists for any SOP of the propagating light. Alternatively speaking, the field pattern of a propagating light in the fixed coordinates will be z-invariant along the whole length of the fast-spun fiber.

It is thus clear that a fast-spun fiber behaves like an isotropic medium in the fixed coordinates. Light of any polarization becomes an eigenmode in the sense of the z-invariance of the respective field pattern. In principle, this "all-eigenmode" property should have been a favorable feature of fast-spun fiber in such application areas as the Faraday rotation devices. Unfortunately, this "all-eigenmode" property is associated with the related property that the fast-spun fiber is exceedingly lo-bi, such that it is incapable of resisting any sort of slight perturbation. In the asymptotic solutions outlined above, any pair of eigenmodes (either x and y aligned linear modes, or right- and left-circular modes, or two orthogonal elliptical modes of any ellipticity) are degenerate in the sense that they have the same phase constant (or zero phase-constant difference, or zero birefringence).

2.8 THE BEAT LENGTH

Beating means the superposition of two modes of different propagation constants (or phase velocities), making an interference pattern (of periodic rises and falls) in the transmission direction. Beat length is defined as the least distance at which the superposed field pattern reproduces itself. The larger the phase-velocity difference of the two modes, the shorter the beat length will be. Thus, shorter beat length implies less power transfer from the desired mode to the undesired mode, and hence stronger resistance to the outside perturbations. In this sense, the parameter beat length is usually taken to be a qualification parameter of the fiber that measures the relative stability of lightwave transmission in fiber.

It is worth noting that this definition of beat length is by no means unambiguous if the chosen coordinates (fixed or local) are not specified. According to the previously derived solutions for lightwave transmission in a fast-spun lo-bi fiber, the phase-velocity difference for any kind of modes (circular, linear, elliptical) vanishes in the fixed coordinates, corresponding to an indefinitely large beat-length. The resistivity of the fiber to external perturbations is therefore vanishingly small. This exactly reflects the physical property of the fast-spun lo-bi fiber, that is,

$$L_b \to \infty \tag{2.53}$$

This expression refers to the fixed coordinates (circumflex $\wedge$ not used for simplicity).

Confusion arises if we take the local coordinates as the reference coordinates. According to Eq. (2.47), an incident light of arbitrary SOP in the local coordinates will reproduce itself under the condition $\tau z = \pi(1 + m)$, where m is zero or a positive integer. The shortest length for an SOP of light to reproduce itself is the beat length (corresponding to $m = 0$), so that the *local* beat length of a fast-spun fiber is given by

$$L_b^l \rightarrow \frac{\pi}{\tau} \tag{2.54}$$

where the superscript denotes that the beat-length expression refers to the local coordinates.

In the special case of linear light, Eq. (2.54) is simply the distance at which the linear light in the local coordinates will reproduce its orientation ($\theta - \tau z$), according to Eq. (2.43). Since τ is a large value for fast-spun fiber, the local beat length for such fiber becomes exceedingly small. So, should the *local* beat length be a measure of the capability of perturbation resistance, then a fast-spun fiber would have been highly resistive to perturbations. This is indeed not the case.

We therefore come to the understanding that while the beat length in the fixed coordinates is a measure of the perturbation resistivity, the local beat length is not. The local beat length is only descriptive of the field-reproducing behavior of the beating local modes, but nevertheless is irrelevant to the perturbation-resistive property of fiber. The reason for this is really fairly simple: the various perturbations are themselves defined with respect to the fixed coordinates, not the rotating local coordinates. It should therefore be the beat length referring to the fixed coordinates that measures the perturbation resistivity.

2.9   RETARDER–ROTATOR FORMULATION OF SPUN FIBER

In the relevant literature [10, 11], a well-known theory is to treat the spun fiber (as well as the post-draw twisted fiber) as a stack of retarder and rotator elements. It is therefore interesting to examine where and how this retarder–rotator theory is connected or related to the method of diagonalization employed in this chapter.

Transfer Matrix by Method of Diagonalization

In Section 2.6 it was shown that under the "fast-spun condition" the transfer matrix in the local coordinates reduces to a rotation matrix. For the sake of more generality, in this section we shall not use this mathematical restriction imposed by the fast-spun condition, and derive the transfer matrix of spun fiber in the local coordinates without introducing early approximations in the derivation. Thus, by Eqs. (2.25), (2.29a), and (2.32a), we are ready to derive a general form of the transfer matrix of spun fiber in the local coordinates:

$$\mathbf{T}_l = \mathbf{O}\tilde{\mathbf{\Lambda}}\mathbf{O}^{-1} \tag{2.55}$$

$$\mathbf{O}^{-1} = \begin{bmatrix} \cos\phi & -j\sin\phi \\ -j\sin\phi & \cos\phi \end{bmatrix}$$

$$\tilde{\mathbf{\Lambda}} = \begin{bmatrix} e^{jgz} & 0 \\ 0 & e^{-jgz} \end{bmatrix} \tag{2.56}$$

$$\mathbf{O} = \begin{bmatrix} \cos\phi & j\sin\phi \\ j\sin\phi & \cos\phi \end{bmatrix}$$

in which

$$\phi = \frac{1}{2}\arctan\left(\frac{2\tau}{\delta\beta}\right)$$

$$g = \left[\tau^2 + \left(\frac{\delta\beta}{2}\right)^2\right]^{1/2} \tag{2.57}$$

Substituting the matrices given by Eq. (2.56) successively into Eq. (2.55) yields:

$$\mathbf{T}_l = \begin{bmatrix} \cos(gz) + j\cos(2\phi)\sin(gz) & \sin(2\phi)\sin(gz) \\ -\sin(2\phi)\sin(gz) & \cos(gz) - j\cos(2\phi)\sin(gz) \end{bmatrix} \tag{2.58}$$

This general form of the transfer matrix of spun fiber in the local coordinates can be transformed to refer to the fixed coordinates by a counterrotation of the coordinates by $(-\tau z)$. The resulting matrix in the fixed coordinates is exactly the Jones matrix:

$$\hat{\mathbf{T}} = \mathbf{R}(-\tau z)\mathbf{T}_l = \mathbf{J}$$

$$\mathbf{J} = \begin{bmatrix} a & -b^* \\ b & a^* \end{bmatrix} \tag{2.59}$$

$$a = \cos(\tau z)\cos(gz) + \sin(\tau z)\sin(gz)\sin(2\phi)$$
$$+ j\cos(\tau z)\sin(gz)\cos(2\phi)$$
$$b = \sin(\tau z)\cos(gz) - \cos(\tau z)\sin(gz)\sin(2\phi) \tag{2.60}$$
$$+ j\sin(\tau z)\sin(gz)\cos(2\phi)$$

where the asterisk (*) in Eq. (2.59) denotes the complex conjugate.

The Retarder–Rotator Formulation of Spun Fiber

We have thus derived Eqs. (2.59) and (2.60) consistently by the method of diagonalization. These equations are exact in the *fixed* coordinates, implying no approximation.

An early work of Kapron et al. [10] showed that, in the Jones calculus scheme, the coupled-mode solution describing the global (macroscopic) behavior of a spun fiber of length L can be cast in the form of conventional optics comprising a uniform linear retarder with phase retardation $R(L)$ and fast-axis angle $\Phi(L)$, followed by (or following) an angular rotation $\Omega(L)$. According to Ref. [10], the ordered product of linear and circular retarder matrices leaves the relation:

$$\begin{bmatrix} A_x(L) \\ A_y(L) \end{bmatrix} = \begin{bmatrix} a & -b^* \\ b & a^* \end{bmatrix} \begin{bmatrix} A_x(0) \\ A_y(0) \end{bmatrix} \tag{2.61}$$

$$a = \cos\frac{R}{2}\cos\Omega + j\sin\frac{R}{2}\cos(2\,\Phi + \Omega)$$
$$b = \cos\frac{R}{2}\sin\Omega + j\sin\frac{R}{2}\sin(2\,\Phi + \Omega) \tag{2.61a}$$

By equating the real parts and the imaginary parts of Eqs. (2.60) and (2.61a), we find that the retardation, the rotation, and the orientation of spun fiber are given, respectively, by the following expressions:

$$R(L) = 2\sin^{-1}[\cos(2\,\phi)\sin(gL)] \tag{2.62}$$

$$\Omega(L) = \tau L - \tan^{-1}[\sin(2\,\phi)\tan(gL)] \tag{2.63}$$

$$\Phi(L) = \frac{1}{2}\tan^{-1}[\sin(2\,\phi)\tan(gL)] \tag{2.64}$$

In order to compare Eqs. (2.62)–(2.64) and Eqs. (1)–(3) given by Barlow et al. [11], it is convenient to introduce a new symbol $\bar{\rho}$ defined by

$$\bar{\rho} = \frac{\delta\beta}{2\tau} \tag{2.65}$$

According to Eq. (2.26), we have

$$\tan(2\,\phi) = \frac{2\tau}{\delta\beta} = \frac{1}{\bar{\rho}}$$

$$\sin(2\,\phi) = \frac{1}{\sqrt{1+\bar{\rho}^2}} \tag{2.66}$$

$$\cos(2\,\phi) = \frac{\bar{\rho}}{\sqrt{1+\bar{\rho}^2}}$$

Substitution of these expressions in Eqs. (2.62)–(2.64) yields:

$$R(L) = 2\sin^{-1}\left(\frac{\bar{\rho}}{\sqrt{1+\bar{\rho}^2}}\sin(gL)\right)$$

$$\Omega(L) = \tau L + \tan^{-1}\left(-\frac{1}{\sqrt{1+\bar{\rho}^2}}\tan(gL)\right) \qquad (2.67)$$

$$\Phi(L) = \frac{\tau L - \Omega(L)}{2} \pm \frac{m\pi}{2} \qquad m = 0, 1, 2\ldots$$

These are exactly Eqs. (1)–(3) formulated by Barlow et al. [11]. The symbols τ and g in the above equations are the symbols ξ and γ in Ref. [11]. Note also that the above Eq. (2.65) does not include α (twist-induced elastooptic coefficient) that appears in Eqs. (4) and (5) of Barlow et al. [11]. This is because the present chapter concerns the spun fiber only. The effect of α will be considered in Chapter 4, where τ in Eq. (2.65) is to be replaced by $(\tau - \alpha)$ (see Section 4.9).

The retarder–rotator formulation describes the inherent property of fiber in the familiar language of conventional bulk optics. In this formulation the basic parameters $R(L)$, $\Omega(L)$, and $\Phi(L)$ are measurable quantities. As described in Ref. [10], a short section of fiber 7.5 cm long was used in the measurement. The fiber was held in a horizontal position at a uniform height by gently taping it onto flat surfaces. Care was taken to avoid inducing any stress in the fiber due to mounting. The polarizer angle was chosen so that the output light was plane polarized (linear SOP of light at output of fiber), as determined by the analyzer. Measured with respect to an "arbitrary fixed" direction, the polarizer angle Φ yielded one or the other principal axis (fast or slow) of the fiber segment. The difference between the analyzer and polarizer angles then represents the rotation Ω. For a more detailed description of the measurement, see Ref. [10].

Geometrical Interpretation of the Jones Matrix of Spun Fiber

If we put the expressions for a, b, a^*, and b^* in the Jones matrix for a spun fiber, the resulting equation is fairly complex looking. By straightforward but slightly tedious matrix–algebraic manipulation, this formulation of the Jones matrix can be arranged to read:

$$\mathbf{J} = \cos\frac{R}{2}\mathbf{R}(-\Omega) + j\sin\frac{R}{2}\mathbf{R}(-\Omega - 2\Phi)\begin{bmatrix} 1 & 0 \\ 0 & -1 \end{bmatrix} \qquad (2.68)$$

where R, Ω, and Φ are given by Eqs. (2.62)–(2.64); $\mathbf{R}(-\Omega)$, $\mathbf{R}(-\Omega - 2\Phi)$ are rotation matrices that perform counterrotations of the coordinate axes by Ω, $\Omega + 2\Phi$, respectively; and the rightmost diagonal matrix performs a reflection that is mirror symmetrical with respect to the x axis. This geometrical interpretation of the Jones matrix for spun fiber slightly simplifies the mathematical formulation.

Retarder–Rotator Description of Fast-Spun Lo-Bi Fiber

Let a conventional fiber of some small inherent linear-birefringence $\delta\beta$ be spun at a high rate τ, such that $\tau \gg \delta\beta$. Then Eqs. (2.62)–(2.64) will reduce to simpler forms. Consider the former equations. If $\tau \gg \delta\beta$, then $\tan(2\,\phi) = 2\,\tau/\delta\beta \gg 1$, $\cos(2\,\phi) = \delta\beta/2g \ll 1$, and $g \to \tau$, such that Eq. (2.62) reduces to

$$R(L) = 2\sin^{-1}[\cos(2\,\phi)\sin(gL)]$$

$$\approx \frac{\delta\beta}{\tau}\sin(\tau L) \qquad (2.69)$$

Meanwhile, under the fast-spun condition $\tan(2\,\phi) \gg 1$, $2\,\phi \approx \pi/2$, and $\sin(2\,\phi) \approx 1$, such that Eqs. (2.63) and (2.64) reduce to

$$\Omega(L) = \tau L - \tan^{-1}[\sin(2\,\phi)\tan(gL)]$$

$$\approx -\frac{(\delta\beta)^2}{8\tau}L \approx 0 \qquad (2.70)$$

$$\Phi(L) = \frac{1}{2}\tan^{-1}[\sin(2\,\phi)\tan(gz)]$$

$$\approx \frac{1}{2}\tau L \qquad (2.71)$$

From Eq. (2.69) it is seen that, for finite spin rate, the residual linear retardation of fiber is oscillatory with tiny amplitudes. From Eq. (2.70), the residual rotation is a negligibly small second-order value. Thus, both retardation and rotation become vanishingly small when the spin rate of fiber tends to be sufficiently fast. The fiber thus behaves as an isotropic waveguiding medium wherein any polarization mode (or light of any SOP) will just go straight without any change.

2.10   NOTES

Summary of Results and Discussions

This chapter aims at a relatively comprehensive study of the fundamental characteristics of the spun lo-bi fiber. The classical method of diagonalization, which has occupied a prominent place in theoretical physics, is consistently used throughout this chapter and also throughout this book. The well-known retarder–rotator approach to spun fiber of Kapron et al. [10] and Barlow et al. [11] is briefly addressed because of its relation to the "method of diagonalization." The connection between the two analytic approaches is illustrated by the theory of coupled modes with constant coefficients.

A great deal of effort has been put into deriving the solutions of the coupled-mode equations, first in the local (precessing) coordinates, and then in the fixed coordinates. Why not adopt one coordinate system, but complicating the task by a rather roundabout way in order to derive the coupled-mode solutions? The answer is: the coupled-mode equations for spun fiber in the local coordinates are readily available in simple and practically tractable form (ordinary differential equations with constant coefficients). Phenomenologically, it should also be possible to establish the coupled-mode equations for spun fiber in the fixed coordinates, but the resulting formulation would be much more complicated, with $\delta\beta$ (descriptive of an elliptical core, for example) no longer denoting a constant but becoming a variable (i.e., function of z). Although the relevant type of differential equations of variable coefficients are not unsolvable, the required mathematics would be too extensive for approximate analytic solutions (see Appendix B).

The main purpose of this chapter is simply to show that spinning a fiber in the "hot" state produces an "averaging effect" (an important finding of the Southampton scientists), such that spun fiber of sufficient spin rate behaves macroscopically just like an isotropic medium. Any polarization mode (or light of any SOP) is an eigenmode of such fiber. But this very favorable "all-eigenmode" property is associated with a lack of polarization-maintaining capability for any polarization mode, that is, all eigenmodes are extremely lo-bi.

It might be asked: Why use such an elaborate mathematical method for such a simple result? The reason is twofold. On the one hand, verification of this result on a rigorous analytic basis needs such a tedious mathematical process. Second, a sufficiently careful discussion of the method of diagonalization is not only useful in the study of the subject of the present chapter (spun lo-bi fiber), but will also prove to be a mathematical prerequisite for studying many of the following chapters.

Derivation of Coupled-Mode Equations by Coordinate Transformation

In Section 2.3 we followed McIntyre and Snyder [9] in deriving the coupled-mode equations. Because these equations and their modified forms are so important in the analytic study of various special fibers, it is of our interest to also include the alternative method of derivation due to Okoshi [3, pp. 192–193, 407–408].

Consider two linear modes along the principal axes x and y of the fiber. When the fiber is unspun, these polarization modes are independent of each other such that they satisfy the uncoupled equations in the fixed coordinates:

$$\frac{d}{dz}\begin{bmatrix} \hat{A}_x \\ \hat{A}_y \end{bmatrix} = \begin{bmatrix} -j\beta_x & 0 \\ 0 & -j\beta_y \end{bmatrix}\begin{bmatrix} \hat{A}_x \\ \hat{A}_y \end{bmatrix} \qquad (2.72)$$

When the fiber is spun, its principal axes are spinning in the transmission direction z. The coordinates that coincide with the spinning principal axes of the fiber are the local coordinates, and the modes defined with respect to these local coordinates are the local modes. The local coordinates and the fixed coordinates are related by a rotation matrix of rotation angle τz:

$$\mathbf{R}(\pm \tau z) = \begin{bmatrix} \cos \tau z & \pm \sin \tau z \\ \mp \sin \tau z & \cos \tau z \end{bmatrix} \qquad (2.73)$$

where the $\pm$ sign refers to the two different senses (right and left handedness, or clockwise and anticlockwise) of spinning.

When the fiber is spun at the rate τ, Eq. (2.72) for an unspun fiber is modified according to the rotation matrix, such that

$$\frac{d}{dz}\begin{bmatrix} \hat{A}_x \\ \hat{A}_y \end{bmatrix} = \mathbf{R}(-\tau z)\begin{bmatrix} -j\beta_x & 0 \\ 0 & -j\beta_y \end{bmatrix}\mathbf{R}(+\tau z)\begin{bmatrix} \hat{A}_x \\ \hat{A}_y \end{bmatrix} \qquad (2.74)$$

where $\mathbf{R}(+\tau z)$ is the rotation of the coordinates by an angle $+\tau z$ in order to follow up the rotating principal axes, so that Eq. (2.72) applies, while, finally $\mathbf{R}(-\tau z)$ counterrotates the coordinates back to their initial "fixed" place.

The modes $\hat{A}_x$, $\hat{A}_y$, which refer to the fixed coordinates (x, y), and the modes A_x, A_y, which refer to the local coordinates (x', y'), are related by

$$\begin{bmatrix} \hat{A}_x \\ \hat{A}_y \end{bmatrix} = \begin{bmatrix} \cos \tau z & -\sin \tau z \\ \sin \tau z & \cos \tau z \end{bmatrix}\begin{bmatrix} A_x \\ A_y \end{bmatrix} \qquad (2.75)$$

Substituting Eq. (2.75) into Eq. (2.74), we have

$$\frac{d}{dz}\begin{bmatrix} \cos \tau z & -\sin \tau z \\ \sin \tau z & \cos \tau z \end{bmatrix}\begin{bmatrix} A_x \\ A_y \end{bmatrix} = \begin{bmatrix} \cos \tau z & -\sin \tau z \\ \sin \tau z & \cos \tau z \end{bmatrix}\begin{bmatrix} -j\beta_x & 0 \\ 0 & -j\beta_y \end{bmatrix}\begin{bmatrix} A_x \\ A_y \end{bmatrix}$$

$$(2.76)$$

Differentiation of the left side yields

$$\begin{bmatrix} \cos \tau z & -\sin \tau z \\ \sin \tau z & \cos \tau z \end{bmatrix}\begin{bmatrix} dA_x/dz \\ dA_y/dz \end{bmatrix} + \begin{bmatrix} -\tau \sin \tau z & -\tau \cos \tau z \\ \tau \cos \tau z & -\tau \sin \tau z \end{bmatrix}\begin{bmatrix} A_x \\ A_y \end{bmatrix}$$

$$= \begin{bmatrix} \cos \tau z & -\sin \tau z \\ \sin \tau z & \cos \tau z \end{bmatrix}\begin{bmatrix} -j\beta_x & 0 \\ 0 & -j\beta_y \end{bmatrix}\begin{bmatrix} A_x \\ A_y \end{bmatrix} \qquad (2.77)$$

Matrix multiplication of both sides of Eq. (2.77) from the left by

$$\begin{bmatrix} \cos \tau z & \sin \tau z \\ -\sin \tau z & \cos \tau z \end{bmatrix}$$

yields the result:

$$\frac{d}{dz}\begin{bmatrix} A_x \\ A_y \end{bmatrix} = -j\begin{bmatrix} B_x & 0 \\ 0 & \beta_y \end{bmatrix}\begin{bmatrix} A_x \\ A_y \end{bmatrix} + \begin{bmatrix} 0 & \tau \\ -\tau & 0 \end{bmatrix}\begin{bmatrix} A_x \\ A_y \end{bmatrix} \qquad (2.78)$$

Rearranging, we have the coupled-mode equations in the local coordinates:

$$\frac{dA_x}{dz} = -j\beta_x A_x + \tau A_y$$

$$\frac{dA_y}{dz} = -\tau A_x - j\beta_y A_y \qquad (2.79)$$

This end result conforms with the previous result, derived by the "stack of plates" model in Section 2.3.

REFERENCES

[1] H. Schneider, H. Harms, A. Papp, and H. Aulich, "Low-birefringence single-mode optical fibers: Preparation and polarization characteristics," *Appl. Opt.*, vol. 17, pp. 3035–3037 (1978).

[2] S. R. Norman, D. N. Payne, M. J. Adams, and A. M. Smith, "Fabrication of single-mode fibres exhibiting extremely low polarization birefringence," *Electron. Lett.*, vol. 15, pp. 309–311 (1979).

[3] T. Okoshi, "Analysis of fiber with an elliptical core," in Section 6.3.2 of *Optical Fibers for Communication* (original in Japanese, 1983), Chinese translation by Post & Telecom Press, Beijing, pp. 172–173 (1989).

[4] J. Sakai and T. Kimura, "Birefringence and polarization characteristics of single-mode optical fibers under elastic deformations," *IEEE J. Quantum Electron.*, vol. QE-17, pp. 1041–1051 (1981).

[5] A. Ashkin, J. M. Dziedzic, and R. H. Stolen, "Outer diameter measurement of low birefringence optical fibers by a new resonant backscatter technique," *Appl. Opt.*, vol. 20, pp. 2299–2303 (1981).

[6] A. J. Barlow, J. J. Ramskov-Hansen, and D. N. Payne, "Anisotropy in spun single-mode fibres," *Electron Lett.*, vol. 18, pp. 200–202 (1981).

[7] A. J. Barlow, D. N. Payne, M. R. Hadley, and R. J. Mansfield, "Production of single-mode fibre with negligible intrinsic birefringence and polarisation mode dispersion," *Electron. Lett.*, vol. 17, pp. 725–726 (1981).

[8] D. N. Payne, A. J. Barlow, and J. J. Ramskov Hansen, "Development of low- and high birefringent optical fibers," *IEEE Trans. J. Quantum Electron.*, vol. QE-18, pp. 477–488 (1982).

[9] P. McIntyre and A. W. Snyder, "Light propagation in twisted anisotropic media: Application to photoreceptors," *J. Opt. Soc. Am.*, vol. 68, pp. 149–157 (1978).

[10] F. P. Kapron, N. F. Borrelli, and D. B. Keck, "Birefringence in dielectric optical waveguides," *IEEE J. Quantum. Electron.*, vol. QE-8, pp. 222–225 (1972).

[11] A. J. Barlow, J. J. Ramskov-Hansen, and D. N. Payne, "Birefringence and polarization mode-dispersion in spun single-mode fibers," *Appl. Opt.*, vol. 20, pp. 2962–2967 (1981).

BIBLIOGRAPHY

R. Dandiker, "Rotational effects of polarization in optical fibers," in *Anisotropic and Nonlinear Optical Waveguides* (C. G. Someda and G. Stegeman, editors), Elsevier, pp. 39–76 (1992)

R. C. Jones, "A new calculus for the treatment of optical systems," *J. Opt. Soc. Am.*, vol. 31, pp. 488–503 (1941).

CHAPTER THREE

Linearly Birefringent Fiber

The pattern of this book, as stated in the Preface, is a nonconventional one with particular emphasis on circularly birefringent fibers. Nevertheless, the class of linearly birefringent fibers is such an important established subject area in fiber optics that a separate chapter wholly devoted to it naturally appears indispensable.

The study of linearly birefringent fiber has a history nearly two decades long, and covers a multitude of fairly diversified topics. Central to the study is the discussion of the two distinctive classes of linearly birefringent fiber.

1. *The polarization-maintaining fiber.* Such fiber is capable of supporting, within a certain operating wavelength range, either of the two orthogonal polarization modes, as long as the desired polarization mode always keeps aligned with the respective principal axis. The terminology varies from laboratory to laboratory, and the fiber is alternatively called polarization-preserving, polarization-retaining, and so on.

2. *The polarizing fiber.* This kind of fiber has a certain low loss region in the wavelength spectrum, in which only one specific polarization mode can propagate while the other cannot. Such a fiber is a real single-mode fiber in the strict sense.

The above two classes of fiber will be treated in Sections 3.4–3.6, following a preliminary survey of the background technology, to be presented in Sections 3.2 and 3.3. The first section will help clarify the meanings of some common terms often used in this technology.

In the present chapter, naturally, only selected topics of close relevance to the polarization property of linearly birefringent fibers will be addressed, and the description can only be concise. A relatively extensive list of references and bibliographic publications is included at the end of the chapter. References [1–5] are review papers addressing optical-fiber polarization in general, and linear polarization in particular.

3.1   LINEARLY BIREFRINGENT FIBER PARAMETERS

In practical application, a linearly birefringent fiber is usually specified by a number of parameters. Some parameters describe the geometry and material contents of the fiber structure. Other parameters describe the transmission characteristics inherent in the specific fiber concerned. In Section 2.1, the birefringence of a fiber was described by the normalized birefringence B, or by the beat length L_b (referring to a specified working wavelength), the latter being used more often in fiber practice because it is a directly measurable quantity. These two parameters are written here in the following form:

$$B = \frac{\lambda}{2\pi}\delta\beta = \delta n = n_y - n_x$$

$$L_b = \frac{2\pi}{\delta\beta} = \frac{\lambda}{B_G + B_S}$$

(3.1)

where n_x and n_y are the effective indices of the two orthogonal polarization modes along the principal axes, x and y, and B_G and B_S denote the geometrical birefringence and the stress-induced birefringence, respectively.

From the coupled-mode viewpoint, the parameter B or L_b alone is not sufficient to specify the polarization-holding capability of the fiber, inasmuch as the process of power conversions and reconversions taking place in fiber between the two orthogonal polarizations depend not only on the phase difference $\delta\beta (= \beta_y - \beta_x)$, but also on the intrinsic imperfections of the fiber medium in an actually unavoidable perturbative environment. The polarization-holding capability is therefore a random process that is not length-invariant. A simple specification of this property is by using the parameter η (the extinction ratio) defined by the power ratio at the output $z = L$:

$$\eta = \frac{P_y(L)}{P_x(L)}$$

(3.2)

where it is assumed that $P_x(L)$ is the output power of the desired polarization mode, and $P_y(L)$ is the output power of the undesired mode. The extinction ratio is generally expressed in negative decibel (dB) units. Since the extinction ratio is not length-invariant, it is necessary for characterization to specify the fiber length to which the extinction ratio refers. A common practice is to give this figure in negative dB over a certain length, say 1 km.

A more exact way to specify the polarization-holding capability of fiber is by the statistical approach. The regular coupled-mode equations involving both amplitude and phase information are too complicated to deal with. We recall that, in the early exploitation of the coupled wave theory and its microwave applications, particularly in using circular metallic waveguide as a communication medium, the problem of extra loss and side flow due to conversions and reconversions of power was studied by a set of coupled power equations. In the present study, P_x and P_y as functions of z are randomly coupled by fiber imperfections and environmental perturbations. Since every length of fiber is different and may vary with time and environment, the best we can do is to compute the ensemble averages. The background of this calculation was provided, for example, by Kaminow [1]. Without resorting to a lengthy derivation, the relevant coupled power equations can be expressed phenomenologically:

$$\frac{dP_x}{dz} = hP_y - hP_x$$

$$\frac{dP_y}{dz} = hP_x - hP_y$$

(3.3)

where P_x and P_y, stand, respectively, for the ensemble averages $\langle P_x \rangle$ and $\langle P_y \rangle$. A single coefficient h (the power-coupling coefficient) is associated with all terms on the right side (see Ref. [4], p. 321).

With monochromatic excitation of the x-mode, the solution of Eq. (3.3) gives the relative cross-coupled power after length L along the fiber as [4]

$$\frac{P_y}{P} = \frac{1}{2}(1 - e^{-2hL})$$

(3.4)

where $P = \langle P_x + P_y \rangle$ stands for the ensemble average of the sum of P_x and P_y. From this equation it is seen that the power-coupling coefficient h describes the average rate at which power is transferred to the cross polarization in an ensemble of birefringent fibers. This coefficient h is therefore called the "polarization-holding parameter."

In most cases of practical interest, $hL \ll 1$, such that

$$\langle \eta \rangle \approx hL$$

(3.5)

where the ensemble average $\langle \eta \rangle$ is directly related to the spatial spectral density of the randomly perturbing birefringence [4]. The above simple approximate equation, which correlates $\langle \eta \rangle$ with h, is valid only in a statistical sense.

In past publications on linear birefringent fiber of different versions, while the perturbation-resistive qualification is always specified by the beat length, the polarization-maintaining (or polarization-holding) qualification is specified alternatively either by the extinction ratio η of a typical fiber over 1 km, or by the measured value of the h parameter.

3.2   AN ELEMENTARY ANALYSIS OF LINEAR Hi-Bi FIBER

A linear hi-bi fiber can be intentionally spun at a specified spin rate during the linear draw (in the "hot" state), giving rise to a special class of elliptically birefringent fibers called the "spun hi-bi fibers." This subject will be dealt with separately in Chapter 6. Besides the kind of intentionally made spun (linearly) hi-bi fibers, it is often required in practice that a length of fiber be wound on a mandrel, or deployed in a natural course, such that small twists always occur somewhere along the fiber. In an analytic study of the effects of such small twists, we consistently adopt the coupled-mode approach with a set of coupled-mode equations given in the same form as Eqs. (2.13) and (2.14):

$$\frac{d\mathbf{A}}{dz} = \mathbf{K}\mathbf{A}$$

$$\mathbf{A} = \begin{bmatrix} A_x \\ A_y \end{bmatrix}$$

$$\mathbf{K} = \begin{bmatrix} j\dfrac{\delta\beta}{2} & \tau_\varepsilon \\ -\tau_\varepsilon & -j\dfrac{\delta\beta}{2} \end{bmatrix}$$

(3.6)

where τ_ε is the coupling coefficient caused by the small twists. Equation (3.6) implies weakly coupling condition, which means that the off-diagonal coupling coefficient is small compared with the difference of the diagonal elements:

$$\frac{\tau_\varepsilon}{\delta\beta} = Q_\varepsilon \ll 1$$

(3.7)

where for simplicity we do not write out the symbol of absolute value for τ_ε and $\delta\beta$, or for the ratio $\tau_\varepsilon/\delta\beta$, and take it for granted that Q_ε is always positive.

Here we observe that the set of coupled-mode equations, Eq. (3.6), like Eqs. (2.13) and (2.14), is valid in the local coordinates, not in the fixed (or laboratory-fixed) coordinates. In the previous case (lo-bi spun fiber), the local coordinates were exactly defined as the rotating coordinates that conform with the principal axes of the spun fiber everywhere. In practice, these principal axes are simply the major and minor axes of the core that retains a slight residual elliptical deformation. Now, we are concerned with a length of linear hi-bi fiber that is not spun, nor intentionally twisted, but laid in a natural course so that it includes some small twists along the length of the fiber. With respect to this kind of problem, how should we define the coordinates?

Apparently, neither the fixed coordinates nor the local coordinates (defined previously as the coordinates that are progressively rotating at the spin rate) will be appropriate in the present analysis. It is therefore a logical step to extend the definition of local coordinates such that at any point (input and output, in particular) these coordinates always refer to the *local* major and minor axes of the fiber cross section at the point. We still call such coordinates the local coordinates, and thenceforward continue our analysis.

The method of diagonalization remains useful for an analytic solution of Eq. (3.6). From Section 2.4 we borrow the following collection of pertinent equations:

$$\mathbf{A}(z) = \mathbf{O}\tilde{\Lambda}\mathbf{O}^{-1}\mathbf{A}(0) \tag{3.8}$$

$$\mathbf{O} = \begin{bmatrix} \cos\phi & j\sin\phi \\ j\sin\phi & \cos\phi \end{bmatrix} \tag{3.9}$$

$$\phi = \frac{1}{2}\arctan\left(\frac{2\tau_\varepsilon}{\delta\beta}\right)$$

$$\tilde{\Lambda} = \begin{bmatrix} e^{jgz} & 0 \\ 0 & e^{-jgz} \end{bmatrix}$$

$$g = \left[\tau_\varepsilon^2 + \left(\frac{\delta\beta}{2}\right)^2\right]^{1/2} \tag{3.10}$$

Equations (3.8)–(3.10) can be simplified with the aid of the small-twist condition, Eq. (3.7). Thus, for $\tau_\varepsilon \ll \delta\beta$, we have $\phi \approx 0$ and $g \approx \delta\beta/2$, such that

$$\mathbf{O} = \mathbf{O}^{-1} \approx \begin{bmatrix} 1 & 0 \\ 0 & 1 \end{bmatrix}$$

$$\tilde{\Lambda} \approx \begin{bmatrix} e^{j(\delta\beta/2)z} & 0 \\ 0 & e^{-j(\delta\beta/2)z} \end{bmatrix} \tag{3.11}$$

By Eq. (3.11), Eq. (3.8) simply takes the form:

$$\mathbf{A}(z) \approx \begin{bmatrix} e^{j(\delta\beta/2)z} & 0 \\ 0 & e^{-j(\delta\beta/2)z} \end{bmatrix}\mathbf{A}(0) \tag{3.12}$$

The result can actually be predicted by intuitive reasoning. Consider a length of linear hi-bi fiber that includes some small twists. For each twist, an input light of any state of polarization (SOP) will follow the twist as if being locked to it. The linear birefringence is preserved in the local coordinates, defined at any point by the local principal axes of the fiber cross section at that point. A linear light aligned with one or the other local principal axis at the input will follow this slightly rotating principal axis of this and other small twists, and maintain its linear SOP all way to the output. The photoelastic rotation due to the small twists is effectively overwhelmed by the linear high birefringence of the fiber.

Here we note in passing that, not only from the analytic viewpoint but also from the measurement viewpoint, it is necessary to extend the meaning of local coordinates, as we did above, such that a fiber laid in a natural course can be readily measured by experiment. Only in rare case is it possible to lay a fiber of very short length precisely straight with each principal axis always keeping parallel, so that we can use the laboratory-fixed axes to describe lightwave transmission. Generally and practically, a length of fiber is seldom all-way straight from input to output. What we measure at the input and output are thus the quantities referring, respectively, to the local principal axes at the input and the output of the fiber.

3.3   GEOMETRICAL BIREFRINGENCE VS. STRESS BIREFRINGENCE

Intuitively it is conceivable that, in order to create a fiber with a sufficiently large phase-velocity difference $\delta\beta(=\beta_y - \beta_x)$, we will have to spoil, as far as possible, the otherwise nominal circular symmetry of a conventional fiber. Generally speaking, there are two approaches to this objective. One is the geometrical approach, which attempts to change the fiber geometry to become as noncircular as practically possible; the other approach is to apply an anisotropic force on the core such that β_y and β_x are separated as much as possible.

In the first approach, it is all obvious that if the core geometry is made to become a long ellipse, or any shape resembling a rectangle, the difference between β_x and β_y can be enhanced with respect to the principal axes of this deformed core. At first thought, one is likely to adopt this simple and intuitively suggestive way for realizing the desired linear birefringence of fiber.

The second (stress-birefringence) approach is likewise intuitively suggestive. One who works a while in a fiber-optics laboratory must have the experience that, by tweezing a piece of nominal single-mode fiber somewhere along its length between fingers, one will not fail to observe an erratic changes in the polarization of light at the output end of the fiber. A simple experience like this suggests that the application of a transverse force on fiber can be utilized in generating in the fiber the desirous birefringence.

While the underlying principle (for either geometrical or stress-induced birefringence) is as simple as just described, determining the sources that contribute to the overall birefringence in an *actual* fiber becomes complicated. Practically, it is by no means simple to quantify what percentage is attributable to the geometrical birefringence, and what percentage to the stress-induced birefringence. Theoretically or mathematically, we are free to idealize the fiber model for analytic simplification, but an actually fabricated fiber usually possesses not simply one kind of birefringence as a result of the fabrication processes adopted. Nevertheless, in many cases, we can consider the overall birefringence of fiber to be of one kind or the other when it is predominantly of that kind. When both geometrical birefringence and stress-induced birefringence coexist, they sometimes add so as to raise the total birefringence, while at other times they subtract so as to lower the total birefringence. Whether the two parts of birefringence add or subtract depends on the specific fiber configuration concerned [3].

The earliest exploratory work on the practicality of the geometrical-birefringence approach was due to Ramaswamy et al. of Bell Labs [6, 7]. It is only natural that the line of approach in fiber-optics research is closely associated with the fabrication technique readily available to the researchers. At Bell Labs, the modified chemical vapor deposition (MCVD) technique was invented early. In their research, they found that the core of preform (and hence of fiber) was considerably deformed if lower pressure was used during the process of preform collapsing. A typical fiber with a noncircular core produced by the MCVD technique involving a "reduced-pressure" process is shown in Figure 3.1.

The cores were usually peanut (or dumbbell) shaped, and could be adapted for investigating the influence of the core shape on polarization effect. At Bell Labs, experiments were performed by injecting linearly polarized light into a few hundred meters of the fiber and analyzing the output polarization. The fiber was laid in a natural course, suffering no extrinsic perturbations or environmental disturbances. The principal axes (i.e., major and minor axes) were determined by successively tuning the input polarizer

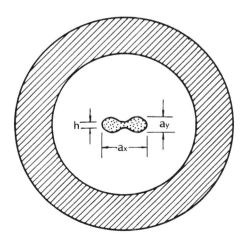

FIGURE 3.1 Geometrical birefringence due to a noncircular core. (From Ref. [7].)

and correspondingly the output analyzer until the output power was minimized as far as possible (by a "minimum-of-minimum" process, so to say).
With the principal axes being determined at the input and output ends of the
fiber, the next step of the experiment was to measure the extinction ratio,
defined as ratio of the magnitudes of the output power analyzed at both
principal axes, for an input linear light oriented along one or the other of the
principal axes. For a core configuration like that shown in Figure 3.1, the
extinction ratio was found to be at a level of -35 dB, indicating a strong
linear polarization at the output.

However, the output polarization was not sufficiently stable, but could be
drastically changed by slightly twisting or bending the fiber, or by changing
the ambient temperature near the fiber. Although the effects were less
pronounced as compared with the case of a conventional single-mode fiber of
nominal circular geometry, the improvement in polarization maintenance was
not substantial. The perturbation resistivity of the peanut-shaped fiber core
can be quantified by measuring the beat length. The result was $L_b = 5.5$ cm.
For a typical circular core, graded-index borosilicate fibers, $L_b \approx 10$ cm. The
experimental data given by Bell Labs thus indicate that a large change in the
core geometry from circular to a fairly flat configuration does not noticeably
improve the birefringence.

For practical fiber-optic system applications, it is desirable that the beat
length of the polarization-maintaining fiber be in the short mm range (2 or
3 mm, for example), or even in the sub-millimeter range. From what has
already been described, we can conclude, as initially stated in Refs. [6] and
[7], that the noncircular geometry alone (with the associated stress-induced
birefringence taken into account) is not sufficient to achieve the practical
objective of polarization-maintaining in single-mode fiber. In other words,
this objective does not seem to be achievable, if not resorting to the
stress-birefringence approach.

During the years when the linear hi-bi fibers were developed, there were
conflicting opinions about the usefulness of the geometrical birefringence
approach. Aside from the above-said negative viewpoint, a proposal was
made that, by deliberately designing the fiber configuration, geometrical
birefringence in fiber could reach a level as high as that of the stress-induced
birefringence. A well-known example is the "side-tunnel" fiber proposed by
Okoshi [2]. An attempt to fabricate this fiber was made at Somitomo,
employing the holed-preform technique. The theoretically predicted geometrical birefringence of the side-tunnel fiber has not been experimentally
demonstrated. In this example, perhaps we can scarcely find a fault in the
analytic-numerical analysis presented in the referenced papers. What one
finds is that a theoretically idealized model is one thing, while the actually
fabricated specimen is not necessarily the same thing. In fact, it was found
that the resulting birefringence in the fabricated "tunnel" fiber specimen was
not purely geometrical.

3.4   INITIAL RESEARCH ON LINEAR BIREFRINGENT FIBER

An initial attempt at generating a higher level of linear birefringence by the stress-inducing approach was begun by scientists at Bell Labs [8–13]. While some early fiber models attempted during late 1970s were not fully developed later, the underlying ideas and working principles did lend perspective to research and development during early 1980s, marked by the advent of the very best-known polarization-maintaining fibers that have been proved to be practical and useful.

Thermal Expansion Mismatch and Strain Birefringence

It was observed during the early experiments that, in a fabricated fiber comprising high thermal-expansion cladding-material like borosilicate, a stress-induced strain is generated in fiber due to the expansion coefficient mismatch between the borosilicate cladding and the pure silica outer jacket [12]. This strain becomes anisotropic when the usual circular or almost circular symmetry of the cladding is made broken. Since such strain is frozen-in during the linear draw of the fiber, it can be a permanent source of birefringence. Additionally, it was observed that such strain birefringence cannot be annealed even after repeated thermal circling.

While this concept is fairly general, there are different ways to break the circular symmetry of the cladding made by high thermal-expansion material. First we make a survey of the research initiated at Bell Labs during late 1970s.

The "Side-Grinding" Technique

Some earliest linear-birefringent fibers were fabricated by the "side-grinding" technique. There are alternative ways to apply this technique to generate linear birefringence in fiber.

One way is to grind flats to the utmost extent on the collapsed preform so that the cladding becomes almost exposed [9]. The associated fabrication method was known as the "exposed-cladding" technique. The fiber drawn from such an unusual flat-sided preform becomes almost circular in outside periphery due to the "rounding effect" in the course of fiber drawing at an appropriately high temperature. The core remains essentially circular, but the borosilicate cladding because of its lower softening temperature will become elliptical-like.

The other way is to grind flats on a thick-walled fused-quartz substrate tube before the borosilicate-cladding and silica-core layers are deposited by the MCVD process [8, 12]. After collapsing, the surface tension causes the

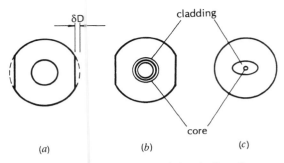

FIGURE 3.2   The "side-grinding" technique: (a) grinding flats on a thick circular substrate-tube; (b) core-clad depositions; (c) collapsed circular preform with an elliptical clad. (From Ref. [8].)

preform to become circular again, resulting in an elliptical cladding and an almost circular core, as illustrated in Figure 3.2.

Some findings reported in Refs. [8–13] are of basic importance. It was found that the ellipticity of the elliptical borosilicate cladding of a fabricated fiber sensitively depends on the amount of the portions of the substrate tube walls being sawed away, to some extent on the substrate wall thickness, the amount of deposited cladding, and the drawing temperature. Usually the germanium-doped core remained essentially round as a consequence of the higher softening temperature of pure silica relative to the borosilicate cladding. But, if a germanium core of lower softening-point is used with a borosilicate cladding and a Pyrex substrate tube, it was found that the core no longer remained circular, but became ribbon-like, and this was considered to be possibly useful for efficient coupling to junction lasers.

It was also found that an inner cladding can be introduced during the fabrication process so as to form a type of "double-cladding" fiber. The inner cladding adjacent to the core is considerably thinner than the outer cladding, but contains a considerably higher level of borosilicate concentration. Because of the more efficient contribution to the strain birefringence, the double-cladding fiber appears to be superior to the single-cladding fiber. It was probably this double-cladding tactic that eventually led to the later invention of the class of flat fiber varieties.

The side-grinding technique was surpassed by other techniques developed later. Cutting or sawing the sides of a preform or substrate tube is a kind of machining process that requires extreme care in order to avoid cracking. The yield, therefore, cannot be high. Especially in the case of longer preforms or substrate tubes, the difficulty associated with the side-grinding processes would be prohibitively augmented.

3.5 PRACTICAL POLARIZATION-MAINTAINING FIBERS REALIZED IN THE EARLY 1980s

The early 1980s witnessed dramatic achievements in the search of *practical* linear birefringent fibers by the strain-birefringence approach. Almost all the best-known fibers of this kind came into being during this period: the flat fiber, the Bow Tie, the Panda, the Corguide, and the Hitachi elliptical fiber, among others. Top laboratories of the world, which had been racing against each other, arrived almost simultaneously at the common goal when they realized one or another type of polarization-maintaining fiber. The specific type of such fiber developed in a laboratory was closely associated with the technical background and tradition of that laboratory.

Also highlighted in this period was the initial achievement of making the type of *real* single-mode fiber. Such fiber is alternatively called the *polarizing* fiber, or the single polarization single mode (SPSM) fiber. Active efforts continued until mid- or late 1980s. Since this research is a fairly involved subject in itself, it will be dealt with separately in Section 3.6.

The Flat Fiber

Regarding the R & D of linear birefringent fiber, Bell Labs seems to have taken a technical approach that more or less kept the flavor of the initially proposed "slab model" of the late 1970s. In an idealized slab model [10], the core, the clad, and the jacket are all rectangular in shape. Such an idealized model can scarcely be achieved in all its details after the fiber drawing. Fabrication processes most likely modify or change the actual fiber in various places. Early work using the exposed-cladding or side-grinding technique yielded a fiber whose clad became elliptical or rectangular-like, but whose core and outer geometry became nearly circular.

Attempts to avoid the preform-machining process led to the "substrate-tube lithography" technique as reported in 1982 [13]. A sequence of preform-making processes was programmed, in which the step of *partial etching* of the deposited borosilicate layer inside the substrate tube was crucial to realizing the desired patterning of an MCVD preform. Different fiber configurations were fabricated by using this technique.

The flat fiber came into being during the early 1980s. Such kind of fiber is featured by reserving the circular shape for the core, while making the claddings and the outer periphery of fiber shaped markedly like a long ellipse or round-cornered rectangle. The flat fiber may well be called a rectangular fiber in view of its overall fiber geometry. Keeping the core round is favorable for connecting or splicing the fiber in-line in a fiber-optic circuitry. But the unique feature of a flat fiber is its outer rectangular shape, which has special advantages that are not to be found in other fibers of circular outer geometry. One apparent advantage is that the principal axes of the flat fiber

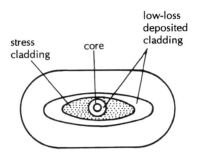

FIGURE 3.3   The flat fiber.

can be determined effortlessly because they are simply the major and minor axes of the cross-sectional geometry of fiber in the normal case. It is a common experience that location of the principal axes of a linear hi-bi fiber of round periphery is often a tedious job, particularly when optical transmission in fiber involves two counterpropagating beams of light. Another related advantage of the flat fiber concerns winding or coiling of the fiber around a form. The flat fiber because of its mechanical property will bend naturally, with its major principal axis all-way aligned with the bending plane (see Fig. 3.3).

The flat fiber is fabricated by the "press-heating" technique [14]. A circular preform is fabricated containing a core and a circular highly doped layer that will become the stress cladding. The preform is then locally heated and squeezed from the sides. The deformed preform has an aspect ratio of about 2:1 to 2.5:1, with a highly elliptical cladding and a round core.

A preform of flat transverse configuration does not mean that the drawn-up fiber necessarily reproduces the flat shape. Drawing a fiber at the usual high temperature produces a "rounding effect," such that it causes the fiber drawn from a flat preform to have an outer periphery not so flat, but more or less round. Clearly, one crucial point to ensure success in making the fiber flat is to avoid the "rounding effect." Thus, the drawing temperature for flat fiber should be lower than the usual drawing temperature for fibers of round outer periphery. Nevertheless, the fiber-drawing temperature cannot be too low when considering other factors, such as the fiber strength and the yield. For flat fiber, therefore, the determination of an optimum drawing temperature remains a difficult research topic of much practical interest.

Fiber drawn from the flat preform at a relatively low drawing temperature, or at a relatively high tension, has nearly the same shape as the preform. The stress-induced linear birefringence is provided by the elliptical boron-doped layer. The inherent absorption of the borosilicate composition of the stress layer increases beyond 1 μm, and therefore low-loss designs for applications beyond 1 μm have a thin barrier layer, usually a fluorine doped composition, which surrounds the core to isolate the guided mode from the absorbing stress material.

It was reported [15] in the mid-1980s that polarization-maintaining flat fiber with a nearly rectangular outer cross section (made by the press-heating technique) achieved an extinction ratio of -43 dB for 5-m lengths, and an h-parameter value of $1.7 \times 10^{-6} \mathrm{m}^{-1}$ for 100-m lengths. At 1.3 and 1.5 μm, typical value of losses was 0.8 dB/km.

In a later report [16] on the polarization properties of AT & T's rectangular polarization-maintaining fiber, the measurement was made at 1320 nm for lengths of fiber equal to 20, 40, 100, 310, and 804 m. The length dependence of the extinction ratio was experimentally demonstrated to obey the random coupling theory. In the measurement, it was due to the rectangular geometry of the fiber that facilitated winding and packaging the fiber with no perturbations. The measured polarization-holding parameter was $h = 2.08 \pm 0.09 \times 10^{-6} \mathrm{m}^{-1}$. The SOP as a function of the input angle of linear light is determined for various lengths. For $L \geq 100$ m, depolarization effects eliminate all other SOPs except the linear on-axis SOP. For shorter lengths of fiber, if the input angle is within 5° of the polarization axis, still only linear on-axis SOPs can propagate. A beat length of 6.78 ± 0.36 mm was directly measured at 1320 nm.

In passing it is noted that the press-heating technique also applies to making the kind of polarizing fiber, a topic that will be discussed in Section 3.6.

The Bow-Tie Fiber by MCVD plus Gas-Phase Partial Etching

Different fabrication methods that were attempted early yielded fiber varieties whose diametrical stress-applying regions resemble a bow tie. However, what is called the Bow-Tie fiber specifically refers to the kind of birefringent fiber first produced (1982) at Southampton, England, using the versatile MCVD method, in which the necessary patterning of the deposited boron-doped stress regions inside a substrate tube is achieved by the *gas-phase* partial etching technique [17] (see Fig. 3.4).

The Southampton's method of fabrication has many advantages. Most prominently, the fabrication processes (essentially consisting of depositing B_2O_3–SiO_2, gas-etching with fluorine, depositing GeO_2–SiO_2, and collapsing) are performed on a glass lathe in one continuous step. Compared with conventional fiber fabrication, the only extra work in making Bow-Tie fiber is the gas-etching process, during which the rotation of the lathe is stopped and two burners positioned diametrically on both sides of the substrate tube provide localized hot zones on the tube walls. These diametrically opposed hot zones travel along the length of the tube while a fluorine-liberating gas is passed down the bore. The remained B_2O_3–SiO_2 depositions then become the stress-applying parts of the fabricated fiber. Since it is not required that during processing the preform be removed from the lathe for machining or other operations, the risk of fracturing is minimized. Moreover, this fabrica-

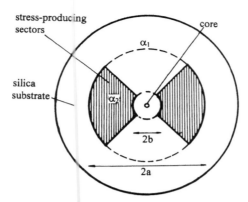

FIGURE 3.4   Transverse geometry of a Bow-Tie fiber. (From Ref. [17].)

tion method, which embodies all the processes in a continuous single step, is believed to be superior in achieving higher precision of the constituent parts of the fiber.

The Bow-Tie fiber is one of the very admirable successes achieved in the area of special fiber optics that the Southampton scientists are ever proud of. Their initial paper [17] reported that beat lengths at 0.6328 μm were routinely less than 1.3 mm. The best result for beat length achieved (1982) at Southampton was 0.55 mm at 0.6328 μm, remarkably better than previously reported data by other fabrication methods. The reported losses were not low for the initial Bow-Tie fiber specimen, but could be reduced by improved design.

In a subsequent analytic study [18], it was theoretically ascertained that the Bow-Tie configuration is optimum in a specified sense. As the present author views it, the outcome of a Bow-Tie figure using the "MCVD plus partial etching" technique was likely an experimental fortune. It can hardly be predicted how the boron-doped regions of fiber will shape themselves after the collapsing procedure. It can be anticipated that a frozen-in high linear birefringence will occur in the resulting preform, inasmuch as before collapse the residual boron-doped regions are positioned diametrically on the two sides of the core. It must be a great surprise to anyone attempting a novel linear birefringent fiber, when he or she for the first time sees through the microscope's eyepiece a nice-looking bow-tie.

The Panda Fiber by VAD

By intuitive reasoning, the idea of making a kind of linear birefringent fiber known as the Panda fiber is actually extremely simple and obvious. In an exhaustive search of all possibly fitting preform structures, one would naturally ask: Why not just put two boron-doped cans into a common round single-mode preform? In principle, the making of a linear birefringent fiber can really be just that simple. In technology, however, the task is not that simple.

The first attempt along this line of thought was made in the NTT Labs, employing the so-called "rods-in-tube" technique [19]. The central rod is a conventional round preform containing along its axis a GeO_2–SiO_2 core. The two side rods are highly boron doped. The other rods are silicate-glass rods without doping, which serve to fill the inside space of the hollow substrate tube. The substrate tube embodying many rods must be considerably larger than the usual MCVD preform. The fiber drawn from a rods-in-tube preform generally looks like that shown in Figure 1 of Ref. [19]. Interestingly, the originally circular-shaped stress rods are deformed into two peculiar-shaped stress-applying regions. The source causing these deformations comes from the residual (air-filled) spaces left inside the rods-in-tube preform. During drawing, the highly boron-doped stress regions of lower viscosity, being surrounded by small irregular air spaces, are bound to be deformed from its original circular shape.

Deformation of the stress parts is not important in itself provided the deformed parts remain mirror-symmetrical with respect to the vertical plane passing the core. But this cannot be ensured in practice, except with a very sophisticated control. Asymmetrical deformations of the two stress parts are the usual case.

About two years later, a new fabrication technique [20], also developed at the NTT Labs, was reported that excelled the initial one described above. In this new technique, a vapor-phase axial deposition (VAD) preform was jacketed into a silica tube (see Fig. 3.5). The preform was so designed that the ratio of the synthesized cladding diameter to the core diameter was chosen to be 5. Two pits of about 10 mm in diameter were bored into this core-cladding preform with an ultrasonic machine, and the stress-applying rods were inserted into the pits. The fiber drawn from such kind of preform

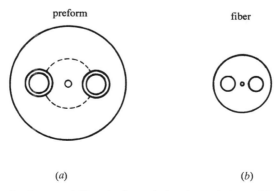

preform   fiber

(a)   (b)

FIGURE 3.5   Preform and fiber in the rods-in-pits technique. (From Ref. [20].)

yields a highly linear-birefringent fiber of long length whose transverse configuration (Fig. 3.5b) accurately reproduces the original geometry of the preform (Fig. 3.5a), and is maintained along the length of fiber. In Ref. [20], over an 8-km-long fiber that was drawn with the "rods-in-pits" techniques, the two circular stress regions maintained the circular shape all the way without deformation and without asymmetrical aligning with respect to the core. Microphotos of the two end faces of the 8-km fiber (not reproduced here) were given in the original report [20]. The geometry at one end face reproduces the geometry at the other end face so well that one cannot view it without surprise and admiration. It is only natural that the "rods-in-pits" technique was initiated and developed at the NTT Labs, where the VAD method is best known. The so-named "Panda fiber" refers specifically to this kind of fiber whose stress parts are circular-shaped, reminiscent of the two black eyes of a panda.

According to Ref. [20], the 8.3-km-long Panda fiber, with relative refractive-index difference of 0.6%, was measured to have a crosstalk of -23 dB at 1.15 μm, where the loss was 1.7 dB/km. In 1987, a 26-km-long polarization-maintaining Panda fiber was reported [21].

The Corguide

The Corguide is the commercial name for the fibers produced by the Corning Glass Works. In Corning, the linear-birefringent fiber is made by the rods-in-tube technique (Fig. 3.6). This linear-birefringent fiber makes no essential difference as compared with other linear-birefringent fibers made by this technique, except that the stress-rod composition is doped with aluminum

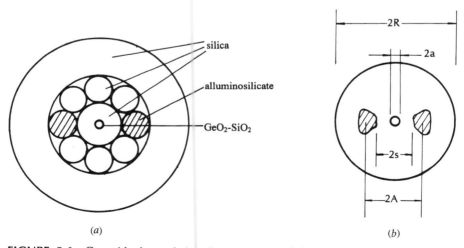

(a)   (b)

FIGURE 3.6 Corguide by rods-in-tube technique. (a) Preform cross section. (b) Resulting fiber cross-section geometry. (From Ref. [23].)

instead of the usual boron. We recall that the initially attempted "Panda" fiber (1981) using the rods-in-tube technique was not as successful as the later attempted Panda fiber (1983) using the rods-in-pits technique. Nevertheless, the Corning technique for polarization-preserving fiber was developed with the rods-in-tube technique [22, 23].

The geometrical aspects of the design of the linear-birefringent Corguide do not appear very special. One point worthy of note is the choices of the aspect ratios of the transverse configuration: the outer aspect ratio was increased over the previous designs to $2R/2A > 2.4$, and the inner aspect ratio was optimized at $s/a = 6$ (typical). But the most strikingly impressive feature of the Corguide is its refractive-index profile across the stress-rod plane, shown in Figure 3.7.

The normalized index difference of the GeO_2 core is $\Delta = 0.6\%$. The hump-shaped portions of the index profile are descriptive of the high level of aluminum doping in the stress rods. Such rods with a very high refractive-index level may form a multimode waveguide, but because they are relatively impure, any light guided along them is attenuated.

The beat length and h-parameter of the aluminum-doped Corguide were reported to be comparable to the corresponding data of the boron-doped fibers. Nevertheless, the attenuation (α-value) of the former was much higher than that of the latter. Because aluminosilicate has a remarkably higher linear thermal expansion coefficient, it was believed that the fiber comprising the stress-regions of this material can be made to achieve a high polarization-holding capability in coiled configurations below 2.0 cm in diameter.

For this type of Corguide (index profile shown in Fig. 3.7), no attempt was made to match the index of the stress rods with that of the claddings, which means that it is not intended to make polarization-preserving fiber devices like coupler and splitter. In such fiber devices composed by two coupled fibers, light in one fiber propagates to the other fiber, and vice versa, such

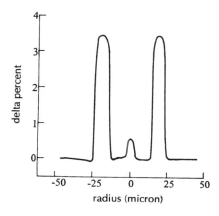

FIGURE 3.7 Index profile across the stress-rod plane. (From Ref. [23].)

that the presence of "index humps" in either fiber becomes a barrier to light transfer. Apparently, should a Corguide be employed in device making, matching the rod index with the cladding index is a necessary design requirement.

The Hitachi's Elliptical Fiber

The Hitachi's elliptical fiber is fabricated using the method of controlled "reduced-pressure" collapse of preform [24–26]. The reduced-pressure approach to core-shaping was initially discovered in 1978 at Bell Labs. This was mentioned briefly in Section 3.3 in connection with Figure 3.1 (see Ref. [7]). In the paper it said: "After completing the deposition steps (of B_2O_3) in this process, the tube was collapsed in the usual manner until the axial core of the preform was about 1 mm in diameter. At this point, the bore was partially evacuated, and the collapse was completed." Nevertheless, this early discovery at Bell Labs was not developed in the same place (Bell Labs) to a level of sophistication. Reference [7] continued: "While application of the vacuum during collapse produced noncircular cores, it did not provide complete dimensional control."

Admirably, at the Hitachi Central Laboratory [26] the reduced-pressure preform-collapsing method was developed afterwards in the early 1980s, and became a sophisticated art that did provide almost complete dimensional control. While this method works in shaping either the core or the clad (or jacket), the Hitachi elliptical fiber discussed here specifically refers to the fiber structure in which the Ge-doped core is kept circular, while the B-doped stress-applying clad is long elliptical, with a thin circular buffer layer introduced between the core and clad to ensure low loss (buffer not shown in Figure 3.8). To avoid possible confusion in terminology, it is noted here that the "buffer-layer" and "clad" in our description were called, respectively, the "clad" and the "jacket" in Ref. [26].

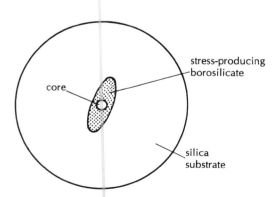

FIGURE 3.8   The Hitachi elliptical fiber.

We know that the kind of elliptical-cladding circular-core fiber had been fabricated earlier at Bell Labs [12]. But there the fiber was fabricated with the aid of the "substrate-tube grinding" technique. This technique does not seem as simple and practical as the reduced-pressure technique developed at Hitachi. More importantly, because the depth of grinding on a substrate tube is very limited, the resulting birefringence cannot be high. Indeed, the Hitachi elliptical fiber makes a distinctive practical fiber version of high birefringence whose configuration is fairly neat, featuring a circular core (and buffer) and long elliptical-cladding within a silica substrate with a circular outer periphery.

In the Hitachi elliptical fiber, the B-doped portion covers a continuous medium. By intuitive reasoning it is conceivable that a tube-shaped region of the stress-applying material will not contribute to the resulting birefringence. This tube-region has an inner periphery coinciding with the outer periphery of the buffer layer, and an outer circular periphery touching the edges of the cladding ellipse along the minor axis. The reason is self-explanatory: a circularly symmetrical region of any material cannot impose on the circular core (and buffer) a linear or other kind of birefringence. This was proved by calculating the stress birefringence in fibers of different configurations using the infinitesimal-element method. Moreover, the calculation provided insights into the birefringence contributions due to continuous as well as discrete regions. In particular, it was shown that a buffer layer around the core will not reduce the birefringence at the core, and that a relatively small stress region of higher doping is preferable to a large stress region of moderate doping.

Nevertheless, according to Ref. [26], the minor dimension of the clad-ellipse cannot be too small in order to ensure circularity of the core (and buffer) in the resulting preform after the reduced-pressure collapsing process. Considering the two factors together (efficient use of stress materials and keeping the core round), the minor dimension of the cladding ellipse was found to be optimum within a certain range. As a design criterion, the inverse ratio of this dimension to the dimension of the core (and buffer) was required to be smaller than a critical value, being typically within the range of 0.3–0.6, as cited in Ref. [5]. This restriction perhaps reflects the not uncommon circumstance that the simplicity and neatness in the reduced-pressure technique is gained at the unavoidable cost of consuming some extra cladding material that does not contribute to the resulting linear birefringence.

The mechanism of the reduced-pressure technique that plays a role in producing a fiber with circular core (and buffer) and elliptical cladding within the silica substrate is described in Ref. [26]. The starting silica tube is relatively thick, such that after being layered inside, the outer side of the layered tube is little influenced by the pressure reduction. Moreover, during preform collapsing the temperature gradient is initially higher at the outer side than it is at the inner side. The outer side of the preform therefore remains circular-shaped due to surface tension. However, at the inner side of the layered tube during collapse, the reduction in pressure plays a dominant role. As the high temperature continues, the inner side is liable to become deformed. This is just like a rubber ball from which air is released. The B-doped cladding layer has a lower softening point than the core (and buffer) layer. Immediately after collapsing, the clad layer becomes elliptical, but the viscosity of this layer gradually lowers down. The core (and buffer) floating in this increasingly less viscose clad are forced by the surface tension to be circular-shaped. Accordingly, the accurate and reproducible yield of the Hitachi elliptical fiber requires careful dimensioning of the fiber geometry and precise control of the collapsing pressure, as well as deliberate selection of the softening points of the core and cladding material.

3.6   THE POLARIZING FIBER

The afore-described birefringent fiber varieties refer to polarization-maintaining fibers. A fiber of this kind supports either an x-polarized light (by an x-oriented excitation) or a y-polarized light (by a y-oriented excitation). We now come to a more challenging topic referred to as the SPSM (Single Polarization Single Mode) fiber, in which only a single desired polarized light is supported. The other undesired polarized light, wherever occurring, will be attenuated and lost due to some kind of "leaky" mechanism called the "polarizing" effect. The SPSM fiber is therefore a real single-mode fiber in the strict sense. Here we note that, in some of the early published papers on linear-birefringent fibers, the relevant terminology may not be precise enough to differentiate a *polarizing* fiber from a polarization-maintaining fiber. In such case, we will therefore have to judge what kind of fiber (whether polarizing or polarization-maintaining) is being discussed, not according to the terminology, but according to how the fiber does behave.

During early 1980s, the attempts to make polarizing fibers achieved initial successes by two approaches, namely, the W-tunneling flat-fiber approach adopted at Bell Labs and the Bow-Tie bent-fiber approach established at Southampton. The polarizing mechanisms associated with these two approaches are distinctively different. They were developed almost simultaneously, though independently of each other.

The Flat-Model Approach to Polarizing Fiber

The structural geometry of a flat polarizing fiber [27] is just like that of a flat polarization-maintaining fiber described in Section 3.5. There are two elliptical-shaped claddings (inner and outer) surrounding the core. The inner cladding is B-doped, and is so highly flattened that the narrow sides almost touch the circular-shaped core (or the thin buffer layer, if existing). The outer cladding is doped with fluorine. The index profiles along the major and minor principal axes are sketched in Figure 3.9.

major axis

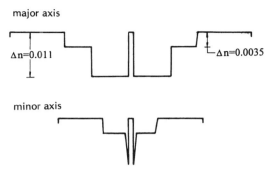

$\Delta n = 0.011$

$\Delta n = 0.0035$

minor axis

FIGURE 3.9 Index profiles of a flat polarizing fiber: (*a*) along major axis and (*b*) along minor axis. (From Ref. [27].)

The two drawings in Figure 3.9 are reminiscent of two types of conventional double-clad fiber of circular geometry, namely the "depressed-cladding" fiber and the "W-type" fiber. The distinction between these two fibers is not clear-cut. Generally, a depressed-cladding fiber refers to one characterized by a wide and shallow depressed-clad, while a W-fiber refers to one characterized by a narrow and deep depressed-clad.

It is therefore reasonable to deal with a flat polarizing fiber as one that behaves like a depressed-cladding fiber for polarization along its major axis, but like a W-fiber with an intentionally exaggerated depression for polarization along its minor axis. To make understanding the flat polarizing fiber easier, a look at the relevant concepts relating to the doubly-clad fibers follows.

From the viewpoint of physics, a mode is said to be guided when its field outside of the fiber core decays exponentially as a function of radial distance. On the other hand, the modal field can form a radial traveling wave in the outer cladding after an initial exponential decay near the core. When this happens, the mode is said to be cut off. Thus, the term "cutoff" defined in a physical model does not imply that power cannot be transmitted along the fiber, but simply means that, because the power is not totally trapped within the light guide, some power can leak out and be lost by the process of radiation. For the sake of convenience in description and analysis, a term called the "modal-effective index," or simply the "effective index," is defined as

$$n_e = \frac{\lambda}{\lambda_g} \tag{3.13}$$

where λ is the plane wavelength of the signal in free space, and λ_g is the modal wavelength of the same signal in the fiber.

In a simple step-index fiber, the wave power does not become cut off even at arbitrarily long wavelengths. This is because the refractive index of the core is greater than that of the surrounding claddings for signals of all wavelengths within the range of the signal wavelength of interest. In contrast, in a depressed cladding fiber the effective index of the fiber tends to decrease as the signal wavelength increases until a value of effective index is reached that is less than that of the outer cladding. When this occurs, the fiber is cut off with respect to that wavelength and longer wavelength signals.

The occurrence of the cutoff is due to the wavelength-dependent attribute of the effective index. At shorter wavelengths, the field is concentrated in the core region, such that the effective index is close to that of the core region and is greater than the index of the outer cladding. For longer wavelengths, the field is not as tightly bound to the core, but has spread into the region of the inner cladding. Because the index of the inner cladding is less than that of both the core and the outer cladding, the effective index at the longer wavelengths tends to decrease and eventually reaches a value that is less than the index of the outer cladding. The signal then radiates or leaks.

It was the understanding of the above-described loss mechanism that led to the discovery that, in a double-clad single-mode fiber with a depressed inner cladding, the high loss region at the longer wavelengths can be made to move out of the useful range by letting the ratio of clad-radius to core-radius be significantly large, say 6.5:1. Thus, if this ratio is made exaggeratively small, the mode is likely to suffer from high leakage losses analogous to the tunneling effect that occurs, through the inner cladding, between the power trapped inside the core and the leakage through the outer cladding region.

The flat polarizing fiber is exactly the desired structure that behaves like a low-loss depressed-cladding fiber for polarization along its major principal axis, but becomes leaky for polarization along its minor principal axis. Such fiber because of its leaky mechanism is therefore called the "W-tunneling" fiber.

In the original design framework the inner cladding in the narrow dimension was made almost touching the core, and was highly doped with boron such that the stress-induced splitting of the two perpendicular polarizations of the fundamental mode was large enough to ensure W-tunneling in one polarization (the one along the narrow dimension), and simultaneously to ensure low-loss propagation in the other polarization along the major axis of the flat-fiber.

At the very early stage, when many factors involved in the polarizing mechanism were still confusing, a fact of basic importance was pointed out by Simpson et al. [27] that, "It is the difference in birefringence between the core and outer cladding that provides the difference in cutoff wavelength for the two polarizations." With respect to the said flat-fiber structure, the calculated differential stress $(\sigma_x - \sigma_y)$ of the core and clad is uniform throughout the core and declines rapidly in the narrow direction. Since the local birefringence is proportional to $(\sigma_x - \sigma_y)$, the requirement of concentration of birefringence in the core is thus satisfied.

According to Simpson et al. [27], the wavelength for useful operation can be tuned by bending the fiber. This is another basic finding that is useful in application. But this bend-induced shifting of the polarizing range does not imply that the W-tunneling effect evidenced in the initial experiment required a curvature for proper operation. As a matter of fact, a W-tunneling polarizing fiber does not have to be bent to cause the leaky losses. That the W-tunneling effect can happen in a really straight flat fiber of specific design is not only well grounded on physical principle, but more importantly, it was experimentally evidenced in the actual realization of a short W-tunneling fiber polarizer (4.8 cm long) mounted in a capillary tube to ensure that there was no curvature [28]. Measurements on this short, straight polarizing fiber showed a -39-dB polarization extinction ratio over a useful bandwidth of 4% centered at 0.6328 μm.

The Bend-Induced Polarizing Mechanism

The bend-induced polarizing effect in Bow-Tie fiber was discovered at Southampton almost simultaneously with the discovery of the flat polarizing fiber at Bell Labs. Deterministic experiments made at Southampton revealed that no polarizing effect could be found in a straight fiber of the Bow-Tie version, while bent Bow-Tie fibers would behave as polarizers when the bend radius was sufficiently reduced [29, 30]. Differential bend loss for the two polarized modes was thus shown to be the predominant polarizing mechanism in Bow-Tie fibers.

The said experimental discovery at Southampton had a theoretical background that deserves to be briefly discussed here, because of its scientific value. The early 1980s witnessed the advent of a multitude of practically useful linear-birefringent fibers by different technological approaches. At the initial stage of this advance when linear-birefringent fibers were mostly of the polarization-maintaining type, Snyder and Ruhl at Canberra [31] made a purely theoretical prediction that, in fibers composed of highly birefringent material, a polarizing mechanism exists due to the leaky mode effect associated with the "small field." Recall that in Section 1.2, the "small field" refers to the tiny curved portion of the electric lines of force of the fundamental mode. The theoretical prediction [31] was based on the presumption that, in a step-index model of a birefringent fiber, the index profile for one polarized mode (say y) can be described by a lowered step-function with respect to that of the other polarized mode x. Consequently, the tiny x-field of the y-mode "sees" a cladding index that is too high to have the field guided, and hence leaks its power.

There was an intense exchange between Canberra and Southampton during that period (1982–1983). Almost immediately after Snyder and Ruhl's prediction, Varnham et al. [29] published their experimental results on the realization of single-polarization operation in Bow-Tie fibers with high levels of stress-induced birefringence. While the waveguide-theory formulation by Snyder and Ruhl was briefly outlined in this paper in explaining the polarizing behavior of Bow-Tie fibers, the Southampton scientists suggested the existence of a bending-related mechanism, which was borne out by their experimental observation that modes leak prematurely at wavelengths shorter than those predicted by the "small-field" leaky-mode theory. They also suggested that the leaky edges could be shifted in wavelength simply by coiling the fiber with a different bending curvature. Elaborate experimentation in a somewhat circuitous way at Southampton eventually led to the conclusion [30] that it was the differential bend loss for the two polarized modes, not the "small field," that was mainly responsible for the polarizing effect in Bow-Tie fibers.

The working principle of the bend-induced polarizing effect in a highly birefringent fiber is attributed to the large internal stress in the fiber that makes one of the two polarized modes a little more strongly guided. When the fiber is bent, a differential bend loss occurs that can be enhanced to the extent that, through a judicious choice of the bend radius and the operating wavelength, one polarized mode becomes leaky while the other polarized mode keeps propagating.

That a birefringent fiber is able to polarize light simply by bending is indeed a very significant finding. Using this method, compact polarizers can be structured by winding Bow-Tie fibers into a close-wound multiturn coil. [32]. A silicone adhesive is applied, which allows the coil to be removed from the form to become a free-standing coiled unit. Measurements showed that such coiled-birefringent-fiber polarizers have well-controlled bending characteristics and permit mode coupling caused by winding tension to be virtually eliminated.

The curves in Figure 3.10 are typical to show the wavelength dependence of the extinction-ratio and the guided-mode loss measured in coiled-fiber polarizers within a fairly wide temperature range (say $-63°$ to $+140°$).

With all the successes achieved at Southampton in understanding the bend-induced polarizing mechanism in Bow-Tie fibers, there were left-over points that still appeared confusing to an analyst. For example, the question of whether the "small field" contributes to the overall polarizing effect remained unanswered.

The technical background for the R & D of polarizing fibers during the early to mid-1980s encouraged Chen and Huang to attempt a relatively comprehensive analytic study of the Bow-Tie fiber, with particular emphasis on quantification of different kinds of loss [33, 34].

Preliminary to this study is the derivation of the mathematical expressions for field components in a Bow-Tie fiber. As expected, the mathematical derivation is fairly lengthy and tedious, and is therefore not fitting to be included in the present text. In brief, the mathematical method employed in

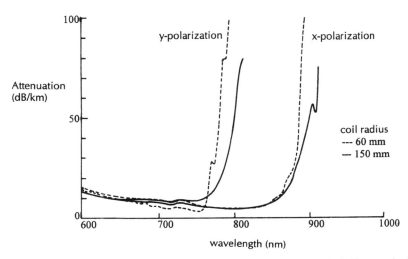

FIGURE 3.10 Wavelength-dependence characteristics of a coiled-fiber polarizer. (From Ref. [32].).

the work is a kind of perturbation-iterative technique that integrates the well-known Green's-function methods due to Snyder et al. in an iterative process. Marcuse's loss formulas are also used in the course of the derivation.

To our satisfaction, the numerical calculations of the analytic solutions yielded results that conform well with the experimental data available at the time of the research. It was found in this analytic study that, in all practical cases of unequal core and cladding anisotropy, losses are predominantly due to the main field, not the "small field." The contribution of the "small-field" losses is negligibly small according to the calculated results. The early prediction by Snyder and Ruhl that the "small field" leaky loss plays a role was confirmed to be *theoretically* correct on the presumption of equal anisotropy in core and clad. Since this presumption is not true in practice, the predicted behavior naturally cannot be found in experiments on actual Bow-Tie fibers.

The analytic study of the bent Bow-Tie fiber showed that, in the actual case of unequal core and cladding anisotropy, there exists a single-polarization window that shifts toward the shorter-wavelength range as the bend radius decreases. Further, the analytic study also showed that it is the inequality of the core anisotropy and the cladding anisotropy that is responsible for the existence of a single-polarization window, and the larger the inequality, the wider the window will be.

View of W-Tunneling and Bend-Induced Polarizing Mechanisms

The key result derived in the analytic study makes the kind of bent birefringent polarizing fibers look very similar to the kind of W-tunneling polarizing

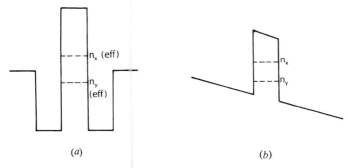

FIGURE 3.11   Sketches of index profiles: (*a*) for W-tunneling and (*b*) for bending. (From an unpublished note by R. H. Stolen [35].)

fibers [27]. In both cases, in order to achieve proper polarizing behavior, it is necessary not to have birefringence in the cladding. Since bending a fiber can be viewed as tilting the index curve, it turns out to be possible to view the bend-inducing polarizing effect in the same way that the W-tunneling polarizing effect is viewed [35]. The index profiles sketched in Figure 3.11 illustrate this point.

In Figure 3.11(*a*), n_1, n_2, and n_3 are the refractive indices of the core, the inner clad, and the outer clad, respectively, in the kind of flat fiber shown in Figure 3.9. At some wavelength, polarized light along the broad dimension of fiber has an effective modal index n_x higher than the outer cladding index so that it propagates in the way of a guided mode. On the other hand, polarized light along the narrow dimension of fiber has an effective modal-index n_y lower than the outer cladding index so that it leaks through W-tunneling. This simple intuitive way of viewing the polarizing effect is analogous to the way that Snyder and Ruhl used in describing the "small field" behavior [27]. But there is a fundamental difference between Ref. [27] and Figure 3.11(*a*) [35]. In the former, the theoretically predicted polarizing effect is attributed to the small field, while in the latter, only the main field needs to be considered.

The tilted index profile shown in Figure 3.11(*b*) is descriptive of a bent Bow-Tie fiber. Because of this tilt of the index profile caused by bending the fiber, the difference between the levels of the effective indices n_x and n_y is enhanced to such an extent that, while one of these effective indices (say n_x) remains higher than the height of the tilt, the other effective index n_y becomes lower than this height. The bend-induced polarizing mechanism is thus self-explanatory. It is also self-explanatory that the strength of the polarizing effect of a bent Bow-Tie fiber can be increased simply by decreasing of the bending radius, which causes a more slanting tilt of the index profile.

With the aid of Figure 3.11, the principle of operation of the two types of polarizing mechanism can be viewed in an integrally consistent way, while recognizing that the W-tunneling mechanism does polarize light in straight fiber not requiring a curvature.

3.7 NOTES

The technology of making linearly birefringent fibers of the "polarization-maintaining" type has been essentially well established since the early 1980s. Relatively long fiber lengths (say from 500 m up, useful in a gyroscope for example) are achievable using different fabrication techniques. For fiber of longer lengths (in terms of tens of km for a transmission line), the VAD-based Panda fiber appears most promising.

So far, practical polarization-maintaining fibers fall exclusively into the stress-induced birefringence category. According to the distribution in the stress region (or regions) in fiber, polarization-maintaining fibers can be grouped according to whether this distribution is continuous or discrete. The latter group (of discrete stress regions) includes the Bow Tie, the Panda, and the Corguide fibers. Example of the former group (of continuous stress regions) is provided by the Hitachi's elliptical fiber. The flat fiber is featured by a stress distribution that is usually continuous, but that can become quasi discrete by making the narrow side of the flat-shaped stress region almost touch the core (or touch the thin layer of buffer surrounding the core). If only the factor of efficient use of stress material is considered, fibers of discrete stress regions are superior. But there are other factors that must be considered in qualifying a specific fiber version.

The unique feature of flat fiber is the ease of principal-axis location, as well as the self-aligning property of the flat shape when being coiled around a form. For other fiber versions of round outer periphery, principal-axes aligning poses a technical problem. In view of easier location and aligning of the principal axes, it is therefore feasible to reshape an otherwise round preform to make it flat, as far as the original structural compositions allow this kind of re-shaping. The flat Panda is a very special fiber version of this kind.

While the fiber versions addressed in this chapter have already achieved high levels in polarization-maintaining capability, further enhancement of the linear birefringence can be conceived using more sophisticated designs of the structural details. Whether or not a design of such kind is practical basically depends not on theory, but on the practicality of the fabrication processes that are required to actually make the fiber.

Almost one and half decades have elapsed since Snyder organized the special Panel on SPSM fibers [36], probably the first international panel devoted to this subject. Despite the remarkable progress that has been achieved throughout the world, the attempted objective (real SPSM fiber, or polarizing fiber) has been approached only half way. The present state of the art is that only short fibers of this kind can be fabricated. Such short polarizing fibers, called "polarizers," have been successively used as lumped devices in various fiber-optic circuitry and systems.

In the course of the R & D of special fiber optics, a number of compact polarizers have been proposed or fabricated by exaggerating the attenuation difference $\delta\alpha = |\alpha_x - \alpha_y|$, rather than the phase-velocity difference $\delta\beta$ of the two polarization modes. But this very specific topic area is beyond our present interest, and so not included in the text.

To conclude, it is worth noting that for the W-tunneling effect to occur in fiber, it is not entirely necessary that the outer geometry of fiber be flat-shaped. What is important is the special index profiles along the two principal axes of the fiber. A fiber of round outer geometry can be made to be W-tunneling, and hence polarizing, if its index profiles are specifically and sophisticatedly designed [37]. Efforts toward achieving long polarizing fiber that behaves as a lightwave transmission line remains an enchanting research direction.

REFERENCES

Review Papers

[1]   I. P. Kaminow, "Polarization in optical fibers," *IEEE J. Quantum Electron.*, vol. QE-17, pp. 15–22 (1981).

[2]   T. Okoshi, "Single-polarization single-mode optical fibers," *IEEE J. Quantum Electron.*, vol. QE-17, pp. 879–884 (1981).

[3]   D. N. Payne, A. J. Barlow, and J. J. R. Hansen, "Development of low- and high-birefringence optical fibers," *IEEE J. Quantum Electron.*, vol. QE-18, pp. 477–488 (1982).

[4]   S. C. Rashleigh, "Origins and control of polarization effects in single-mode fibers," *IEEE J. Lightwave Technol.*, vol. LT-1, pp. 312–331 (1983).

[5]   J. Noda, K. Okamoto, and Y. Sasaki, "Polarization-maintaining fibers and their applications," *IEEE J. Lightwave Technol.*, vol. LT-4, pp. 1071–1089 (1986).

Geometrical Birefringence

[6]   V. Ramaswamy and W. G. French, "Influence of noncircular core on the polarisation performance of single mode fibers," *Electron. Lett.*, vol. 14, pp. 143–144 (1978).

[7]   V. Ramaswamy, W. G. French, and R. D. Standley, "Polarization characteristics of noncircular core single-mode fibers," *Appl. Opt.*, vol. 17, pp. 3014–3017 (1978).

Stress Birefringence

[8]   R. H. Stolen, V. Ramaswamy, P. Kaiser, and W. Pleibel, "Linear polarization in birefringent single-mode fibers," *Appl. Phys. Lett.*, vol. 33, pp. 699–701 (1978).

[9]   V. Ramaswamy, I. P. Kaminow, P. Kaiser, and W. G. French, "Single polarization optical fibers: Exposed cladding technique," *Appl. Phys. Lett.*, vol. 33, pp. 814–815 (1978).

[10] I. P. Kaminow and V. Ramaswamy, "Single-polarization optical fibers: Slab model," *Appl. Phys. Lett.*, vol. 34, pp. 268–270 (1979).

[11] I. P. Kaminow, J. R. Simpson, H. M. Presby, and J. B. MacChesney, "Strain birefringence in single-polarisation germanosilicate optical fibres," *Electron. Lett.*, vol. 15, pp. 677–679 (1979).

[12] V. Ramaswamy, R. H. Stolen, M. D. Divino, and W. Pleibel, "Birefringence in elliptically clad borosilicate single-mode fibers," *Appl. Opt.*, vol. 18, pp. 4080–4084 (1979).

[13] R. H. Stolen, R. E. Howard, and W. Pleibel, "Substrate-tube lithography for optical fibres," *Electron. Lett.*, vol. 18, pp. 764–765 (1982).

The Flat Fiber

[14] R. H. Stolen, W. Pleibel, and J. R. Simpson, "High-birefringence optical fibers by preform deformation," *IEEE J. Lightwave Technol.*, vol. LT-2, pp. 639–641 (1984).

[15] J. R. Simpson, R. H. Stolen, A. J. Ritger and, H. T. Shang, "Properties of rectangular polarizing and polarization-maintaining fiber," *SPIE*, vol. 719, pp. 220–225 (1986).

[16] M. W. Shute, C. S. Brown, and A. J. Ritger, "A study of the polarization properties of AT & T's rectangular polarization-maintaining fiber," *SPIE*, vol. 841, pp. 358–366 (1988).

The Bow Tie

[17] R. D. Birch, D. N. Payne, and M. P. Varnham, "Fabrication of polarisation-maintaining fibres using gas-phase etching," *Electron. Lett.*, vol. 18, pp. 1036–1038 (1982).

[18] M. P. Varnham, D. N. Payne, A. J. Barlow, and R. D. Birch, "Analytic solution for the birefringence produced by thermal stress in polarization-maintaining optical fibers," *IEEE J. Lightwave Technol.*, vol. LT-1, pp. 332–339 (1983).

The Panda

[19] T. Hosaka, K. Okamoto, T. Miya, Y. Sasaki, and T. Edahiro, "Low-loss single polarisation fibres with asymmetrical strain birefringence," *Electron. Lett.*, vol. 17, pp. 530–531 (1981).

[20] Y. Sasaki, T. Hosaka, K. Takada, J. Noda, "8 km-long polarization-maintaining fibre with highly stable polarisation state," *Electron. Lett.*, vol. 19, pp. 792–794 (1983).

[21] Y. Sasaki, K. Tajima, and S. Seikai, "26 km-long polarization-maintaining optical fibre," *Electron. Lett.*, vol. 23, pp. 127–128 (1987).

The Corguide

[22] P. E. Blaszyk, R. M. Hawk, and M. J. Marrone, "Polarization-retaining single-mode fiber with improved coil performance," *SPIE*, vol. 566, pp. 381–386 (1985).

[23] M. J. Marrone, S. C. Rashleigh, and P. E. Blaszyk, "Polarization properties of birefringent fibers with stress rods in the cladding," *IEEE J. Lightwave Tech.*, vol. LT-2, pp. 155–160 (1984).

The Hitachi's Elliptical Fiber

[24] T. Katsuyama, H. Matsumura, and T. Suganuma, "Low-loss single-polarization fibres," *Electron. Lett.*, vol. 17, pp. 473–474 (1981).

[25] T. Katsuyama, H. Matsumura, and T. Suganuma, "Low-loss single polarization fibers," *Appl. Opt.*, vol. 22, pp. 1741–1747 (1983).

[26] T. Katsuyama, H. Matsumura, and T. Suganuma, "Reduced pressure collapsing MCVD method for single polarization optical fibers," *IEEE J. Lightwave Technol.*, vol. LT-2, pp. 634–639 (1984)

Polarizing Fiber and Analytic Topics

[27] J. R. Simpson, R. H. Stolen, F. M. Sears, W. Pleibel, J. B. MacChesney, and R. E. Howard, "A single polarization fiber," *IEEE J. Lightwave Technol.*, vol. LT-1, pp. 370–374 (1983).

[28] R. H. Stolen, W. Pleibel, J. R. Simpson, W. A. Reed, and G. Mitchell, "Short W- tunnelling fibre polarizers," *Electron. Lett.*, vol. 24, pp. 524–525 (1988).

[29] M. P. Varnham, D. N. Payne, R. D. Birch, and E. J. Tarbox, "Single polarization of highly birefringent bow-tie optical fibres," *Electron. Lett.*, vol. 19, pp. 246–247 (1983).

[30] M. P. Varnham, D. N. Payne, R. D. Birch, and E. J. Tarbox, "Bend Behaviour of polarizing optical fibres," *Electron. Lett.*, vol. 19, pp. 679–680 (1983).

[31] A. W. Snyder and F. Ruhl, "New single-mode single-polarization optical fibre," *Electron. Lett.*, vol. 19, pp. 185–186 (1983).

[32] M. P. Varnham, D. N. Payne, A. J. Barlow, and E. J. Tarbox, "Coiled-birefringent-fiber polarizers," *Opt. Lett.*, vol. 9, pp. 306–308 (1984).

[33] Y. Chen and H. C. Huang, "Analytic solution of bow-tie fibres," *Electron. Lett.*, vol. 22, pp. 713–715 (1986).

[34] Y. Chen and H. C. Huang, "Macrobends of bow-tie fibres with constant curvature," *Electron. Lett.*, vol. 23, pp. 159–160 (1987).

[35] R. H. Stolen, "View of W-tunneling and bent-induced polarizing mechanisms," (unpublished, 1987).

[36] A. W. Snyder (organizer), "Panel Discussion on SPSM fibers," in Int. Symp. Opt. Waveguide Sciences held 1983 in Guilin, China, published in double special-issue of *Applied Scientific Research*, Martinus Nijhoff, pp. 365–368 (1984).

[37] M. J. Messerly, J. R. Onstott, and R. C. Mikkelson, "A broad-band single polarization optical fiber," *IEEE J. Lightwave Technol.*, vol. LT-9, 817–820 (1991).

BIBLIOGRAPHY

M. J. Adams, D. N. Payne, and C. M. Ragdale, "Birefringence in optical fibres with elliptical cross-section," *Electron. Lett.*, vol. 15, pp. 298–299 (1979).

C. G. Askins and M. J. Marrone, "Technique for controlling the internal rotation of principal axes in the fabrication of birefringent fibers," *IEEE J. Lightwave Technol.*, vol. LT-6, pp. 1402–1405 (1988).

A. J. Barlow and D. N. Payne, "The stress-optic effect in optical fibers," *IEEE J. Quantum Electron.*, vol. QE-19, pp. 834–839 (1983).

Y. Chen and H. C. Huang, "Anisotropic fiber with a depressed-cladding profile," *Opt. Lett.*, vol. 12, pp. 279–280 (1987).

Y. Chen and H. C. Huang, "Microbending losses of bow-tie and similar anisotropic fibres," *Electron. Lett.*, vol. 23, pp. 157–159 (1987).

Y. Chen and H. C. Huang, "Microbending and macrobending behavior of bow-tie fibers and similar anisotropic optical fibers," *J. Opt. Soc. Am. A*, vol. 5, pp. 380–386 (1988).

P. L. Chu and R. A. Sammut, "Analytical method for calculation of stresses and material birefringence in polarization-maintaining optical fiber," *IEEE J. Lightwave Technol.*, vol. LT-1, pp. 650–662 (1983).

R. B. Dyott, J. R. Cozens, and D. G. Morris, "Preservation of polarisation in optical-fibre waveguides with elliptical cores," *Electron. Lett.*, vol. 15, pp. 380–382 (1979).

W. Eicknoff, "Stress-induced single-polarization single-mode fiber," *Opt. Lett.*, vol. 7, pp. 629–631 (1982)

M. Fontaine, B. Wu, V. P. Tzolov, W. J. Bock, and W. Urbanczyk, "Theoretical and experimental analysis of thermal stress birefringent optical fiber," *IEEE J. Lightwave Technol.*, vol. LT-14, pp. 585–591 (1996).

T. Hinata, S. Furukawa, N. Namatame, and S. Nakajima, "A single-polarization optical fiber of hollow pit type with zero total dispersion at wavelength of 1.55 μm," *IEEE J. Lightwave Technol.*, vol. LT-12, pp. 1921–1925 (1994).

T. Hosaka, K. Okamoto, Y. Sasaki, and T. Edahiro, "Single mode fibres with asymmetrical refractive index pits on both sides of core," *Electron. Lett.*, vol. 17, pp. 191–193 (1981).

Kin Seng Chiang, "Stress-induced birefringence fibers designed for single-polarization single-mode operation," *IEEE J. Lightwave Technol.*, vol. LT-7, pp. 436–441 (1989).

F. Kitamura and Y. Sasaki, "Optimum design for polarization crosstalk reduction in a polarization maintaining optical fiber of Panda profile," *Electronics and Communications in Japan*, Pt. 2 (*Electronics*), vol. 78, pp. 19–27 (1995).

J. D. Love, R. A. Sammut, and A. W. Snyder, "Birefringence in elliptically deformed optical fibres," *Electron. Lett.*, vol. 15, pp. 615–616 (1979).

D. Marcuse, "Simplified analysis of a polarizing optical fiber," *IEEE J. Quantum Electron.*, vol. QE-26, pp. 550–557 (1990).

M. J. Marrone and M. A. Davis, "Low-temperature behaviour of high-birefringence fibres," *Electron. Lett.*, vol. 21, pp. 703–704 (1985).

K. Mochizuki, Y. Namihira, and Y. Ejiri, "Birefringence variation with temperature in elliptically cladded single-mode fibers," *Appl. Opt.*, vol. 21, pp. 4223–4228 (1982).

K. Okamoto, "Single-polarization operation in highly birefringent optical fibers," *Appl. Opt.*, vol. 23, pp. 2638–2642 (1984).

K. Okamoto, T. Hosaka, and J. Noda, "High-birefringence polarizing fiber with flat cladding," *IEEE J. Lightwave Technol.*, vol. LT-3, pp. 758–762 (1985).

T. Okoshi and K. Oyamada, "Single-polarization single-mode optical fibre with refractive-index pits on both sides of core," *Electron. Lett.*, vol. 16, pp. 712–713 (1980).

T. Okoshi, K. Oyamada, M. Nishimura, and H. Yokota, "Side-tunnel fobre: an approach to polarization-maintaining optical waveguiding scheme," *Electron. Lett.*, vol. 18, pp. 824–826 (1983).

S. C. Rashleigh, "Wavelength dependence of birefringence in highly birefringent fibers," *Opt. Lett.*, vol. 7, pp. 294–296 (1982).

S. C. Rashleigh, W. K. Burns, and R. P. Moeller, "Polarization holding in birefringent single-mode fibers," *Opt. Lett.*, vol. 7, pp. 40–42 (1982).

S. C. Rashleigh and M. J. Marrone, "Temperature dependence of stress birefringence in an elliptically clad fiber," *Opt. Lett.*, vol. 8, pp. 127–129 (1983).

S. C. Rashleigh and M. J. Marrone, "Polarisation holding in coiled high-birefringence fibres," *Electron. Lett.*, vol. 19, pp. 850–851 (1983).

S. C. Rashleigh and R. Ulrich, "High birefringence in tension-coiled single-mode fibers," *Opt. Lett.*, vol. 5, pp. 354–356 (1980).

F. F. Ruhl and D. Wong, "True single polarization design for bow-tie optical fibers," *Opt. Lett.*, vol. 14, pp. 648–650 (1989).

Shangyuan Huang and Zongqi Lin, "Measuring the birefringence of single-mode fibers with short beat length of nonuniformity: a new method," *Appl. Opt.*, vol. 24, pp. 2355–2361 (1985).

Chao-Xiang Shi and Rong-Qing Hui, "Polarization coupling in single-mode single-polarization optical fibers," *Opt. Lett.*, vol. 13, pp. 1120–1122 (1988).

N. Shibata, K. Okamoto, and Y. Sasaki, "Structure design for polarization-maintaining and absortion-reducing optical fibers," *Review of the Electrical Comm. Labs.*, vol. 31, No. 3, pp. 393–399 (1983).

A. W. Snyder and F. Ruhl, "Practical single-polarization anisotropic fibres," *Electron. Lett.*, vol. 19, pp. 687–688 (1983).

R. H. Stolen, "Calculation of stress birefringence in fibers by an infinitesimal element method," *IEEE J. Lightwave Technol.*, vol. LT-1, pp. 297–301 (1983).

K. Tajima, M. Okoshi, and Y. Sasaki, "Polarization-maintaining optical fibres with AL_2O_3 stress-applying parts," *Electron. Lett.*, vol. 24, pp. 634–635 (1988).

Xing Ma and P. L. Chu, "New type of temperature-compensated birefringent optical fiber," *Tech. Digest CLEO*'94 (Conference on Lasers and Electro-optics), vol. 8, pp. 16–17 (1994).

CHAPTER FOUR

Twist-Induced Circular Birefringence in Fiber

Twisting a fiber induces circular birefringence in the fiber proportional to the twist rate. Nevertheless, a twisted fiber generally does not behave circularly birefringent, but is more likely to behave elliptically birefringent. This is because the fiber being twisted also has an inherent linear birefringence superimposed on the twist-induced circular birefringence, so that the resultant birefringence is elliptical. Only when the twist rate is sufficiently strong that the twist-induced circular birefringence far exceeds the inherent linear birefringence does the twisted fiber practically become a circularly birefringent fiber. Therefore, when dealing with the topic of twisted fiber we are actually concerned with a fairly wide range of birefringence in fiber. For convenience, two types of problem will be treated. One is the strongly twisted fiber, which approximates a fiber of circular birefringence; the other is the slightly twisted fiber, which is elliptically birefringent with different elliptical birefringence.

This chapter aims at a relatively comprehensive study of the circular birefringence produced by twisting a fiber in its "cool" state after the linear draw. In particular, we emphasize the fundamental difference between the effect of postdraw twisting (in the cool state) and the effect of on-drawing spinning (in the hot state), the latter having been discussed in Chapter 2.

Actually, the process of twisting does not make a "permanent" waveguiding structure that meets the usual qualifications of a special fiber version. The twisted state is not a built-in state, but can persist only under constraint. Once this constraint is removed, and the fiber is freed, the fiber will immediately spring back to its original untwisted state. Although such fiber is not a special kind of fiber in the strict sense, it is still called "twisted-fiber" for convenience.

Coupled-mode formulation in terms of *circular* base modes will be introduced in this chapter. However, before we get to circular base-mode formulation, we shall continue using the more familiar linear base mode in formulating coupled-mode equations of twisted-fiber problems. Generally, the choice of linear or circular base modes will make no difference in describing the transmission characteristics of fiber. The circular base-mode formulation is useful only for those problems in which the dominating mode in fiber is essentially a circular light.

4.1 TECHNICAL BACKGROUND

During the late 1970s or early 1980s, when the technology of making conventional optical fiber was fairly well established, attempts were begun to make polarization-maintaining fibers because of the need for a stable mode, or a stable polarization, during the entire course of transmission. While linear light (linearly polarized light) has been used almost exclusively in fiber-optic practice, the advantages of optical transmission using circular light (circularly polarized light) were recognized early by scientists engaged in this art. For the past 15 years or more, efforts along these two lines (one toward searching the linear polarization-maintaining fibers, and the other toward searching the circular polarization-maintaining fibers) were made almost in parallel.

During this time the research and development (R & D) of special fiber optics came to favor the linear polarization-maintaining fiber, as evidenced by the advent of a multitude of special fibers of this kind: the Panda, the Bow Tie, the elliptical-cladding fiber, and the flat fiber. The preceding chapter gave an account of this most fruitful research area.

On the other hand, in spite of the assiduous efforts made in devising specialized fiber structures for circular polarization-maintenance, achievements in this parallel area were far less fruitful. From the practical viewpoint, the only means capable of producing circular birefringence in fiber remains the early way of twisting a drawn-up conventional fiber in its cool state [1]. Only recently (by September 1995) was a U.S. patent of invention granted to me on a practical circular polarization-maintaining fiber. This specific subject will be dealt with separately and in adequate detail in Chapter 5.

The contents of this chapter are exclusively concerned with the twisted fiber with both strong and slight twisting. A careful and serious discussion on the major aspects of this kind of special fiber is an inseparable part of this book. Aside from its archival value, an understanding of the twist-induced characteristics is relevant to the analysis of any special fiber problem involving some kind of rotation in the fiber structure.

In retrospect, one major impetus for the study of twisted fiber came from the understanding that a fiber-optic system supporting circular light would be particularly attractive for coherent optical transmission. During the early 1980s, while the exploitation of linear hi-bi fibers achieved remarkable initial successes, the question was raised about the strict requirement of the accurate alignment of the polarization axes (i.e., the principal axes) of the linear hi-bi fibers at a joint. A misalignment of the axes at a joint degrades the extinction ratio of lightwave transmission in fiber. Successive connectors further degrade the extinction ratio in a complicated way due to the accumulated effect of mismatches at the joints. It is thus likely that the very long links comprising a number of joints will be unable to preserve a stable polarization state. In contrast, a fiber with circularly polarized eigenstates appears to be an ideal medium for transmitting light of a stable polarization. A scheme of circular polarization-maintaining single-mode fiber cable design was proposed in 1980 by Jeunhomme and Monerie [2]. Although practical circular hi-bi fiber was not available at that time, it was thought that such fiber could be well approximated by sufficiently twisting a fiber in which the initial linear birefringence is small.

At the NTT Labs, a relatively large-scale experiment was carried out on twisted fiber over a dozen kilometers long during the early 1980s. This experiment explored whether this kind of fiber really proves useful in coherent optical communication [3]. Serious study of the transmission behaviors of twisted fiber continued until the mid-1980s. While the twisted fiber was not proved to be practical in the realm of coherent optical communication after all, this research on twisted fiber did accumulate a wealth of scientific data (theoretical and experimental) that should prove to be of fundamental importance in the study of twist-related fiber transmission characteristics [3, 4].

Another major impetus for twisted-fiber research came from the prospective use of this fiber in the area of fiber gyroscope. It was apparently due to this impetus that a special "twist-and-draw" machine was devised that attempted "to freeze the twist in line while drawing the fiber" [5]. According to this paper, "the twist is transmitted to the molten cone of glass by the elasticity of the fiber." Nevertheless, this and other similar attempts did not succeed in preserving circular birefringence in fiber. Any attempt to fix up the twisting state of fiber at a "hot" condition tends to diminish the birefringence of any kind that initially exists in the fiber.

The preceding description is a sketchy survey of the technical background in the earlier study of twisted fiber on a relatively large scale. Also noteworthy was the initial experiment on Faraday rotation sensing, in which disturbing linear birefringence is suppressed by twisting the fiber [6]. Despite all the efforts, the twisted fiber was not a feasible fiber version for essentially two reasons. First, the mechanical strength of glass fiber limits the allowable twist rate, such that the fiber will crack before a sufficient circular birefringence is reached [7]. Second, a twisted fiber, being in a constrained state, is not easy to handle. The use of twisted fiber is therefore rather restrictive from the standpoint of system application, though in laboratory such fiber furnishes a useful medium for some experiments of fundamental value or academic interest.

4.2   PHENOMENOLOGICAL ANALYSIS OF TWISTED FIBER

The Twist-Induced Optical Rotation

The effect of twisting is to introduce a torsional stress in the fiber. Under the simplifying assumption of uniform elastic and uniform elastooptic properties throughout the fiber, and in absence of linear birefringence, it was shown by theory and experiment that a uniform twist produces a total optical rotation Ω in a length z of fiber in proportion to the twist rate:

$$\Omega = \alpha z = (\varsigma\tau)z \qquad (4.1)$$

where α is the twist-induced optical rotation per unit length and τ is the twist-rate (or mechanical rotation per unit length). The proportionality factor ς is the elastooptic coefficient, which is the ratio of the twist-induced optical rotation to the mechanical rotation of the twist, that is,

$$\varsigma = \frac{\alpha}{\tau} \qquad (4.2)$$

The sense of the optical rotation α is the same as the sense of the mechanical twist τ. For doped silica, the value ς for this ratio is found to be fairly small. Rashleigh and Ulrich [8] gave a calculated value of $\varsigma \approx 0.08$. Ulrich and Simon [1] reported their measured value of $\varsigma \approx 0.065 \pm 0.005$ (in Ref. [1] it was also stated that M. Johnson did an independent and accurate determination of this figure). Elsewhere, Okoshi [9] reported the experimental result that one turn (360°) of mechanical twisting of fiber produces an optical rotation of about 26°, so that $\varsigma \approx 26°/360° = 0.072$. The coefficient ς depends to a small extent on the wavelength, the doping of the core, as well as the waveguiding boundary effect. Nevertheless, the range of $\varsigma \approx 0.065-0.08$ is believed to be universally valid for weakly guiding silica fibers.

Phenomenological Approach to Local Coupled-Mode Equations

In the general case where the residual linear birefringence is not negligibly small compared with the twist-induced optical rotation, a rigorous formulation of the coupled-mode equations for twisted fiber would be difficult to derive, either in local coordinates or in fixed coordinates.

If the twist is sufficiently strong, we expect that the coupled-mode equations can be formulated approximately by a simple phenomenological approach. We recall that in the coupled-mode analysis of a fast-spun fiber, an observer in the local coordinates "sees" a linear light rotating in a sense counter to the sense of spinning. Now the sense of optical rotation $\alpha = \varsigma\tau$ is the same as the sense of the mechanical twist τ. Consequently, an observer in the local coordinates will see a linear light rotating at a rate $-\tau + \varsigma\tau = -\tau(1 - \varsigma)$ in the strongly twisted fiber. Such a phenomenological view is useful in deriving the coupled-mode equations for a twisted fiber in local coordinates from the corresponding equations for a fast-spun fiber simply by substituting $(1 - \varsigma)\tau$ for τ in the latter equations. Thus, from Eq. (2.12), we have

$$
\begin{aligned}
\frac{dA_x}{dz} &= \frac{j\delta\beta}{2}A_x + (1 - \varsigma)\tau A_y \\
\frac{dA_y}{dz} &= -(1 - \varsigma)\tau A_x - \frac{j\delta\beta}{2}A_y
\end{aligned}
\tag{4.3}
$$

By analogy with Eq. (2.33) for spun fiber, the twist rate of fiber is considered fast in the local coordinates if the following condition is satisfied:

$$
\frac{(1 - \varsigma)\tau}{\delta\beta} \gg 1
\tag{4.4}
$$

where $\varsigma \approx 0.065 - 0.08$. Exactly speaking, Eq. (4.4) refers to a condition of "strong twist." Sometimes, however, we keep using the earlier term "fast-spun" condition to refer to this equation. Equation (4.4) is similar to Eq. (2.33) in form. However, the twist rate achievable in the cool state is far lower than the commonly used spin rate in the hot state. The former cannot be very high because of the mechanical strength limitation of silica glass material. Coarsely speaking, 50 to 100 turns/meter poses an upper bound for the twist rate. Consequently, in order to satisfy the strong-twist condition given by Eq. (4.4), it becomes a prerequisite that the residual linear birefringence $\delta\beta$ be sufficiently small.

The physical properties of twisted fiber and spun fiber are entirely different. As illustrated in Chapter 2, a spun fiber behaves like an isotropic medium in the macroscopic sense, so that such fiber is extremely low birefringent. On the other hand, in a twisted fiber an azymuthally varying stress acts on the waveguiding medium, thereby inducing a circular birefringence in the twisted fiber structure.

4.3 PERTURBATION ANALYSIS OF TWISTED FIBER

The coupled-mode equations for twisted fiber in the local coordinates can be differently derived by a perturbation approach with the aid of Ref. [1]. Consider two linear base modes in the fixed coordinates. The coupled-mode equations can be put in the form:

$$\frac{d\hat{\mathbf{A}}}{dz} = j \begin{bmatrix} k_{11} & k_{12} \\ k_{21} & k_{22} \end{bmatrix} \hat{\mathbf{A}}$$

where, as in Section 2.7, the circumflex ($\wedge$) denotes the fixed coordinates. In the matrix the diagonal coefficients k_{11}, k_{22} are propagation constants, and the off-diagonal coefficients k_{12}, k_{21} are the coupling coefficients.

Twisting a Fiber of Round Geometry

In the case of an idealized fiber model with a round core, twisting a fiber will not affect the propagation constants of the polarization modes, but will introduce coupling between the modes such that [1]

$$k_{11} = k_{22} = 0$$
$$k_{12} = -k_{21} = j\alpha \tag{4.5}$$

where $\alpha = \varsigma\tau$, as afore-defined, with τ denoting the spin rate and $\varsigma = \alpha/\tau$ denoting the ratio of optical rotation to mechanical rotation due to twisting.

Effect of Core Deformation

In real fibers, the cross section of the core cannot be perfectly round. One of the simplest deviations is the elliptical deformation of the fiber core. The resulting perturbation is then to cause a linear birefringence described by the following coefficients [1]:

$$k_{11} = -k_{22} = \frac{\delta\beta}{2} \cos 2\phi_B$$
$$k_{12} = k_{21} = \frac{\delta\beta}{2} \sin 2\phi_B \tag{4.6}$$

where ϕ_B is the inclination angle of the deformed core from the fixed coordinate axes. The first expression represents a detuning of the two polarization modes, and the second expression represents coupling. By "detuning" we simply mean the generation of a difference of the propagation constants of the two modes concerned. When $\phi_B = 0$, there is only detuning, but no coupling. When $\phi_B = 45°$, there is no detuning, but maximum coupling.

When a real fiber with a deformed core is twisted, the overall effect is additive in the sense of perturbation, such that the coupled mode equations in the fixed coordinates can be written as

$$\frac{d\hat{\mathbf{A}}}{dz} = \hat{\mathbf{K}}\hat{\mathbf{A}}$$

$$\hat{\mathbf{K}} = \begin{bmatrix} j\dfrac{\delta\beta}{2}\cos 2\phi_B & -\alpha + j\dfrac{\delta\beta}{2}\sin 2\phi_B \\[2ex] \alpha + j\dfrac{\delta\beta}{2}\sin 2\phi_B & -j\dfrac{\delta\beta}{2}\cos 2\phi_B \end{bmatrix} \tag{4.7}$$

where $\alpha = \varsigma\tau$ as before, and $\phi_B = \tau z$. In the fixed coordinates, Eq. (4.7) is valid under the condition of perturbation approximation, that is

$$\delta\beta \ll \varsigma\tau \tag{4.7a}$$

by order of magnitude.

The Perturbation Approach to Local Coupled-Mode Equations

Equation (4.7) is difficult to solve because it represents a set of simultaneous differential equations with variable coefficients (functions of z). We therefore try to change the equations to refer to the local coordinates by rotating the coordinate axes:

$$\mathbf{A} = \mathbf{R}(\tau z)\hat{\mathbf{A}}$$

$$\hat{\mathbf{A}} = \mathbf{R}(-\tau z)\mathbf{A}$$

where the expression for the rotation matrix is given by Eq. (2.6).
    Substituting the two preceding equations into Eq. (4.7) yields

$$\frac{d\mathbf{R}(-\tau z)}{dz}\mathbf{A} + \mathbf{R}(-\tau z)\frac{d\mathbf{A}}{dz} = \hat{\mathbf{K}}\mathbf{R}(-\tau z)\mathbf{A}$$

$$\frac{d\mathbf{A}}{dz} = \mathbf{K}\mathbf{A}$$

$$\mathbf{K} = \mathbf{R}^{-1}(-\tau z)\hat{\mathbf{K}}\mathbf{R}(-\tau z) - \mathbf{R}^{-1}(-\tau z)\frac{d\mathbf{R}(-\tau z)}{dz}$$

    It is satisfying that straightforward matrix-algebraic manipulation of the last expression for $\mathbf{K}$ yields a very simple result in exact conformity with the formulation by the previous phenomenological approach:

$$
\mathbf{K} =
\begin{bmatrix}
j\dfrac{\delta\beta}{2} & (1-s)\tau \\[2mm]
-(1-s)\tau & -j\dfrac{\delta\beta}{2}
\end{bmatrix}
\tag{4.8}
$$

Note that this equation is derived on the basis of perturbation approximation.

4.4 SOLUTION OF COUPLED-MODE EQUATIONS FOR TWISTED FIBER

We have thus derived by two different approaches the same coupled-mode equations, Eq. (4.3) and Eq. (4.8), for a twisted fiber in the local coordinates, which look fairly simple and neat in form. The lines of thought implied in the two derivations are entirely different. Both are *not* exact, but are approximate. For more discussion on the range of validity of Eq. (4.8), see the last subsection of Section 4.9.

While recognizing the inexact nature of the above derivations, it is presumable that the coupled-mode equations written in the form of Eq. (4.3) or its equivalent Eq. (4.8), are sufficiently accurate from a practical point of view. In the following we therefore feel free to use these equations for an analytic study of the twisted fiber.

Comparing the coupled-mode formulation, Eq. (4.3), for twisted fiber and the coupled-mode formulation, Eq. (2.12), for spun fiber, we can see that a "short-cut" approach to the solution of twisted fiber is possible by borrowing the corresponding solution for spun fiber, simply with τ (where it occurs) replaced by $(\tau - \alpha) = \tau(1 - s)$. Although it is easy to reach the final solution using this short-cut method, most physical meanings are lost in this approach. Here we would prefer to follow consistently the method of diagonalization when solving problems of the twisted fiber. It is by this classical method that the underlying physical meanings can be made clear in the course of a step-by-step derivation.

To begin with, we put the coupled-mode equations in the local coordinates in matrix form:

$$
\frac{d\mathbf{A}}{dz} = \mathbf{K}\mathbf{A}
$$

$$
\mathbf{K} =
\begin{bmatrix}
j\dfrac{\delta\beta}{2} & (1-s)\tau \\[2mm]
-(1-s)\tau & -j\dfrac{\delta\beta}{2}
\end{bmatrix}
\tag{4.9}
$$

by Eq. (4.8) or Eq. (4.3).

The diagonalizing matrix is introduced by the transformation:

$$\mathbf{A} = \mathbf{OW}$$

$$\mathbf{O} = \begin{bmatrix} \cos\phi & j\sin\phi \\ j\sin\phi & \cos\phi \end{bmatrix} \tag{4.10}$$

$$\phi = \frac{1}{2}\arctan\frac{2(1-s)\tau}{\delta\beta}$$

Substituting Eq. (4.10) into Eq. (4.9) yields independent differential equations for the elements W_x and W_y of the column matrix $\mathbf{W}$, whose exponential solutions are

$$\mathbf{W}(z) = \tilde{\Lambda}\mathbf{W}(0)$$

$$\tilde{\Lambda} = \begin{bmatrix} e^{\lambda z} & 0 \\ 0 & e^{-\lambda z} \end{bmatrix} \tag{4.11}$$

$$\lambda = j\left\{[(1-s)\tau]^2 + (\delta\beta/2)^2\right\}^{1/2}$$

where λ and $-\lambda$ are the eigenvalues. Correspondingly, the exponential functions $e^{\lambda z}$ and $e^{-\lambda z}$ are the eigenfunctions.

Transfer Matrix of Twisted Fiber in the Local Coordinates

By Eq. (4.10), $\mathbf{W}(0) = \mathbf{O}^{-1}\mathbf{A}(0)$. Then by Eq. (4.11), $\mathbf{W}(z) = \tilde{\Lambda}\mathbf{W}(0)$. Again by Eq. (4.10), $\mathbf{A}(z) = \mathbf{OW}(z)$. Thus a formal solution for the twisted fiber is given by

$$\mathbf{A}(z) = \mathbf{O}\tilde{\Lambda}\mathbf{O}^{-1}\mathbf{A}(0)$$

$$\mathbf{T}_l = \mathbf{O}\tilde{\Lambda}\mathbf{O}^{-1} \tag{4.11a}$$

where $\mathbf{T}_l$ is the transfer matrix of the twisted fiber with respect to the local coordinates.

Asymptotic Approximation under the Strong-Twist Condition

Under the strong-twist condition of Eq. (4.4):

$$\mathbf{O} \rightarrow \frac{1}{\sqrt{2}}\begin{bmatrix} 1 & j \\ j & 1 \end{bmatrix}$$

$$\mathbf{O}^{-1} \rightarrow \frac{1}{\sqrt{2}}\begin{bmatrix} 1 & -j \\ -j & 1 \end{bmatrix} \tag{4.12}$$

$$\tilde{\Lambda} \rightarrow \begin{bmatrix} e^{j(1-s)\tau z} & 0 \\ 0 & e^{-j(1-s)\tau z} \end{bmatrix}$$

Equations (4.9)–(4.11a) provide a complete mathematical framework for twisted-fiber analysis in the local coordinates. Under the strong-twist condition, the mathematical derivation is greatly simplified by using the asymptotic expressions of Eq. (4.12). Substituting these asymptotic expressions into Eq. (4.11a) yields, in the local coordinates:

$$\mathbf{A}(z) = \mathbf{T}_l(z)\mathbf{A}(0)$$

$$\mathbf{T}_l(z) = \begin{bmatrix} \cos(1 - \varsigma)\tau z & \sin(1 - \varsigma)\tau z \\ -\sin(1 - \varsigma)\tau z & \cos(1 - \varsigma)\tau z \end{bmatrix} \qquad (4.12a)$$

where $\mathbf{T}_l(z)$ is the transfer matrix of the strongly twisted fiber in the local coordinates. Under the condition of strong twisting, this transfer matrix represents a rotation of the field pattern by an angle $(1 - \varsigma)\tau z$, in agreement with the result by intuition.

Transfer Matrix of Strongly Twisted Fiber in Fixed Coordinates

The above result refers to the local coordinates with linear modes as the base modes. When the solution in the local coordinates is obtained, the corresponding solution in the fixed coordinates can be derived by turning the coordinates back to the original fixed position by counterrotating the coordinates with the matrix:

$$\mathbf{R}(-\tau z) = \begin{bmatrix} \cos(\tau z) & -\sin(\tau z) \\ \sin(\tau z) & \cos(\tau z) \end{bmatrix} \qquad (4.13)$$

At the initial point $z = 0$, $\mathbf{R}(-\tau z)$ reduces to a unit matrix, such that the local coordinates and the fixed coordinates coincide. The solution referring to the local coordinates and that referring to the fixed coordinates are therefore identical at $z = 0$, that is, $\hat{\mathbf{A}}(0) = \mathbf{A}(0)$, where (as defined earlier) the circumflex ($\wedge$) denotes a quantity in the fixed coordinates.

By Eqs. (4.13) and (4.12a), the solution in the fixed coordinates is given by

$$\hat{\mathbf{A}}(z) = \mathbf{R}(-\tau z)\mathbf{A}(z) = \mathbf{R}(-\tau z)\mathbf{T}_l(z)\mathbf{A}(0)$$

$$= \hat{\mathbf{T}}(z)\hat{\mathbf{A}}(0)$$

$$\hat{\mathbf{T}}(z) = \begin{bmatrix} \cos(\varsigma\tau z) & -\sin(\varsigma\tau z) \\ \sin(\varsigma\tau z) & \cos(\varsigma\tau z) \end{bmatrix} \qquad (4.14)$$

Both Eq. (4.12a) for $\mathbf{T}_l(z)$ and Eq. (4.14) for $\hat{\mathbf{T}}(z)$ are descriptive of the overall property of the twisted-fiber structure itself, regardless of what the input light is. The way in which the lightwave actually propagates in the fiber depends on the kind of input light that excites the fiber at $z = 0$. For practical purposes, circular light excitation and linear light excitation are considered first.

4.5   TRANSMISSION BEHAVIORS OF TWISTED FIBER

Circular Light in Strongly Twisted Fiber

Let the input light incident to the fiber be a circular light (say, right circular):

$$\mathbf{A}(0) = \hat{\mathbf{A}}(0) = \frac{1}{\sqrt{2}} \begin{bmatrix} 1 \\ j \end{bmatrix} \tag{4.15}$$

By Eqs. (4.12a) and (4.14), lightwave transmission in the local coordinates and in the fixed coordinates are respectively given by:

$$\mathbf{A}(z) = \frac{1}{\sqrt{2}} \begin{bmatrix} 1 \\ j \end{bmatrix} e^{j(1-s)\tau z} \tag{4.16a}$$

$$\hat{\mathbf{A}}(z) = \frac{1}{\sqrt{2}} \begin{bmatrix} 1 \\ j \end{bmatrix} e^{-js\tau z} \tag{4.16b}$$

When a left circular light excites the fiber, the result is

$$\mathbf{A}(0) = \hat{\mathbf{A}}(0) = \frac{1}{\sqrt{2}} \begin{bmatrix} 1 \\ -j \end{bmatrix}$$

$$\mathbf{A}(z) = \frac{1}{\sqrt{2}} \begin{bmatrix} 1 \\ -j \end{bmatrix} e^{-j(1-s)\tau z} \tag{4.17a}$$

$$\hat{\mathbf{A}}(z) = \frac{1}{\sqrt{2}} \begin{bmatrix} 1 \\ -j \end{bmatrix} e^{js\tau z} \tag{4.17b}$$

Thus, a circular light (right or left) keeps the circular state of polarization (SOP) and its handedness all the way along the fiber. The right circular light undergoes a phase shift of $(1 - s)\tau z$ and $- s\tau z$, respectively, in the local and fixed coordinates. The phase shifts change sign in the case of the left circular light.

Linear Light in Strongly Twisted Fiber

Let the input light incident to a strongly twisted fiber be a linear light expressed by

$$\mathbf{A}(0) = \hat{\mathbf{A}}(0) = \begin{bmatrix} \cos \theta \\ \sin \theta \end{bmatrix}$$

By Eqs. (4.12a) and (4.14), the polarization-mode solution in the local coordinates and in the fixed coordinates are, respectively:

$$\mathbf{A}(z) = \begin{bmatrix} \cos[\theta - (1 - s)\tau z] \\ \sin[\theta - (1 - s)\tau z] \end{bmatrix} \qquad (4.18a)$$

$$\hat{\mathbf{A}}(z) = \begin{bmatrix} \cos(\theta + s\tau z) \\ \sin(\theta + s\tau z) \end{bmatrix} \qquad (4.18b)$$

The geometrical interpretation of the preceding expressions is fairly obvious. In a strongly twisted fiber, a linear light after traversing a distance z in the local coordinates will counterrotate (in the opposite sense to the twist) with an optical rotation equal to $(1 - s)\tau z$. On the other hand, in the fixed coordinates, this linear light will undergo an optical rotation equal to $s\tau z$ (in the same sense as the twist).

We understand that linear light of any initial orientation is *not* an eigenmode of the strong-twist fiber. For a linear incident light, although the emergent light keeps linear, the orientation angle changes by $-(1 - s)\tau z$ in the local coordinates, and by $(s\tau z)$ in the fixed coordinates. A careful examination of Eqs. (4.18a) and (4.18b) reveals that linear light transmission in a strong-twist fiber actually involves the eigenfunctions of both the positive and negative exponents. The coexistence of two eigenfunctions produces a beating pattern in the transmission direction. It happens that, in a strong-twist fiber, this beating pattern appears as a precessing (or spinning) linear light.

Arbitrary Elliptical Light in Strongly Twisted Fiber

The (ε, ψ)-representation in the form of Eq. (2.45) for elliptical light is used in the present analysis. Thus, let the incident light be an arbitrary elliptical light given by

$$\mathbf{A}(0) = \hat{\mathbf{A}}(0) = \frac{1}{\sqrt{1 + \varepsilon^2}} \begin{bmatrix} \cos\psi - j\varepsilon \sin\psi \\ \sin\psi + j\varepsilon \cos\psi \end{bmatrix} \qquad (4.19)$$

By Eqs. (4.12a) and (4.14), the solutions of light transmission in the local and fixed coordinates are given respectively by

$$\mathbf{A}(z) = \frac{1}{\sqrt{1 + \varepsilon^2}} \begin{bmatrix} \cos[\psi - (1 - s)\tau z] - j\varepsilon \sin[\psi - (1 - s)\tau z] \\ \sin[\psi - (1 - s)\tau z] + j\varepsilon \cos[\psi - (1 - s)\tau z] \end{bmatrix} \qquad (4.20)$$

$$\hat{\mathbf{A}}(z) = \frac{1}{\sqrt{1 + \varepsilon^2}} \begin{bmatrix} \cos(\psi + s\tau z) - j\varepsilon \sin(\psi + s\tau z) \\ \sin(\psi + s\tau z) + j\varepsilon \cos(\psi + s\tau z) \end{bmatrix} \qquad (4.21)$$

The results show that the ellipticity of the input light remains unchanged in a strongly twisted fiber in either the local coordinates or the fixed coordinates. According to Eqs. (4.20) and (4.21), the polarization ellipse counterrotates in the local coordinates at the rate $-(1 - \varsigma)\tau$. In the fixed coordinates, the polarization ellipse rotates at the rate $\varsigma\tau$ in the same sense as the twist.

Thus, in the case of an arbitrary elliptical light, it is again exactly the small term ς that makes a strong-twist fiber qualitatively different from a fast-spun fiber. For a fast-spun fiber in the fixed coordinates, arbitrarily polarized light simply goes straight just like the case of lightwave propagation in an idealized isotropic medium. In contrast to the fast-spun fiber, for a strong-twist fiber in the fixed coordinates, light of an arbitrary polarization undergoes a rotation. This SOP rotation in the *fixed* coordinates is due solely to the twist-induced torsional stress in fiber. The sense of this rotation is the same as the sense of the twisting.

Equation (4.21) for an arbitrary elliptical light naturally applies to the two limiting cases: linear light and circular light. In the case of linear light, for which $\varepsilon = 0$ and $\psi = \theta$, Eq. (4.21) becomes a linear light with an orientation $(\theta + \varsigma\tau z)$ in the fixed coordinates. For $\varepsilon = 1$, $\psi = 0$ and $\varepsilon = -1$, $\psi = 0$ (corresponding to right circular and left circular light), Eq. (4.21) reduces to Eq. (4.16b) and Eq. (4.17b), respectively.

Beat Length of Strongly Twisted Fiber

As defined, the beat length is the shortest distance at which the superposed pattern of the two polarization modes in the fiber reproduces itself. Accordingly, two different beat lengths should be distinguished, one referring to the local coordinates and one referring to the fixed coordinates. These two beat lengths are fundamentally two different parameters that specify a fiber. The only thing they have in common is that either beat length is descriptive of the field repetitive property of the beating pattern in the respective coordinates.

More importantly from the practical viewpoint, the parameter beat length specifies the perturbation resistivity of fiber, or the capability of fiber to preserve the light power of the desired mode from loss to the undesired mode. In this regard, only the beat length in the *fixed* coordinates is of interest. The beat length in the local coordinates is of no relevance in the perturbation-resistive description, because the perturbation effect of different sorts (bending, transverse force, etc.) is itself quantified in the fixed coordinates, not in the local coordinates. In practice, therefore, when we speak of the term "beat length," we mean the beat length in the fixed coordinate, if not specified otherwise.

For the fast-spun lo-bi fiber discussed in Chapter 2, the beat length in the fixed coordinates is indefinitely large, inasmuch as the fiber behaves like an isotropic medium, such that any sort of birefringence tends to zero in these coordinates. In the case of a strong-twist fiber, we have shown that the right and left circular modes are the eigenmodes. In the fixed coordinates, these two circular eigenmodes have different phase velocities so that they beat. By Eqs. (4.16b) and (4.17b), which refer to the fixed coordinates, the phase-velocity difference of the two circular eigenmodes is $2\varsigma\tau z$, such that the beat length in the fixed coordinates can be determined by the equality $2\varsigma\tau L_b = 2\pi$, yielding

$$L_b = \frac{\pi}{\varsigma\tau} \tag{4.22}$$

which refers to the fixed coordinates.

Since the twist rate τ cannot be sufficiently high because of the mechanical strength limitation, the circular birefringence generated by the twisting technique unfortunately cannot be strong enough. By Eq. (4.22), therefore, the beat length cannot be short enough to satisfy the specification for the perturbation-resistive requirement.

The beat-length formula, Eq. (4.22), is derived from Eqs. (4.16b) and (4.17b) for the circular eigenmodes in a strongly twisted fiber. This equation can also be derived from the transmission behavior of a linear light, or generally of an arbitrary elliptical light. Thus, by Eq. (4.18b), the beat length is equal to the distance at which the orientation of a linear light reproduces itself. By Eq. (4.21), the same is true for an arbitrary elliptical light.

Coupled-Mode Equations in the Fixed Coordinates

In the foregoing treatment of twisted fiber, we adopted a method of approach that essentially comprises two mathematical techniques. The first technique is to solve the coupled-mode equations in the local coordinates. The second technique is to transform the local-mode solutions to solutions in the fixed coordinates by a geometrical rotation of the coordinate axes. Such a method seems rather roundabout. In the coupled-mode analysis for problems involving a rotation (spin or twist) of the fiber structure, we are forced to go this way in order to obtain the desired solution.

Naturally, were it possible to deal with coupled-mode equations in the fixed coordinates at the beginning of the analysis, the analytic framework for coupled-mode solutions would become simpler and cleaner. Then, all we would need to do is solve the coupled-mode equations, without the extra work of coordinate transformation. Unfortunately, this is scarcely possible in the general case.

In the special case of *strong-twist* fiber, we can approximately put the coupled-mode equations in the fixed coordinates in a phenomenological way. Thus, from the solutions (in fixed coordinates) derived by the method of diagonalization, we find that we can derive these solutions from a set of coupled-mode equations wherein the coupling coefficient $(1 - \varsigma)\tau$ with respect to the local coordinates is replaced by $\varsigma\tau$. Phenomenologically, therefore, coupled-mode equations for strong-twist fiber in the fixed coordinates can be written in the form:

$$\frac{d\hat{A}_x}{dz} = j\frac{\delta\beta}{2}\hat{A}_x - s\tau\hat{A}_y$$

$$\frac{d\hat{A}_y}{dz} = s\tau\hat{A}_x - j\frac{\delta\beta}{2}\hat{A}_y$$

(4.23)

where, as before, the circumflex ($\wedge$) designates the fixed coordinates. Clearly, if we solve the above coupled-mode equations for different SOPs of light at the input, we shall obtain the same solutions as those previously derived by the method of diagonalization and the coordinate transformation.

4.6   COUPLED-MODE THEORY WITH CIRCULAR BASE MODES

Previous coupled-mode analysis was made by using linear base modes, either in local coordinates or in fixed coordinates. The choice of linear base modes became fairly familiar in the early development of the coupled-mode theory. Nevertheless, the choice of linear base modes is by no means unique. Besides linear base modes, we can equally well choose circular base modes to treat the same coupled-mode problem, with more or less mathematical complexity depending on the type of problem treated. In principle, it is also possible to choose other kinds of base modes besides linear and circular such as a pair of orthogonal elliptical modes. But in practice this does not appear to be necessary.

Intuitively, it is conceivable that for a fiber problem wherein a linear light predominates, the choice of linear base modes is advantageous. On the other hand, if a circular light predominates, then the choice of circular base modes may be more appropriate. In the next section, we shall deal with a strong-twist fiber that is bent, and we wish to determine the minimum tolerable bending radius. In this problem of much practical interest, the choice of circular base modes will prove preferable in view of mathematical simplicity. As a first step in this analysis, we need to establish the coupled circular-mode formulation on a sound analytic basis.

Transformation of Base Modes from Linear to Circular and Vice Versa

We expand a general lightwave as the sum of right and left circular modes:

$$\mathbf{E} = \mathbf{E}_R + \mathbf{E}_L = \frac{1}{\sqrt{2}}\begin{bmatrix}1\\j\end{bmatrix}E_R + \frac{1}{\sqrt{2}}\begin{bmatrix}1\\-j\end{bmatrix}E_L$$

$$= \frac{1}{\sqrt{2}}\begin{bmatrix}E_R + E_L\\j(E_R - E_L)\end{bmatrix}$$

(4.24)

Note that the column matrices $\mathbf{E}_R$ and $\mathbf{E}_L$ (bold letters) denote lightwaves whose SOPs are right and left circular, respectively. The numeral $1/\sqrt{2}$ is a normalization factor. The "light" letters E_R and E_L are *complex amplitudes* (functions of z) of the respective right and left circular modes. What we are concerned with at this point is that E_R and E_L themselves are *not* circular modes of whatever sense.

To elucidate the implication of the above expansion in terms of circular base modes, we similarly expand a general lightwave in terms of linear modes for comparison:

$$\mathbf{A} = \mathbf{A}_x + \mathbf{A}_y = \begin{bmatrix} 1 \\ 0 \end{bmatrix} A_x + \begin{bmatrix} 0 \\ 1 \end{bmatrix} A_y$$
$$= \begin{bmatrix} A_x \\ A_y \end{bmatrix} \tag{4.25}$$

In this expansion, elements of the column matrix for linear mode x happen to be 1 and 0, and those for linear mode y happen to be 0 and 1. Incidentally, therefore, the linear modes $\mathbf{A}_x$ and $\mathbf{A}_y$ become the same as their complex amplitudes A_x and A_y, respectively. Thus, when we use linear base modes, there is no need to differentiate the modal matrixes and the respective complex amplitudes. But when we use circular base modes, we do have to make this distinction.

In comparing Eq. (4.24) and Eq. (4.25), we have the following transformation that relates the expansion in circular modes and that in linear modes:

$$A_x = \frac{1}{\sqrt{2}} (E_R + E_L)$$
$$A_y = j \frac{1}{\sqrt{2}} (E_R - E_L) \tag{4.26}$$

$$E_R = \frac{1}{\sqrt{2}} (A_x - jA_y)$$
$$E_L = \frac{1}{\sqrt{2}} (A_x + jA_y) \tag{4.27}$$

where, as said, either E_R or E_L denotes a complex amplitude, not a circular mode in itself.

When the pair of circular modes (right and left) are chosen as base modes in an initial-value problem, the expressions for the initial values must be changed accordingly. These initial values for circular-light base modes are related to the initial values for linear-light base modes by Eqs. (4.26) and (4.27).

TABLE 4.1   Initial Conditions for Linear Base Modes and Circular Base Modes

Initial Value	Right Circular	Left Circular	Linear
$\mathbf{A}(0) = \begin{bmatrix} A_x(0) \\ A_y(0) \end{bmatrix}$	$\dfrac{1}{\sqrt{2}} \begin{bmatrix} 1 \\ j \end{bmatrix}$	$\dfrac{1}{\sqrt{2}} \begin{bmatrix} 1 \\ -j \end{bmatrix}$	$\begin{bmatrix} \cos\theta \\ \sin\theta \end{bmatrix}$
$\mathbf{E}(0) = \begin{bmatrix} E_R(0) \\ E_L(0) \end{bmatrix}$	$\begin{bmatrix} 1 \\ 0 \end{bmatrix}$	$\begin{bmatrix} 0 \\ 1 \end{bmatrix}$	$\dfrac{1}{\sqrt{2}} \begin{bmatrix} e^{-j\theta} \\ e^{j\theta} \end{bmatrix}$

With the aid of Eqs. (4.26) and (4.27), it is easy to obtain the initial conditions for circular base modes from the corresponding initial conditions for linear base modes, or vice versa. For ready reference, Table 4.1 lists the corresponding matrix expressions for the initial conditions for linear base modes and circular base modes.

Coupled-Mode Equations Using Circular Base Modes

The above equations, which transform linear modes to circular modes, or vice versa, are valid irrespective of whether the coordinates are local or fixed. For mathematical simplicity, we take the fixed coordinates, wherein the coupled-mode equations with linear base modes are given by Eq. (4.23). Substituting Eq. (4.26) into Eq. (4.23), we obtain the coupled-mode equations with circular base modes in fixed coordinates:

$$\frac{dE_R}{dz} = -js\tau E_R + j\frac{\delta\beta}{2}E_L$$
$$\frac{dE_L}{dz} = j\frac{\delta\beta}{2}E_R + js\tau E_L \tag{4.28}$$

To simplify the notation, we dispense with the circumflex ($\wedge$) over the variables in the equations, while recognizing E_R and E_L now refer to the fixed coordinates. Comparing Eq. (4.28) with Eq. (4.23), we see that the off-diagonal elements of the above square matrix are both imaginary (of the same signs), different from the coupled-mode formulation using linear base modes wherein the corresponding off-diagonal elements are both real (of opposite signs). Both formulations (in terms of linear base modes or circular base modes) are legitimate or canonical in the light of coupled-mode theory. As indicated by Eqs. (1.26)–(1.28), the principle of conservation of power requires that the off-diagonal terms (i.e., the coupling coefficients) of the coupled-mode formulation be "negative conjugate." It can be seen immediately that both Eq. (4.28) and Eq. (4.23) satisfy this negative-conjugate requirement.

Diagonalization of Coupled Circular-Mode Equations

In spirit, the method of diagonalization for coupled circular-mode equations is in no way different from that for coupled linear-mode equations. Equation (4.28) is written as

$$\frac{d\mathbf{E}}{dz} = \mathbf{K}_c\mathbf{E}$$

$$\mathbf{E} = \begin{bmatrix} E_R \\ E_L \end{bmatrix}$$

$$\mathbf{K}_c = \begin{bmatrix} -js\tau & j\dfrac{\delta\beta}{2} \\ j\dfrac{\delta\beta}{2} & js\tau \end{bmatrix}$$

$$(4.29)$$

To diagonalize the matrix equation, we introduce the transformation:

$$\mathbf{E} = \mathbf{O}_c\mathbf{W} \qquad (4.30)$$

where $\mathbf{O}_c$ is a diagonalizing matrix for the coupled circular-mode equations. Substituting Eq. (4.30) into Eq. (4.28) yields

$$\frac{d\mathbf{W}}{dz} = \left(\mathbf{O}_c^{-1}\mathbf{K}\mathbf{O}_c\right)\mathbf{W} = \Lambda_c\mathbf{W}$$

$$\Lambda_c = \mathbf{O}_c^{-1}\mathbf{K}\mathbf{O}_c = \begin{bmatrix} -\lambda & 0 \\ 0 & \lambda \end{bmatrix}$$

$$(4.31)$$

$$\lambda = j\left[(s\tau)^2 + \left(\frac{\delta\beta}{2}\right)^2\right]^{1/2}$$

where $-\lambda$ and λ are the eigenvalues, and

$$\mathbf{O}_c = \begin{bmatrix} \cos\varphi & \sin\varphi \\ -\sin\varphi & \cos\varphi \end{bmatrix}$$

$$\varphi = \frac{1}{2}\arctan\frac{\delta\beta}{2s\tau}$$

$$(4.32)$$

The normal mode solutions are given by

$$\mathbf{W}(z) = \tilde{\Lambda}_c\mathbf{W}(0)$$

$$\tilde{\Lambda}_c = \begin{bmatrix} e^{-\lambda z} & 0 \\ 0 & e^{\lambda z} \end{bmatrix}$$

$$(4.33)$$

where the two exponential functions (diagonal elements) are the eigenfunctions.

The mathematics can be enormously simplified if in the coupled circular-mode equations the off-diagonal coefficient (residual linear-bi) is small as compared to the difference of diagonal coefficients (twist-induced optical rotation), that is,

$$\frac{\delta\beta}{s\tau} \ll 1 \qquad (4.34)$$

Under this condition, we have $\varphi \to 0$ in Eq. (4.32), such that the orthogonalizing matrix and its inverse are both reduced to a unit matrix:

$$\mathbf{O}_c = \mathbf{O}_c^{-1} \to \begin{bmatrix} 1 & 0 \\ 0 & 1 \end{bmatrix} \qquad (4.35)$$

The eigenvalues become:

$$\pm\lambda \to \pm js\tau \qquad (4.36)$$

Clearly, Eq. (4.34), which refers to the fixed coordinates, corresponds to the inverse expression of Eq. (4.4), which refers to the local coordinates. The physical meanings of both equations are exactly the same. Equation (4.34) is actually an alternative expression for the strong-twist condition. Obviously, for the given residual linear-bi, Eq. (4.34) implies a more restrictive requirement for the strong twist.

To illustrate the use of the above-listed formulas, let a circular light (say right-circular) be launched onto a strongly twisted fiber. The initial condition is

$$\mathbf{E}(0) = \begin{bmatrix} 1 \\ 0 \end{bmatrix} \qquad (4.37)$$

It is then simple to derive the desired solution. Because the diagonalizing matrix and its inverse are both equal to a unit matrix for strong twisting, by Eq. (4.35), $\mathbf{E}$ and $\mathbf{W}$ are identical for all z, such that

$$\mathbf{E}(z) = \mathbf{W}(z) = \tilde{\Lambda}\mathbf{W}(0) = \begin{bmatrix} 1 \\ 0 \end{bmatrix} e^{-js\tau z} \qquad (4.38)$$

For left circular light excitation, the solution is

$$\mathbf{E}(0) = \begin{bmatrix} 0 \\ 1 \end{bmatrix}$$

$$\mathbf{E}(z) = \begin{bmatrix} 0 \\ 1 \end{bmatrix} e^{js\tau z} \qquad (4.39)$$

We therefore see that right circular light and left circular light are the two eigenmodes of a strongly twisted fiber.

Light of any polarization that is not circular will excite both circular eigenmodes, so that they beat along the fiber. The most important special case is the linear light excitation. The relevant initial condition is given in Table 4.1. A simple and straightforward matrix-algebraic derivation yields:

$$\mathbf{W}(0) = \mathbf{E}(0) = \frac{1}{\sqrt{2}} \begin{bmatrix} e^{-j\theta} \\ e^{j\theta} \end{bmatrix}$$

$$\mathbf{W}(z) = \tilde{\Lambda}_c \mathbf{W}(0) \qquad\qquad (4.40)$$

$$\mathbf{E}(z) = \mathbf{W}(z) = \frac{1}{\sqrt{2}} \begin{bmatrix} e^{-j(\theta + s\tau z)} \\ e^{j(\theta + s\tau z)} \end{bmatrix}$$

In a circularly birefringent fiber, linear light transmission actually involves both circular eigenmodes so that they beat. The beating pattern is a precessing linear light in the transmission direction. This general physical property of the circular hi-bi fiber is true irrespective of the choice of coordinates (local or fixed), or the choice of base modes (linear or circular).

4.7 TOLERANCE OF STRONG-TWIST FIBER TO BENDING

We have seen that a twisted fiber behaves approximately circularly birefringent if the twist rate far exceeds the residue linear birefringence. One practical problem is to see the capability of such strongly twisted fiber in tolerating macroscopic perturbations, notably the bending of fiber. When a circular light (say right circular) is incident onto the strongly twisted fiber, the problem is to calculate the amount of power loss due to conversion of power from the desired right circular light to the undesired left circular light. In other words, we are interested in estimating the extinction ratio of a strongly twisted fiber that is bent with a radius of curvature that is likely to occur in practice.

Weak Coupling Theory Using Circular Base Modes

The kind of problem under consideration is one of weak coupling between the modes. The power transfer from the desired mode to the undesired mode is some tens of negative decibels (dB) (loss is 1% for -20 dB, 0.1% for -30 dB, for example).

The advantage of using circular base modes is obvious in problems of circular hi-bi fiber in which lightwave power is essentially carried by one circular mode (the desired mode), with only slight power transfer to the other undesired circular mode due to certain kind of perturbation. Here, the

choice of circular reference modes allows mathematical simplification because of weak coupling between the circular modes during transmission.
Should we choose linear modes as the base modes in our coupled-mode
analysis, the weak coupling condition, and hence the mathematical simplification due to this condition, would become invalid. For the current problem,
the choice of circular base modes is therefore obviously advantageous.

Consider the coupled-mode equations, Eqs. (4.28), wherein $\delta\beta$ is now the
overall linear birefringence, that is, $\delta\beta = \delta\beta_{in} + \delta\beta_{ex}$ ($\delta\beta_{in}$ is the built-in or
intrinsic linear-bi, and $\delta\beta_{ex}$ is the extrinsic linear-bi due to the perturbation).
In a circular birefringent fiber, wherein the intrinsic linear-bi is small in
comparison with the extrinsic linear-bi, we have $\delta\beta \approx \delta\beta_{ex}$.

An intentional macroscopic bending of fiber provides a particularly important application example of the weak coupling theory of circular hi-bi fiber.
In order to restrict the transfer of power from the desired to the undesired
mode at a prescribed low level, the bending radius of fiber cannot be too
small. The problem then concerns an estimation of the allowable minimum
bending radius at a specified maximum allowable power loss in $-dB$. Or
conversely, it is required to calculate the power loss in $-dB$ for a given
bending radius of fiber. Since this power loss cannot be large (some tens of
$-dB$ by order of magnitude), it is legitimate to assume that the depletion of
power of the desired mode (say, right circular E_R) is negligibly small during
transmission along the fiber, such that

$$E_R(z) \approx e^{-js\tau z} \tag{4.41}$$

Putting this approximate solution of E_R into Eq. (4.28), we obtain an
independent differential equation for E_L in the following form:

$$\frac{dE_L}{dz} = j\frac{\delta\beta}{2}e^{-js\tau z} + js\tau E_L \tag{4.42}$$

The solution is

$$E_L(z) = j\frac{\delta\beta}{2s\tau}\sin(s\tau z) \tag{4.43}$$

The maximum power conversion from the desired E_R mode to the undesired
E_L mode is

$$|E_L(z)|^2 = \left(\frac{\delta\beta}{2s\tau}\right)^2 \tag{4.44}$$

where $\delta\beta \approx \delta\beta_{ex}$.

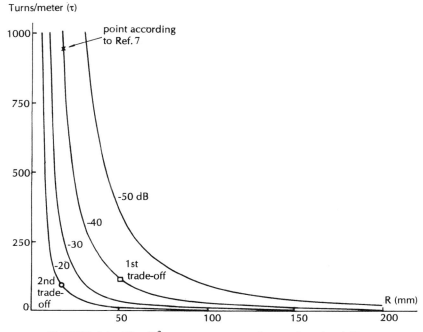

FIGURE 4.1   The $R^2\tau$ = const. curves of strongly twisted fiber.

The $R^2\tau$ = const. Relation

Since the power in the desired mode is normalized to be unity, $|E_L|^2$ is simply the extinction ratio. For a macroscopic bending, it is known that the fast axis of the bend-induced linear-bi is on the plane of the bend, and the magnitude of this linear-bi is inversely proportional to the square of the radius of curvature R of the bend:

$$\delta\beta_{ex} \approx 0.85(r^2/\lambda)R^{-2}$$

where r and λ denote the fiber radius and the working wavelength, respectively [10, 11]. Substituting this equation into Eq. (4.44) yields

$$R^2\tau = \text{const.} \tag{4.45}$$

$$\text{const.} = \frac{0.85}{2\varsigma}\frac{r^2}{\lambda\eta^{1/2}} \approx 6.07(r^2/\lambda)\eta^{-1/2} \tag{4.45a}$$

where η denotes the extinction ratio $|E_L/E_R|^2$.

Theoretical curves for Eq. (4.45) are shown in Figure 4.1 for different extinction ratios. Equation (4.45) reveals that, for the prescribed extinction ratio, the required twist rate τ is inversely proportional to the square of the radius R allowed in a fiber bend. Consequently, a small relaxation of the allowed value of R will result in a more substantial relaxation of the required twist rate τ. The bottom left portion of the family of curves shows the critical region in which vast changes of τ occur for only slight changes of R.

From Figure 4.1 we find that, if the maximum allowable extinction ratio is prescribed to be -40 dB, then a bend of radius 17.5 mm would require a twist rate as high as 1000 turns per meter (see the point marked by a cross in Figure 4.1), which is clearly not achievable within the limit of the mechanical strength of glass fiber. The data for the cross in the figure agree exactly with those of the Southampton scientists [7].

Note that the Southampton data (cross in figure) lie in the critical region where τ changes drastically if R is allowed to relax only slightly. Thus, Figure 4.1 shows that trade-offs can readily be made in different ways. Suppose the maximum allowable extinction ratio remains -40 dB, but the radius of curvature R of the bend is relaxed to be around 50 mm, then the required twist rate τ is reduced to about 100 turns/m, and this is a twist rate that can be realized without breaking the glass fiber. Or, if the radius of curvature R is kept at 17.5 mm, but the maximum allowable extinction ratio is relaxed to -20 dB, then the required twist rate is again at the level of 100 turns/m.

As can be deduced from the previous arguments, twisted fiber can still be used in an actual system where the specified tolerances can be less strict. From the technological viewpoint, one major disadvantage of twisted fiber is that it is not easy to handle. Continuously fixing the twisting state is not impossible, but needs some very sophisticated means, which does not appear to be cost effective. This may be another important factor that, besides the mechanical strength consideration of the fiber material, explains why early attempts at using twisted fiber for circular polarization maintaining in long length (as in coherent optical transmission) were virtually abandoned years ago.

4.8   SLIGHT TWISTING OF FIBER

Previous sections mostly concern strongly twisted fibers that behave more or less circularly birefringent, depending on the achievable twist-rate relative to the overall linear (intrinsic or extrinsic) birefringence of the fiber. In this section we are interested in exploring how a twisted fiber will behave if the twist rate is only slight.

our mathematical treatment is again based on the afore-derived coupled-mode equations for twisted fiber [Eq. (4.9)]. These equations are inexact, implying the fast-spun condition or the perturbation approximation. But nothing else is available that can be used as the analytic framework of our current subject. Apparently, when studying slight twisting, it is of special importance to examine the range of validity for the analytic results based on

Eq. (4.9). We shall discuss this point in Section 4.9. At the moment, we can only begin our study on slight twisting from the coupled-mode equations.

In the case of slight twisting, the coefficients (diagonal and off-diagonal) involved in the coupled-mode equations are of comparable magnitudes. Under such circumstances, it is generally not possible to adopt mathematical simplifications during derivation as we did in the case of strong twisting. In this case, we will have to choose the local coordinates for the coupled-mode formulation. The choice of base modes (circular or linear), is no longer important, because neither circular light nor linear light is predominant in slightly twisted fiber.

We use linear base modes in the local coordinates. Let the incident light be a right circular mode:

$$\mathbf{A}(0) = \frac{1}{\sqrt{2}} \begin{bmatrix} 1 \\ j \end{bmatrix} \qquad (4.46)$$

By the method of diagonalization, we derive $\mathbf{W}(0)$, $\mathbf{W}(z)$ from Eqs. (4.10) and (4.11), successively. Then once more by Eq. (4.10), we have

$$\mathbf{A}(z) = \frac{1}{\sqrt{2}} \begin{bmatrix} \cos gz + j(\cos 2\phi + \sin 2\phi)\sin gz \\ (\cos 2\phi - \sin 2\phi)\sin gz + j\cos gz \end{bmatrix}$$

$$\phi = \frac{1}{2}\arctan\frac{2(1-\varsigma)\tau}{\delta\beta} \qquad (4.47)$$

$$\lambda = j\left\{[\tau(1-\varsigma)]^2 + \left(\frac{\delta\beta}{2}\right)^2\right\}^{1/2}$$

Numerical calculation of Eq. (4.47) yields the curves in Figure 4.2. We can see that the SOP of the emergent light from a slightly twisted fiber pulsates in a violent manner. The magnitude of the SOP pulsation subsides for stronger twisting, until a fairly circular SOP is asymptotically approached at a sufficiently high rate of twisting.

The fairly complex-looking curves in Figure 4.2 are actually rather simple to explain, in view of the underlying physical principle. For slight twisting, the twist-induced circular birefringence is of comparable strength with the intrinsic linear birefringence. Both therefore act on the fiber to produce an elliptical SOP, which varies in a complicated oscillatory way. The SOP can be made to approach circular either by twisting the fiber at a high rate, or reducing the intrinsic linear birefringence until it becomes sufficiently low.

For a slightly twisted fiber, Figure 4.2 shows that the oscillatory variations of the SOP subside to an insignificant amplitude when the intrinsic linear birefringence is diminished to a level of a few degrees per meter. Such low intrinsic linear birefringence can be readily achieved by the spinning of fiber.

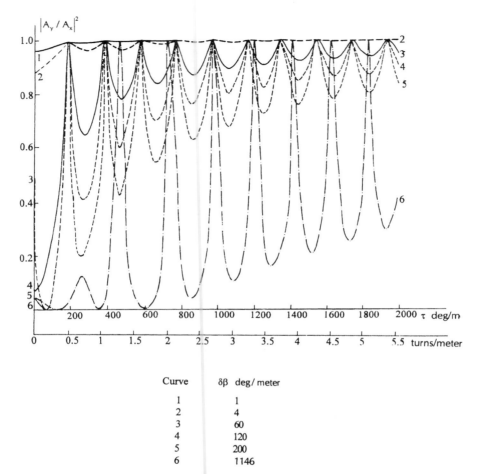

Curve	δβ deg/ meter
1	1
2	4
3	60
4	120
5	200
6	1146

FIGURE 4.2   Polarization behavior due to slight to strong twisting.

It is therefore a simple and practical way to realize a relatively stable circular SOP of light in a conventional fiber if the fiber is first spun during drawing, and then twisted afterward to form a twisted spun fiber.

The experimental data of a twisted spun-fiber specimen obtained in our laboratory are shown in Figure 4.3, wherein the data of a regular twisted fiber (not having been spun beforehand) are also shown for comparison. For the simple twisted-fiber specimen, dark and almost dark points occurred from slight twisting of the fiber (see sequence of +). For the twisted spun-fiber specimen of nearly the same length, no dark or almost dark points were found (see sequence of □) in a large number of repetitive testings [12].

Figure 4.4 shows the laboratory setup for making the fiber-twisting experiment. The fastener is so structured that the fiber passing through it is

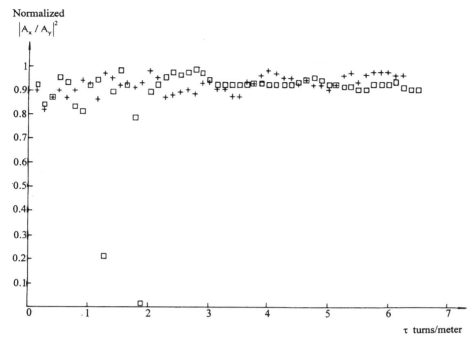

FIGURE 4.3 Experimental data of twisted fiber. $+$, twisted spun fiber; $\Box$, regular twisted fiber.

subjected to circumferentially symmetrical pressure, without suffering pressure in some preferred direction. Thus, the introduction of such fasteners into the experimental setup will not perturb this kind of fiber-twisting experiment [12].

4.9 NOTES

Some Views on Spun and Twisted Fibers

Concerning a uniformly spun fiber, it is known that the process of spinning a conventional fiber (with slight residual birefringence) causes this fiber to become lo-bi with respect to any kind of birefringence. The mechanism responsible for this lo-bi property of spun conventional fiber is attributed to the "averaging effect" of the spinning process. Indeed, this mechanism works not only in lo-bi fiber, but in principle at least, it also works in hi-bi nonconventional fibers, provided the respective spin-rate is sufficiently fast. Thus, as far as the fast-spinning technology will allow, currently available linear hi-bi fibers can be made to behave lo-bi by an exceedingly fast spin rate. In mathematical language, the required ratio of spin rate and the linear-bi, or $(\tau/\delta\beta)$, should be at least 10 or larger.

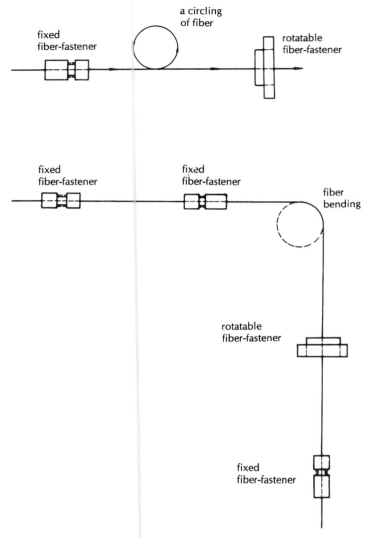

FIGURE 4.4 Setup for fiber twisting.

Meanwhile, in the utmost limit of fast spin, any noncircularity existing before the spin will be averaged so that the resulting fiber becomes circularly isotropic in the macroscopic sense. In a uniformly fast-spun (isotropic) fiber, any polarization mode (linear, circular, or elliptical) becomes an eigenmode. Such an "all-eigenmode" characteristic would be extraordinarily appealing, were it not for the fact that none of these theoretical eigenmodes ensures a stable transmission in a practical fiber line that includes even a very slight perturbation. As a matter of fact, these "all-eigenmode" and "all lo-bi" aspects are only the twofold inherence of a uniformly fast-spun fiber.

If a noticeable linear birefringence exists in fiber, it is natural and preferable to take its principal axes as the reference coordinates. This is the case even for a fiber that is not categorized as a spun-fiber, but is deployed in a natural way, such that the orientation of the set of principal axes is not invariant of the transmission direction z. Strictly speaking, the coordinate axes everywhere defined as coincident with the principal axes of fiber are not laboratory-fixed axes, but are local axes. In the presence of linear birefringence, except for a very short section of fiber in a careful layout, it is only practical to choose local axes. The inclusion of the laboratory-fixed axes is thus of significance only as an idealized model for mathematical analysis.

In the fast spin limit, an adequate choice of the reference coordinates for the fiber becomes a problem. The prespin principal axes of fiber, originally suitable for acting as the axes of reference, become a mere mathematical concept for the fiber after the fast spin. Practically speaking, if the spin tends to be very fast, one can scarcely define over a fiber cross-section an orthogonal pair of reference axes (or preferred directions) with respect to which measurements can be made without ambiguity. In this case, one would rather use the laboratory-fixed coordinates for analysis and measurements.

Some confusing points concerning spun fibers can be related to the choice of reference coordinates. For example, by plain intuition, a linear light of any orientation traversing a conventional spun fiber will just go straight because the fiber is isotropic or nearly so. Yet, in the local coordinates, which rotate in conformity with the spinning of fiber, a mathematical solution shows that a linear light traversing the fiber appears to counterrotate at an angular rate $-\tau$. This slightly deceptive view can be easily cleared up, if we rotate the local coordinates by an angle $-\tau z$ back to the laboratory-fixed coordinates, such that the final mathematical solution reappears as a straight-going linear light, in agreement with intuition.

The deceptive viewpoint becomes more intriguing if we go one step further to include the (postdraw) twist effect in a more general study of the rotational-type of fiber. For this composite case, which involves both spinning and twisting of fiber, the original coupling coefficient τ is replaced by $(1 - \varsigma)\tau$ or $(\tau - \alpha)$, where $\alpha = \varsigma\tau \approx (0.065-0.08)\tau$ is the twist-induced photoelastic optical rotation. Since α is small in comparison with τ (not 10% as large), purely mathematical reasoning might lead one to suppose that the effect of α is only a modification of the effect of τ. One would therefore expect that the end results derived from Eq. (2.12) and from Eq. (4.3) could not differ much, because $\varsigma \ll 1$. Also, when carrying out a usual mathematical derivation, it is not unreasonable to disregard a relatively small term in a step of simplification. Viewing it purely from the mathematical angle, it might therefore even be questioned if it is legitimate to neglect the small coefficient ς in the coupled-mode equation, Eq. (4.3).

But the preceding argument is conceptually misleading. To see this, it is simple, for example, to examine Eqs. (4.18a) and (4.18b), which describe the mathematical solutions of a strongly twisted fiber in the local coordinates and the fixed coordinates, respectively. The solutions show that the linear light appears to counterrotate by $-(\tau - \alpha)L = -\tau(1 - \varsigma)L$ after traversing the fiber of length L in the local coordinates, but appears to rotate by $\alpha L = \varsigma \tau L$ in the fixed coordinates. Thus, in the fixed coordinates, α (small as it is in comparison with τ) is the only factor that affects the transmission of light. We therefore see that different choices of the reference coordinates may lead to seemingly controversial interpretations of the mathematical results: while the effect of the geometrical spin seems to dominate (mathematically, $\tau \gg \alpha$) in the local coordinates, the elastooptic effect α becomes the only factor left and shows dominance in the fixed coordinates.

It is therefore my view that for fiber problems involving a small nongeometrical optical rotation (due to twist, Faraday effect, etc.) superposed on a large geometrical spin rate, it is feasible to choose the fixed coordinates, wherein the high spin rate produces an averaging effect on fiber at a macroscopic scale, thus allowing the small nongeometrical optical rotation to show its dominance. On the other hand, for the rotational type of fiber problems that involve geometrical spinning only, it is simple and natural to choose the local coordinates provided these coordinates are definable by the principal axes of the fiber. The adoption of local coordinates, wherever legitimate, is advantageous from the theoretical viewpoint, as well as from the experimental viewpoint.

Retarder–Rotator Formulation of Twisted Fiber

The retarder–rotator formulation of spun fiber is intuitively appealing in view of the connection it makes between fiber optics and the more familiar conventional optics. Nevertheless, because this formulation necessarily refers to the fixed coordinates, the required mathematical manipulation becomes much more laborious, even to speak of the general solution.

We recall that in Chapter 2, a theoretical framework for spun fiber was formulated by the "method of diagonalization." Meanwhile, following the early work of Kapron et al. [13] and Barlow et al. [14], the retarder–rotator formulation was also introduced in that chapter. As shown, the two formulations conform exactly. The present chapter is wholly devoted to the twisted fiber whose analysis has so far been based on the method of diagonalization. Before closing this chapter, we would like also to describe the twisted fiber in the language of the retarder–rotator formulation.

We start with the coupled-mode equations for twisted fiber in the local coordinates. Comparing these equations and those for spun fiber, we see that the difference is only in the off-diagonal coupling coefficients, with $(1 - \varsigma)\tau$ for the twisted fiber corresponding to τ for the spun fiber. We therefore take over the previous Eqs. (2.62)–(2.64) in our present derivation for twisted fiber:

$$R(z) = 2\sin^{-1}[\cos(2\phi)\sin(gz)] \tag{4.48}$$

$$\Omega(z) = \tau z - \tan^{-1}[\sin(2\phi)\tan(gz)] \tag{4.49}$$

$$\Phi(z) = \frac{1}{2}\tan^{-1}[\sin(2\phi)\tan(gz)] \tag{4.50}$$

in which the parameters ϕ and g are modified to take the following forms:

$$\phi = \frac{1}{2}\arctan\left[\frac{(1-\varsigma)\tau}{\delta\beta/2}\right]$$

$$g = \left\{[(1-\varsigma)\tau]^2 + \left(\frac{\delta\beta}{2}\right)^2\right\}^{1/2} \tag{4.51}$$

The above equations conform exactly with the retarder–rotator formulation by Barlow et al. [14]. Equations (4.48)–(4.51) apply to the more general case, where the geometrical spin τ and the elastooptic effect due to the postdraw twist $\varsigma\tau$ coexist. If no postdraw twist is present, Eqs. (4.48)–(4.51) reduce to the form of Eqs. (2.62)–(2.64) and (2.57).

For a highly twisted fiber ($\tau \gg \delta\beta$), we have from Eqs. (4.48) and (4.49):

$$R(z) \approx \frac{\delta\beta}{(1-\varsigma)\tau}\sin[(1-\varsigma)\tau z] \ll 1$$

$$\Omega(z) \approx \varsigma\tau z \tag{4.52}$$

Thus, the linear retardation becomes small, while the rotation becomes dominant.

Range of Validity of Coupled-Mode Equations for Twisted Fiber

In Sections 4.3 and 4.4, we derived the coupled-mode equations for twisted fiber whose standard matrix form in the local coordinates is given by Eq. (4.9). We adopted two different approaches to the formulation of this equation, and made the remark that both approaches are not exact, but approximate.

We recall that, methodologically, the said two approaches are entirely different. In the first approach, the coupled-mode equations for twisted fiber (in the local coordinates) are simply borrowed from the coupled-mode equations for spun fiber (also in the local coordinates). Without any derivation, the equations for twisted fiber are written out at the very beginning in the form of Eq. (4.3). This approach is approximate (see Section 4.3), inasmuch as it implies the "fast-spun" condition, so that the elastooptic effect of twisting can be viewed as a rotation, in the same sense as the twist, of the modal field-configuration by an angle αz about the z-axis.

In the second approach, the coupled-mode equations for twisted fiber are derived on the basis of the analytic results of Ref. [1]. Differing from the first approach, this one begins with establishing the coupled-mode equations in the fixed coordinates in the form of Eq. (4.7), and arrives at the coupled-mode equations in the local coordinates by a coordinate transformation. This second approach is also approximate (see Section 4.4), inasmuch as the starting matrix formulation of Eq. (4.7) is valid only under the condition of perturbation approximation, Eq. (4.7a), which refers to the fixed coordinates. It is worth noting that the perturbation approximation by Eq. (4.7a) in the fixed coordinates is exactly identical to the fast-spun condition by Eq. (4.4) in the local coordinates. This can be clearly seen if we simply transform the coupled-mode equations from one coordinate system to the other (i.e., from local to fixed, or vice versa). The coupled-mode equations for twisted fiber, in the form of Eq. (4.9), are therefore perfectly fitting to be applied to strong-twist problems, for which the fast-spun condition or the perturbation approximation is necessarily satisfied.

For slight-twisting problems, the polarization characteristics become greatly complicated. For convenience, we make reference to Section 4.8 entitled "Slight Twisting of Fiber." Note that in Figure 4.2, a set of curves were drawn according to Eq. (4.9). The twist rate τ varies from zero to about 5.5 turns/m—a range commonly understood to refer to slight twisting. The inherent linear-bi $\delta\beta$, however, are assumed to cover a fairly wide range—from 1 deg/m to 1146 deg/m. Thus, the so-called "fast-spun" condition can still be satisfied for τ taking a relatively large value on the abscissa, with $\delta\beta$ as a parameter taking a small value. Thus, for Eq. (4.9) to apply to the case of slight twisting, the inherent linear birefringence is strictly required to be exceedingly small. With respect to the curves in Figure 4.2, the ranges of validity obviously lie around the upper right region.

Even though the ranges of validity of the curves in Figure 4.2 are restricted by the "fast-spun" condition, these curves based on Eq. (4.9) do provide an overall picture of the SOP patterns caused by slight twisting. They are especially useful in illustrating the importance of reducing the intrinsic linear birefringence of the fiber in order to achieve a higher degree of circular birefringence in twisted fiber.

At this point it appears to be of much theoretical interest to ask: How about formulating a set of coupled-mode equations for a twisted fiber whose inherent residual linear-bi is not small, but fairly large? In other words: How do you treat analytically the kind of fiber that may well be called the "twisted hi-bi" fiber? In Chapter 6, we shall deal exclusively with the kind of special fiber called the "spun hi-bi" fiber. We shall observe that the coupled-mode equations derived initially in the form of Eq. (2.12) or Eq. (2.13) apply equally well to spun hi-bi fiber as they did earlier to spun lo-bi fiber. Now, in the case of twisted fiber, does Eq. (4.7) remain valid in the hi-bi case? The answer is in the negative. When k_{12} and k_{21} are not small compared to α, and so cannot be considered as perturbations terms, it becomes illegitimate to form, in a simple additive way, either of the off-diagonal elements in Eq. (4.7). Fortunately, it is clear enough that the kind of "twisted hi-bi" fiber cannot be useful anyway. Should it be useful in some way—and one wishes to analyze it purely out of theoretical or academic interest—one would find that there are no suitable coupled-mode equations available for the purpose. The mathematics would be formidably difficult, should one attempt such a task.

REFERENCES

[1] R. Ulrich and A. Simon, "Polarization optics of twisted single-mode fibers," *Appl. Opt.*, vol. 18, pp. 2241–2251 (1979).

[2] L. Jeunhomme and M. Monerie, "Polarization-maintaining single-mode fibre cable design," *Electron. Lett.*, vol. 16, pp. 921–922 (1980).

[3] S. Machida, J. Sakai, and T. Kimura, "Polarization preservation in long-length twisted single-mode optical fibers," *Trans. IECE Jpn.*, vol. E65, no. 11, pp. 642–647 (1982).

[4] J. Sakai, S. Machida, and T. Kimura, "Existence of eigen polarization modes in anisotropic single-mode optical fibers," *Opt. Lett.*, vol. 6, pp. 496–498 (1981).

[5] F. Gauthier, J. Dubos, S. Blaison, Ph. Graindorge, and H. J. Addity, "Attempt to draw a circular polarization conserving fiber," *Proc. Int. Congr. Fiberoptic Rotation Sensors*, MIT, Cambridge, pp. 196–200 (Nov. 1981).

[6] S. C. Rashleigh and R. Ulrich, "Magneto-optic current sensing with birefringent fibers," *Appl. Phys. Lett.*, vol. 34, pp. 768–770 (1979).

[7] A. J. Barlow and D. N. Payne, "Polarization maintenance in circularly birefringent fibers," *Electron. Lett.*, vol. 17, pp. 388–389 (1981).

[8] S. C. Rashleigh and R. Ulrich, "Polarization mode dispersion in single mode fibers," *Opt. Lett.*, vol. 3, pp. 60–62 (1978).

[9] T. Okoshi, "Coupled-mode formulation of twist-induced elasto-optic rotation," in *Optical Fibers for Communication*, sec. 6.6.4 (original in Japanese, 1983), Chinese translation by Post & Telecom Press, Beijing, p. 194 (1989).

[10] H. C. Lefevre, "Single-mode fibre fractional wave devices and polarisation controllers," *Electron. Lett.*, vol. 16, pp. 778–780 (1980).

[11] R. Ulrich, S. C. Rashleigh, and W. Eickhoff, "Bending induced birefringence in single mode fiber," *Opt. Lett.*, vol. 5, pp. 273–275 (1980).

[12] H. C. Huang and Y. C. He, "Polarization behavior of spun fiber versus conventional fiber under strong and slight twisting," *Microwave Opt. Tech. Lett.*, vol. 9, pp. 37–41 (1995).

[13] F. P. Kapron, N. F. Borrelli, and D. B. Keck, "Birefringence in dielectric optical waveguides," *IEEE J. Quantum Electron.*, vol. QE-8, pp. 222–225 (1972).

[14] A. J. Barlow, J. J. Ramskov-Hansen, and D. N. Payne, "Birefringence and polarization mode-dispersion in spun single-mode fibers," *Appl. Opt.*, vol. 20, pp. 2962–2967 (1981).

BIBLIOGRAPHY

Y. Fujii and K. Sano, "Polarization coupling in twisted elliptical optical fiber," *Appl. Opt.*, vol. 19, pp. 2602–2605 (1980).

K. Hotate and T. Ito, "Fibre ring resonator with stable eigenstate of polarisation using twisted single-mode optical fibre," *Electron. Lett.*, vol. 32, pp. 923–924 (1996).

N. R. Sadykov, "Propagation of circularly polarised radiation along a twisted path," *IEEE J. Quantum Electron.*, vol. QE-26, pp. 271–274 (1996).

R. E. Schuh, J. G. Ellison, L. M. Gleeson, S. S. R. Sikora, A. S. Siddigui, N. G. Walker, and D. H. O. Bebbington, "Theoretical analysis and measurement of the effect of fiber twist on the polarization OTDR of optical fibers," *OFC'96, Opt. Fiber Commun.*, vol. 2, pp. 297–298 (1996).

Chao-Xiang Shi, "A novel twisted Er-doped fiber ring laser: proposal, theory and experiment," *Opt. Commun.* (Elsevier), vol. 125, pp. 349–358 (1996).

Y. Yen and R. Ulrich, "Birefringence measurement in fiber optic devices," *Appl. Opt.*, vol. 20, pp. 2721–2725 (1981).

Dong Xiaopeng, Hu Hao, and Qian Jingren, "Measurement of fiber Verdet constant with twist method," *SPIE*, vol. 1572, pp. 56–60 (1991).

CHAPTER FIVE

Practical Circular-Polarization-Maintaining Optical Fiber

Currently, when one speaks of a highly birefringent (hi-bi) fiber, it is almost always taken for granted that hi-bi means high *linear*-birefringence (*linear*-bi). This is a natural reflection of the circumstance that in the exploitation of (linear and circular) hi-bi fibers over the past one and half decades or more, only linear hi-bi fibers have been successful, as notably evidenced by the advent of the Panda, the Bow Tie, the elliptical-cladding, and the flat fibers. Despite assiduous efforts being also made in the research and development (R & D) in the area of circular hi-bi fiber, nearly in parallel with the R & D of linear hi-bi fiber, the advancement of the former has been far less fruitful. Until now, the only practical means available for producing circular birefringence in fiber is still the early-known method of postdraw twisting. The previous chapter addressed the subject of the twist-induced circular birefringence in fiber.

The present chapter aims at an exposition, in a style that fits the book, of a new invention entitled "Practical circular-polarization-maintaining optical fiber," for which I own a U.S. patent [1]. For convenience, this novel fiber is given the name the "Screw fiber." A paper of mine also of the same title as the invention is just published [2].

5.1 DESCRIPTION OF RELATED ART

There have been several major impetuses for applied scientists to search for circular hi-bi fiber. One impetus came from the idea that using circular light in coherent communications will make it possible to eliminate the need of orienting fiber sections at splices [3–5]. Another impetus was due to the prospective applicability of such fiber in all-fiber gyroscope architecture. Still another impetus came from the demand for some kind of fiber capable of sensing the Faraday effect, and meanwhile capable of maintaining the sensed Faraday effect.

In the preceding chapter, we studied the postdraw twisted fiber [6]. Such fiber was shown to be impractical essentially on two counts. First, the mechanical strength of glass fiber limits the allowable twist rate, inasmuch as the fiber will crack before a sufficient circular birefringence is reached [7]. Second, a twisted fiber, being in a constrained state, is not easy to handle. Therefore, the twisted fiber is not practical in application, though in laboratory such fiber furnishes a useful medium for certain experiments of fundamental value or academic interest.

Primarily for use in gyroscope [8], a special "spin and draw" machine was devised early that attempted "to freeze the twist in line while drawing the fiber." According to Ref. [8], "the twist is transmitted to the molten cone of glass by the elasticity of the fiber." Nevertheless, this and other similar attempts did not succeed in preserving circular birefringence. Any attempt to fix up the twisting state of fiber at a "hot" condition would tend to diminish the otherwise existing birefringence of any kind.

A more recent approach to lightwave transmission in circular state of polarization (SOP) is essentially to make use of the geometrical effect of a helical core, or of the geometrical and torsional effects due to a helical winding [9–15]. This kind of fiber did arouse a great deal of interest, and hence many articles and papers, in the fiber-optic community worldwide. However, there are several inherent difficulties that make such fiber impractical. For one thing, the fiber must be of an unusually larger dimension. More serious are the problems of launching of light into the helical core, the radiative nature of the mode, and the joining or splicing of this type of fiber segments.

But the effort of searching for circular hi-bi fiber has never discontinued. Nearly about the same time, there also evolved in the literature and patent documents a family of so-called circularly "form-birefringent" fibers whose common working principle is to generate a circular birefringence by spinning an intentionally made azymuthal-dependent index distribution of the core [16–19]. Such fibers are alternatively called the twisted "multilobe-core" fibers, and include a variety of fiber versions with different index patterns in the core, and have such names as "Spiral" for fiber with a core of one lobe, "Twisted Cross" or "Clover-Leaf" for fiber with a core of four lobes, and "Octopus" for fiber with a core of eight lobes [20, 21], and so on. Of all these proposed fibers, the only published information about an actually fabricated specimen is the one-lobe "Spiral" fiber, which closely resembles a helix fiber. From the standpoint of fabrication, the core configurations are so sophisticated that these fibers would be extremely difficult, if not impossible, to fabricate. From the standpoint of transmission theory, launching the incident light onto the fiber and splicing the fiber sections of this class, among other things, inevitably pose serious technological problems.

5.2 PRACTICAL CIRCULAR-POLARIZATION-MAINTAINING OPTICAL FIBER

The thought of using circular light instead of linear light in certain fiber-optic circuitry or transmission system is appealing at least on two counts. One is the ease of splicing, inasmuch as circular light has no preferred orientations over the fiber cross section, such that no principal-axes aligning is required. The other is the use of circular hi-bi fiber to detect the Faraday effect and also to hold it. In addition to these special advantages inherent to circular light, a practical circular hi-bi fiber must also possess some other qualifications in common with a practical fiber of any kind. These include, for example, relative ease of fabrication without too much extra need of fabrication facilities, low cost, and compatibility with the existing single-mode fiber technology.

From experience, we know that birefringence can generally be induced in fiber by two different mechanisms: the geometry-induced birefringence and the stress-induced birefringence. The former surpasses the latter in regard of essential immunity from the effect of temperature. Nevertheless, the early study of linear-bi fibers [22] revealed that in practice (not in principle) the geometry-induced birefringence is orders of magnitude too small as compared with the stress-induced birefringence. With this technical background in mind, we had to give up attempting to find a geometry-induced circular hi-bi fiber, and direct our whole attention on the search for a stress-induced circular hi-bi fiber.

The aim of the present chapter is to reveal a new approach to practical circular hi-bi fiber. The structure comprises an on-axis straight core and an off-axis stress filament that helically whirls around the core (Fig. 5.1a). By intuitive reasoning, the azymuthally varying stress acting on the core is expected to produce a stress-induced circular birefringence. The method that is suitable for making such fiber structure is apparent enough. We use a very peculiar preform comprising an on-axis core rod, and *one* off-axis stress rod (Fig. 5.1b). The preform is spun in its hot state during the linear draw, yielding the kind of fiber shown in Fig. 5.1a. Being spun during the fiber-drawing process, the residual "noncircular" birefringence in the core is further diminished due to the averaging effect, which is equivalent to enhancing the effective circular birefringence induced by the off-axis whirling stress filament.

Here is the secret of making a circular hi-bi fiber. From Figures 5.1a and 5.1b it is seen that the cross-sectional configuration of the fiber (or of the preform) is asymmetrical. That is, the mirror-symmetrical geometry, being customarily utilized in the making of linear hi-bi fibers, is intentionally ruined in the present case. The linear-bi associated with such mirror-symmetry is always a hindrance to the realization of circular-bi. The rotational technique will not help if this intrinsic linear-bi is not effectively eliminated. That is why spun Bow-Tie and other spun hi-bi fibers can never become circular-bi,

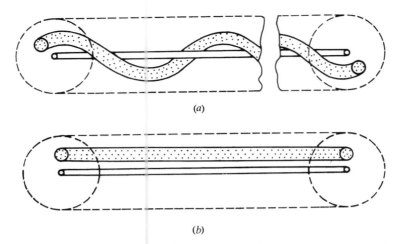

(a)

(b)

FIGURE 5.1 Fiber and preform. (*a*) Structure of a Screw fiber: core (on-axis straight), stress-filament (off-axis helical). (*b*) Preform: core rod (on-axis straight), stress rod (off-axis straight).

however fast the spin rate is. Actually, spinning linear hi-bi fibers very fast during the "hot" state results in lo-bi fibers due to the "averaging" effect.

At this point, it is interesting to compare, from the angle of design and fabrication, the circular hi-bi Screw fiber and the existing varieties of linear hi-bi fiber. For linear hi-bi fibers, the two stress regions are strictly required to be mirror-symmetrical with respect to the fast axis passing the core center. Any slight deviation of this geometrical mirror-symmetry, or slight deviation of the two constituent stress regions from being exactly identical in configuration and doping distribution, will degrade the resulting linear birefringence of the fiber. On the other hand, for the circular hi-bi Screw fiber the core and stress-filament configuration over any cross section of the fiber is intentionally made *asymmetrical*, such that the only geometrical parameter involved is the off-axis distance of the stress filament, whose tolerance is rather loose. It is therefore conceivable that the circular hi-bi Screw fiber should be easier to make than the various kinds of linear hi-bi fiber.

Using our small homemade fiber-drawing tower (with a d.c. motor incorporated on top), we fabricated our first several specimens of Screw fiber in 1994. Figure 5.2 is a microphoto of a cross section of a typical Screw fiber specimen.

In principle, it is also possible to make circular hi-bi fibers comprising more stress filaments [2] (see also Figures C2*a* and C2*b* in Appendix C). The placement of the multiple stress filaments allows the residual "noncircular" birefringence to be further reduced, thus further enhancing the circular birefringence. Nevertheless, using more stress filaments increases the complexity of the design and fabrication. Very likely, the simplest is the most practical. In this chapter we focus on the simple Screw fiber with one stress filament only.

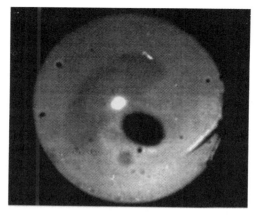

FIGURE 5.2   Microphoto of Screw fiber cross section (initial specimen fabricated in 1994).

5.3   COUPLED-MODE ANALYSIS OF SCREW FIBER

Coupled-Mode Equations in Local Coordinates

Like the strongly twisted fiber, the coupled-mode equations for the Screw fiber in the *local* coordinates can be formulated in a phenomenological way as follows:

$$\frac{d\mathbf{A}}{dz} = \mathbf{K}\mathbf{A}$$

$$\mathbf{A} = \begin{bmatrix} A_x \\ A_y \end{bmatrix}$$

$$\mathbf{K} = \begin{bmatrix} j\dfrac{\delta\beta}{2} & (1 - \kappa)\tau \\ -(1 - \kappa)\tau & -j\dfrac{\delta\beta}{2} \end{bmatrix}$$

(5.1)

where A_x and A_y are the local modes. The expressions of the elements of $\mathbf{K}$ depend on the choice of the reference coordinates (local or fixed), as well as on the choice of the base modes (linear or circular). The above form of $\mathbf{K}$ applies to the case of local coordinates and linear base modes, wherein $\delta\beta$ is the phase-velocity difference, τ is the built-in spin rate of the fiber, and κ is a characteristic figure descriptive of the stress-induced optical rotation in the Screw fiber.

The mathematical form of Eq. (5.1) for Screw fiber closely resembles the well-established coupled-mode formulation for a twisted fiber, with the characteristic figure κ of the former corresponding to the characteristic figure ς (≈ 0.08) of the latter. In the mathematical formulation, one can scarcely detect an essential difference between a Screw fiber and a strongly twisted fiber. Both behave circularly birefringent. In physical and technical particulars, however, there is a crucial difference between the two fibers that is fundamentally important, that is, in the former, the rotation of fiber and hence the formation of the azymuthally varying stress acting on the core is *built in* during the fabrication process, while in the latter, the elastooptic rotation is formed in an "impermanent" constrained state.

Specifying the local coordinates for a Screw fiber poses some analytic complexity. This can be seen when it is compared with the previous case of a twisted fiber. For a twisted fiber, this specification is simple. Practically, the core is always slightly elliptical, which provides a mirror-symmetry with respect to a certain pair of axes (i.e., the principal axes), such that the local coordinates are defined naturally as being identical to these principal axes. In the case of a Screw fiber, however, things are complicated because of the very unusual transverse structure of the fiber. The cross section of a Screw fiber is asymmetrical. While the core locally keeps the small ellipticity, the usual mirror-symmetry of fiber geometry is completely lost. It therefore becomes difficult to ascertain quantitatively the proportion of the residual linear birefringence relative to the effective circular birefringence. Regarding the local cross-sectional configuration of the fiber shown in Figure 5.2, the strength of the residual linear birefringence cannot be large, because the single stress filament applies only a one-sided push or pull to the core (depending on the kind of dopant in the stress filament). On the other side of the core, the jacket is relatively remote and flat so that the stress pattern around the core is likely to be scattered rather than concentrated. Intuitively, this kind of one-sided stress is expected to be much smaller than in the case of a linear hi-bi fiber, where the the core is acted upon by a two-sided force due to the two stress filaments placed symmetrically. Because of this apparent complication, an exact coupled-mode formulation of the Screw fiber is mathematically difficult. It is therefore desirable to note that, while the coupled-mode equations for Screw fiber, Eq. (5.1), are closely analogous to those for twisted fiber, the former are more phenomenological in nature. Here, we make no further attempt to formulate a sophisticated and (seemingly) more rigorous treatment of the relevant transverse boundary-value problem, because the mathematics would be too complicated and we would stray too far from the subject under study. From a practical point of view, it seems prudent to regard the coefficient κ in Eq. (5.1) as a characteristic figure that can be readily determined by, for example, the beat-length measurement [see Eq. (5.28a)].

The Method of Diagonalization

We consistently use the method of diagonalization for solutions of the coupled-mode equations, Eq. (5.1). We introduce the transformation [23]

$$\mathbf{A} = \mathbf{OW} \qquad (5.2)$$

where

$$\mathbf{W} = \begin{bmatrix} W_x \\ W_y \end{bmatrix}$$

$$\mathbf{O} = \begin{bmatrix} \cos\phi & j\sin\phi \\ j\sin\phi & \cos\phi \end{bmatrix}$$
(5.2a)

$$\phi = \frac{1}{2} artan\left[\frac{2(1-\kappa)\tau}{\delta\beta}\right]$$
(5.2b)

Substituting Eq. (5.2) into Eq. (5.1) yields:

$$\frac{d\mathbf{W}}{dz} = \Lambda\mathbf{W}$$
(5.3)

$$\Lambda = \mathbf{O}^{-1}\mathbf{KO} = \begin{bmatrix} jg & 0 \\ 0 & -jg \end{bmatrix}$$
(5.3a)

$$g = \left\{ \left(\frac{\delta\beta}{2}\right)^2 + [(1-\kappa)\tau]^2 \right\}^{\frac{1}{2}}$$

Integrating Eq. (5.3) yields

$$\mathbf{W}(z) = \tilde{\Lambda}\mathbf{W}(0)$$

$$\tilde{\Lambda} = \begin{bmatrix} e^{jgz} & 0 \\ 0 & e^{-jgz} \end{bmatrix}$$
(5.4)

where jg and $-jg$ are the eigenvalues, and e^{jgz} and e^{-jgz} are the corresponding eigenfunctions. If the common wave factor $e^{j(\omega t - \beta z)}$ is written out, we can see that e^{jgz} and e^{-jgz} represent a fast mode and a slow mode, respectively. By the transformation Eq. (5.2), the local-mode solutions for A_x and A_y are obtained by the normal-mode solutions for W_x and W_y. See Appendix B for a more complete description of the normal modes, or eigenmodes, in the normal coordinates.

The Transfer Matrix

The above equations provide a complete theoretical framework for solving the coupled-mode equations by the method of diagonalization. The method is simple and natural by intuitive reasoning. By the transformation Eq. (5.2), the simultaneous differential equations become independent differential equations, and hence readily solvable. Thus, by Eqs. (5.2)–(5.4), the general solution of a Screw fiber in the local coordinates is given by

$$A(z) = T_l A(0)$$
$$T_l = O\tilde{\Lambda}O^{-1}$$

$$(5.5)$$

The solution given by Eq. (5.5) is the general solution, where the initial condition $A(0)$ is considered arbitrary. The local transfer matrix T_l in Eq. (5.5) is an inherent characteristics of the fiber, regardless of the kind of input light.

While the general solution to the problem is put in as simple and neat a form as Eq. (5.5), actual calculation of this equation in the general case will involve very complicated mathematical manipulations. We therefore make recourse to asymptotic approximation before proceeding to the analytic study of various practical transmission problems.

5.4   ASYMPTOTIC APPROXIMATIONS

As we did in Eq. (4.4), which defines the strong-twist condition for twisted fiber, we can similarly define a fast-spun condition for the Screw fiber in the following mathematical form:

$$\frac{(1 - \kappa)\tau}{\delta\beta} \gg 1$$

$$(5.6)$$

There is a fundamental difference between Eq. (4.4) and Eq. (5.6) in respect to technological practice. For the former, the strong-twist (in the cool state of fiber) is a severe condition that is not easy to achieve within the mechanical strength of glass. For the latter, the fast-spinning (in the hot state of fiber) is always satisfied by the existing fabrication technique. Equation (5.6) therefore no longer poses a condition to technology, but simply expresses a matter of fact.

Note that, like Eq. (4.4) for twisted fiber, the fast-spun condition, Eq. (5.6), for Screw fiber, is a mathematical condition that admits the following asymptotic approximations for solutions of the coupled-mode equations in the local coordinates. From the standpoint of the theory of differential equations, the ratio on the left side of Eq. (5.6) is, with reference to the coupled-mode equations, Eqs. (5.1), the ratio of the off-diagonal coefficient $(1 - \kappa)\tau$ to the difference between the diagonal coefficients $(\delta\beta/2) - (-\delta\beta/2) = \delta\beta$. Note that $\delta\beta$ is not exactly interpretable as residual "linear" birefringence in the current case. As afore-described, the local transverse structure of Screw fiber is asymmetrical, resulting in a distribution of complex (not simply linear) birefringence. To be exact, it is therefore appropriate to view $\delta\beta$ as a mathematical coefficient in the coupled-mode formulation that is descriptive of the overall *noncircular* birefringence components, or simply as an "equivalent" linear birefringence.

Under the fast-spun condition, Eq. (5.6), $\phi \approx \pi/4$ such that

$$\mathbf{O} \to \frac{1}{\sqrt{2}} \begin{bmatrix} 1 & j \\ j & 1 \end{bmatrix} \qquad (5.7)$$

$$\mathbf{O}^{-1} \to \frac{1}{\sqrt{2}} \begin{bmatrix} 1 & -j \\ -j & 1 \end{bmatrix} \qquad (5.8)$$

In the above two expressions, it is tacitly assumed that the fiber spins in the specified $(+)$ sense. If the fiber spins in the $(-)$ sense, the corresponding expressions become

$$\mathbf{O} \to \frac{1}{\sqrt{2}} \begin{bmatrix} 1 & -j \\ -j & 1 \end{bmatrix} \qquad (5.9a)$$

$$\mathbf{O}^{-1} \to \frac{1}{\sqrt{2}} \begin{bmatrix} 1 & j \\ j & 1 \end{bmatrix} \qquad (5.9b)$$

Under the same fast-spun condition, g in Eq. (5.3a) is simplified

$$g \to (1 - \kappa)\tau \qquad (5.10)$$

Equations (5.7)–(5.10) are asymptotic in the sense that the approximation involved is legitimate under the fast-spun condition, Eq. (5.6).

5.5   GENERAL SOLUTION OF SCREW FIBER

General Solution of Screw Fiber in the Local Coordinates

With the aid of the preceding asymptotic approximations, the general solution, Eq. (5.5), which is descriptive of the inherent fiber property, can be treated analytically and expressed in a simple and useful form. Thus, substituting Eqs. (5.7)–(5.10) into Eq. (5.5) yields

$$\mathbf{A}(z) = \mathbf{T}_l \mathbf{A}(0)$$

$$\mathbf{T}_l \approx \begin{bmatrix} \cos[(1 - \kappa)\tau] & \sin[(1 - \kappa)\tau] \\ -\sin[(1 - \kappa)\tau] & \cos[(1 - \kappa)\tau] \end{bmatrix} \qquad (5.11)$$

The general solution, Eq. (5.11), refers to the local coordinates. It is approximate in the sense of asymptotic approximations, but applies generally without regard to the kind of input light exciting the fiber.

General Solution of Screw Fiber in the Fixed Coordinates

For circular hi-bi Screw fiber of our current concern, it is possible and feasible to carry out the analysis in the fixed coordinates. The local coordinates deviate from the fixed coordinates progressively by a rotational angle (τz), that is, the angular deviation increases indefinitely along the transmission direction z at a rate τ. Thus, for any solution in the local coordinates, we shall have to counterrotate the local coordinates by ($-\tau z$) in order to obtain the corresponding solution in the fixed coordinates. The matrix that effects this counterrotation is

$$\mathbf{R}(-\tau z) = \begin{bmatrix} \cos(\tau z) & -\sin(\tau z) \\ \sin(\tau z) & \cos(\tau z) \end{bmatrix} \tag{5.12}$$

The transfer matrix in the fixed coordinates, identifiable to the Jones matrix, is thus derivable from Eqs. (5.12) and (5.11):

$$\hat{\mathbf{T}} = \mathbf{R}(-\tau z)\mathbf{T}_l$$

$$= \begin{bmatrix} \cos(\kappa\tau z) & -\sin(\kappa\tau z) \\ \sin(\kappa\tau z) & \cos(\kappa\tau z) \end{bmatrix} \tag{5.13}$$

where the circumflex ($\wedge$) refers to the fixed coordinates. The general solution of the Screw fiber in the fixed coordinates is then given by

$$\hat{\mathbf{A}}(z) = \hat{\mathbf{T}}\hat{\mathbf{A}}(0)$$

$$= \begin{bmatrix} \cos(\kappa\tau z) & -\sin(\kappa\tau z) \\ \sin(\kappa\tau z) & \cos(\kappa\tau z) \end{bmatrix}\hat{\mathbf{A}}(0) \tag{5.14}$$

where the initial condition $\hat{\mathbf{A}}(0)$ is yet taken to be arbitrary.

A particular solution is defined as a solution that satisfies the coupled-mode equations as well as the specified initial condition. As the initial conditions can be indefinitely many, the corresponding particular solutions are likewise indefinitely many. Putting the initial condition $\mathbf{A}(0) = \hat{\mathbf{A}}(0)$ into Eqs. (5.11) and (5.14) yields, respectively, a particular solution in the local coordinates and in the fixed coordinates. The transfer-matrix calculation yields the final results without describing the transmission process involved. In other words, the underlying physical meaning is not revealed when this short-cut is used.

It is therefore desirable to classify the numerous particular solutions in a simple and practically useful way. The classification depends on whether a single eigenmode is involved, or both eigenmodes are involved, in the course of transmission. To distinguish the single eigenmode scheme from the dual-eigenmode (or bi-eigenmode) scheme, we need to derive the respective solutions using a step-by-step procedure instead of using Eq. (5.11) or Eq. (5.14) directly.

5.6 SINGLE-EIGENMODE TRANSMISSION IN SCREW FIBER

We follow the method of diagonalization as outlined in Section 5.3. Let it be required that light transmission in a circular hi-bi fiber involves one and only one eigenmode, say the fast eigenmode $W_x(z) = e^{jgz}$, implying $W_y(z) \equiv 0$ for all z, then the initial condition for **W** should be

$$\mathbf{W}(0) = \begin{bmatrix} 1 \\ 0 \end{bmatrix} \tag{5.15}$$

By Eqs. (5.7) and (5.2), the initial condition for **A** is given by

$$\mathbf{A}(0) = \mathbf{OW}(0)$$

$$= \frac{1}{\sqrt{2}} \begin{bmatrix} 1 \\ j \end{bmatrix} \tag{5.16}$$

This is a circular light, specified as right circular for definiteness.

Under the fast-spun condition, $g \approx (1 - \kappa)\tau$, such that, by Eq. (5.4),

$$\mathbf{W}(z) = \tilde{\Lambda}\mathbf{W}(0)$$

$$= \begin{bmatrix} e^{j(1-\kappa)\tau z} & 0 \\ 0 & e^{-j(1-\kappa)\tau z} \end{bmatrix} \begin{bmatrix} 1 \\ 0 \end{bmatrix} \tag{5.17}$$

$$= \begin{bmatrix} 1 \\ 0 \end{bmatrix} e^{j(1-\kappa)\tau z}$$

By Eq. (5.17), it is easy to derive the particular solution for the initial condition, Eq. (5.16):

$$\mathbf{A}(z) = \mathbf{OW}(z)$$

$$= \frac{1}{\sqrt{2}} \begin{bmatrix} 1 \\ j \end{bmatrix} e^{j(1-\kappa)\tau z} \tag{5.18}$$

The above derivations all refer to the local coordinates. In the fixed coordinates, formulation of the corresponding particular solution of the same initial-value problem can be derived from Eq. (5.12):

$$\hat{\mathbf{A}}(0) = \mathbf{A}(0) = \frac{1}{\sqrt{2}}\begin{bmatrix} 1 \\ j \end{bmatrix}$$

$$\hat{\mathbf{A}}(z) = \mathbf{R}(-\tau z)\mathbf{A}(z) \tag{5.19}$$

$$= \frac{1}{\sqrt{2}}\begin{bmatrix} 1 \\ j \end{bmatrix}e^{-j\kappa\tau z}$$

Comparing Eq. (5.18) with Eq. (5.19), we see that the only difference between the solution in the local coordinates and that in the fixed coordinates is in the exponential phase factor, not in the SOP of light.

Similarly, if light transmission is required in the slow eigenmode $W_y(z) = e^{-j g z}$, implying that $W_x(z) \equiv 0$ for all z, then a step-by-step derivation yields:

$$\mathbf{W}(0) = \begin{bmatrix} 0 \\ 1 \end{bmatrix}$$

$$\mathbf{A}(0) = \mathbf{OW}(0) = \frac{j}{\sqrt{2}}\begin{bmatrix} 1 \\ -j \end{bmatrix} \tag{5.20}$$

$$\mathbf{W}(z) = \tilde{\Lambda}\mathbf{W}(0)$$

$$= \begin{bmatrix} e^{j(1-\kappa)\tau z} & 0 \\ 0 & e^{-j(1-\kappa)\tau z} \end{bmatrix}\begin{bmatrix} 0 \\ 1 \end{bmatrix} = \begin{bmatrix} 0 \\ 1 \end{bmatrix}e^{-j(1-\kappa)\tau z}$$

$$\mathbf{A}(z) = \mathbf{OW}(z) \tag{5.21}$$

$$= \frac{j}{\sqrt{2}}\begin{bmatrix} 1 \\ -j \end{bmatrix}e^{-j(1-\kappa)\tau z}$$

This result refers to the local coordinates. In the fixed coordinates, we again use Eq. (5.12) to derive the corresponding particular solution of the initial-value problem wherein a left circular light is launched to excite the fiber:

$$\hat{\mathbf{A}}(0) = \mathbf{A}(0) \tag{5.22}$$

$$\hat{\mathbf{A}}(z) = \mathbf{R}(-\tau z)\mathbf{A}(z)$$

$$= \frac{j}{\sqrt{2}}\begin{bmatrix} 1 \\ -j \end{bmatrix}e^{j\kappa\tau z} \tag{5.23}$$

Comparing Eq. (5.21) with Eq. (5.23) again reveals the simple fact that in single-eigenmode transmission, the particular solution in the local coordinates and that in the fixed coordinates differ by an exponential phase factor only.

The preceding step-by-step derivation yields the following fundamental results:

1. Circular modes of both senses (right and left) are the eigenmodes of the kind of Screw fiber concerned.
2. If the fiber is excited by a circular light (either right or left), the propagating lightwave along the fiber will be this and only this circular light during the entire course of transmission.

These analytic results agree exactly with intuition. The single eigenmode transmission is realized when the incident light matches with either eigenmode (right circular or left circular) of the fiber at the input end $z = 0$. The condition for single-eigenmode transmission is therefore simply the "matching condition" of the exciting light and the eigenmode of the fiber.

5.7 DUAL-EIGENMODE TRANSMISSION IN SCREW FIBER

If the initial condition (or excitation condition) does not match one or the other eigenmode of the fiber, then both eigenmodes will be excited at the input and propagate along the fiber. In fact, circular light excitation is the only case that satisfies the matching condition at the input, thereby causing single-eigenmode transmission in the circular hi-bi Screw fiber. For input light of all other noncircular SOPs, both eigenmodes will be excited, thereby causing dual-eigenmode transmission in the fiber. We consider linear light excitation first because of its particular importance in practical application.

Linear Light Excitation

Let the incident linear light be given by

$$A(0) = \begin{bmatrix} \cos\theta \\ \sin\theta \end{bmatrix} \tag{5.24}$$

where θ is the input orientation angle. By Eqs. (5.24), (5.8), (5.4), and (5.7), successively:

$$W(0) = O^{-1}A(0) = \begin{bmatrix} e^{-j\theta} \\ -je^{j\theta} \end{bmatrix}$$

$$W(z) = \tilde{\Lambda}W(0) = \begin{bmatrix} e^{-j[\theta-(1-\kappa)\tau z]} \\ -je^{j[\theta-(1-\kappa)\tau z]} \end{bmatrix} \tag{5.25}$$

$$A(z) = OW(z)$$

$$= \begin{bmatrix} \cos[\theta-(1-\kappa)\tau z] \\ \sin[\theta-(1-\kappa)\tau z] \end{bmatrix}$$

The above particular solution for a linear light excitation refers to the local coordinates. This solution can be readily transformed to refer to the fixed coordinates by the coordinate transformation, Eq. (5.12). In the fixed coordinates the particular solution of the relevant initial-value problem can be formulated as

$$\hat{\mathbf{A}}(0) = \mathbf{A}(0) = \begin{bmatrix} \cos\theta \\ \sin\theta \end{bmatrix}$$

$$\hat{\mathbf{A}}(z) = \mathbf{R}(-\tau z)\mathbf{A}(z) \tag{5.26}$$

$$= \begin{bmatrix} \cos(\theta + \kappa\tau z) \\ \sin(\theta + \kappa\tau z) \end{bmatrix}$$

By Eqs. (5.25) and (5.26), with a linear light exciting the fiber at the input, the output light keeps the linear SOP, but not the orientation, either in the local coordinates or in the fixed coordinates. In the local coordinates, the linear light precesses around the z axis at a rate $-(1 - \kappa)\tau$ such that the orientation of the linear light at the output makes a total angle of $-(1 - \kappa)\tau L$, where L is the length of the fiber, from the initial orientation θ at the input. In the fixed coordinates, the linear light precesses around the z axis at a rate $(\kappa\tau)$ such that the orientation of the linear light at the output makes a total angle of $(\kappa\tau L)$ from the initial orientation θ at the input.

For a Screw fiber with a reversed sense of spin, the linear light will counter-precess around the z axis at a rate $(1 - \kappa)\tau$ in the local coordinates. In the fixed coordinates, the linear light will precess at a rate of $(-\kappa\tau)$.

We recall that in our previous treatment of twisted fiber with a high twist rate, the fiber behaves circularly hi-bi, such that linear light precesses in the fiber. In fact, precession of linear light is a basic property of circular hi-bi fiber of any variety. As we said, the precessing linear light is *not* an eigenmode of the circular hi-bi fiber, even though the linear polarization is maintained during light transmission. The qualifications of an eigenmode imply not only the invariance of the field pattern, but also the invariance of the orientation, in the transmission direction z. Nothing changes during eigenmode transmission, except for an exponential phase shift descriptive of the wave motion.

When making an experiment, it is not improbable to mistake linear light for an eigenmode of a circular hi-bi fiber. This is because, when we launch a linear light onto the input of the fiber, we will always detect a linear light at the output. When we change the orientation of the input light by a certain angle, the orientation of the output light will change by the same angle. According to the previous analytic study, the output light actually undergoes a precession angle equal to $(\kappa\tau L)$ in the fixed coordinates, but this is a constant angle always involved in the experiment, and hence easily slips the attention of an observer. What one observes is simply the phenomenon of "linear light in, linear light out," as well as a subsequent change in orientation at the output exactly equal to the change in orientation at the input.

Here we therefore wish to put sufficient emphasis on the analytic result that a precessing linear light is not an eigenmode in the circular hi-bi Screw fiber of our current concern. In this task, we decompose the linear light at $z = 0$ into the two circular constituents (exactly the two eigenmodes of the circular hi-bi fiber):

$$\hat{\mathbf{A}}(0) = \mathbf{A}(0) = \begin{bmatrix} \cos \theta \\ \sin \theta \end{bmatrix}$$

$$= \frac{1}{2} \begin{bmatrix} 1 \\ j \end{bmatrix} e^{-j\theta} + \frac{1}{2} \begin{bmatrix} 1 \\ -j \end{bmatrix} e^{j\theta} \tag{5.27}$$

We use the fixed coordinates, while recognizing that use of the local coordinates will lead to the same conclusion. Since the two circular eigenmodes have phase velocities equal in magnitude but opposite in sign, in the fixed coordinates they recombine at the output to yield

$$\hat{\mathbf{A}}(z) = \frac{1}{2} \begin{bmatrix} 1 \\ j \end{bmatrix} e^{-j(\theta + \kappa\tau z)} + \frac{1}{2} \begin{bmatrix} 1 \\ -j \end{bmatrix} e^{j(\theta + \kappa\tau z)} \tag{5.27a}$$

$$= \begin{bmatrix} \cos (\theta + \kappa\tau z) \\ \sin (\theta + \kappa\tau z) \end{bmatrix} \tag{5.27b}$$

Naturally we arrive at the same result as Eq. (5.26). The precessing linear light is not to be confused with the single-eigenmode transmission in fiber. Instead, both eigenmodes (right and left circular) are involved. By the above derivation, the incident linear light is seen as superposition of the two circular eigenmodes (right and left) of equal amplitudes, but opposite senses. These circular eigenmodes have different phase velocities (in the sense of equal magnitudes but opposite signs), so that they beat along the fiber. It happens that the resulting beating pattern appears in the form of a precessing linear light, that is, the two circular light components at any point always combine to become a linear light, but with an orientation that varies continuously in the transmission direction.

The precessing linear light in circular hi-bi fiber is actually a special form of beating. For linear light excitation and transmission, the beat length L_b is the period of the precessing linear light. As known, the beat length in the fixed coordinates is descriptive of the perturbation resistivity of the fiber. Thus, by Eq. (5.27a), the phase velocity difference of the two circular eigenmodes is $2\kappa\tau$, such that the beat length of the Screw fiber in the fixed coordinates is determined from the equality $2\kappa\tau L_b = 2\pi$:

$$L_b = \frac{\pi}{\kappa\tau} \tag{5.28}$$

It is noted here that in fiber optics publications, the beat length is defined in two ways, leading to two formulas that differ by a factor of 2. Consider Eq. (5.27b). The SOP of light is reproduced when $\kappa\tau z = \pi(1 + m)$, $m = 0, 1, 2 \ldots$. The beat length is the shortest distance in which the SOP of light reproduces itself. When $m = 0$, a negative sign appears before the matrix on the right of Eq. (5.27b). In waveguide practice, we are often more interested in the pattern of power (square of the field strength), such that the said negative sign does not need to be considered. Thus, for $m = 0$, we have $L_b = \pi/(\kappa\tau)$, as given by Eq. (5.28). Some authors prefer to define the beat length by letting $m = 1$, so that Eq. (2.27b) will be completely reproduced. The beat length so defined is therefore twice as large as Eq. (5.28), that is, $L_b = 2\pi/(\kappa\tau)$.

By Eq. (5.28), we have

$$\kappa = \frac{\pi}{\tau L_b} \tag{5.28a}$$

Since L_b (in the fixed coordinates) is a readily measurable quantity, and the spin rate τ (a structural parameter) is known a priori, the fundamental parameter κ of the Screw fiber can be determined by Eq. (5.28a) if L_b is available. The above equation thus suggests a simple and practical way for the determination of κ.

Elliptical Light Excitation

Using the (ε, ψ)-representation given by Eq. (2.45), an arbitrary elliptical light incident onto the fiber is expressed as

$$\mathbf{A}(0) = \frac{1}{\sqrt{1 + \varepsilon^2}} \begin{bmatrix} \cos\psi - j\varepsilon\sin\psi \\ \sin\psi + j\varepsilon\cos\psi \end{bmatrix} \tag{5.29}$$

where $\varepsilon = b/a$ (ratio of minor and major axes of the polarization ellipse), and ψ is the orientation angle.

In the local coordinates, the output light can be derived by using the transfer matrix, Eq. (5.11). Straightforward matrix-algebraic manipulation leads to:

$$\mathbf{A}(z) = \mathbf{O}\tilde{\mathbf{\Lambda}}\mathbf{O}^{-1}\mathbf{A}(0)$$

$$= \frac{1}{\sqrt{1 + \varepsilon^2}} \begin{bmatrix} \cos\{\psi - (1 - \kappa)\tau z\} - j\varepsilon\sin\{\psi - (1 - \kappa)\tau z\} \\ \sin\{\psi - (1 - \kappa)\tau z\} + j\varepsilon\cos\{\psi - (1 - \kappa)\tau z\} \end{bmatrix} \tag{5.30}$$

By Eqs. (5.29) and (5.14), the output light in the fixed coordinates is

$$\hat{A}(z) = R(-\tau z)A(z)$$

$$= \frac{1}{\sqrt{1 + \varepsilon^2}} \begin{bmatrix} \cos(\psi + \kappa\tau z) - j\varepsilon\sin(\psi + \kappa\tau z) \\ \sin(\psi + \kappa\tau z) + j\varepsilon\cos(\psi + \kappa\tau z) \end{bmatrix} \quad (5.31)$$

This result explicitly shows that the ellipticity of an arbitrary elliptical light is preserved in a circular hi-bi fiber, with the orientation of the polarization ellipse continuously precessing in the transmission direction.

Clearly, Eq. (5.31) includes the two special cases treated previously: the behaviors of circular light and of linear light in circular hi-bi fiber. In the former case, $\varepsilon = 1$, so that Eqs. (5.29) and (5.31) reduce to:

$$\hat{A}(0) = \frac{1}{\sqrt{2}} \begin{bmatrix} 1 \\ j \end{bmatrix} e^{-j\psi}$$

$$\hat{A}(z) = \frac{1}{\sqrt{2}} \begin{bmatrix} 1 \\ j \end{bmatrix} e^{-j(\psi + \kappa\tau z)} \quad (5.32)$$

Equation (5.32) is the same as Eq. (5.19), except for the appearance of a constant phase factor $e^{j\psi}$ both in the input light and in the output light. This factor simply defines the initial phase of the circular light, and is trivial in regard to the circular polarization-maintaining property of the Screw fiber.

The other special case is linear light incidence, for which $\varepsilon = 0$ in Eqs. (5.29) and (5.31):

$$\hat{A}(0) = \begin{bmatrix} \cos\psi \\ \sin\psi \end{bmatrix}$$

$$\hat{A}(z) = \begin{bmatrix} \cos(\psi + \kappa\tau z) \\ \sin(\psi + \kappa\tau z) \end{bmatrix} \quad (5.33)$$

The result agrees with Eq. (5.27a), which describes the precessing linear light in a circular hi-bi fiber.

5.8   PRACTICAL ASPECTS OF SCREW FIBER

Offset Joint of Two Circular Hi-Bi Fiber Sections

Jointing two fiber sections is always required in any practical application. It is known to all fiber specialists that, in the case of linear hi-bi fiber, jointing of two fiber sections requires strict aligning of the principal axes. The linear polarization-maintaining behavior of a linear hi-bi fiber can be severely deteriorated if the principal axes aligning on both sides of a joint is not accurately ensured within a few degrees.

Consider the case of the circular hi-bi Screw fiber. Analytically, offset at a joint can be described by a small rotation of the principal axes of one fiber section with respect to the principal axes of the other fiber section

$$\mathbf{R}(\sigma) = \begin{bmatrix} \cos\sigma & \sin\sigma \\ -\sin\sigma & \cos\sigma \end{bmatrix} \tag{5.34}$$

To avoid circuitous mathematical manipulation, we use the afore-derived particular solutions in the fixed coordinates directly. Let the output light from the first fiber section be a right circular light given by Eq. (5.19):

$$\hat{\mathbf{A}}(L_1) = \frac{1}{\sqrt{2}}\begin{bmatrix} 1 \\ j \end{bmatrix} e^{-j\kappa\tau L_1} \tag{5.35}$$

where L_1 is the length of the first fiber section. Let the sense of spin of the two fiber sections be the same. Because of the offset at the joint, the input light to the second fiber section becomes

$$\mathbf{R}(\sigma)\hat{\mathbf{A}}(L_1) = \frac{1}{\sqrt{2}}\begin{bmatrix} 1 \\ j \end{bmatrix} e^{j(\sigma-\kappa\tau L_1)} \tag{5.36}$$

The result can be anticipated, inasmuch as a rotation of the principal axes that conform with the coordinate axes will not affect the circular SOP of light, but will only introduce a phase factor. The step-by-step derivation shows that light transmission in the second fiber section keeps maintaining the single eigenmode transmission in right circular SOP throughout the entire fiber, yielding a final output light of the form:

$$\hat{\mathbf{A}}(L_1 + L_2) = \frac{1}{\sqrt{2}}\begin{bmatrix} 1 \\ j \end{bmatrix} e^{j[\sigma-\kappa\tau(L_1+L_2)]} \tag{5.37}$$

Thus, the task of jointing two circular hi-bi fiber sections is just as simple as that of jointing two conventional fiber sections, requiring an aligning of the cores only. In the above expressions, the same value of $\kappa\tau$ is assumed for both fiber sections. In practice, if the $\kappa\tau$-values are different for different fiber sections (say $\kappa_1\tau_1$ and $\kappa_2\tau_2$ for the two sections), what happens is just a modification of the phase factor in Eq. (5.37), with $\kappa\tau(L_1 + L_2)$ replaced by $\kappa_1\tau_1 L_1 + \kappa_2\tau_2 L_2$. Still, the two-section piece of fiber will keep maintaining the circular light.

Jointing of Two Fiber Sections of Opposite Senses of Spin

Consider the case where the two circular hi-bi fiber sections to be jointed are of opposite senses of spin. According to Eqs. (5.7), (5.8), and (5.9a, b), the transfer matrices (in the local coordinates) of the two-section piece of fiber are given by

$$\mathbf{T}_1 = \mathbf{O}_1 \tilde{\Lambda} \mathbf{O}_1^{-1} \approx \begin{bmatrix} 1 & j \\ j & 1 \end{bmatrix} \begin{bmatrix} e^{j(1-\kappa_1)\tau_1 L_1} & 0 \\ 0 & e^{-j(1-\kappa_1)\tau_1 L_1} \end{bmatrix} \begin{bmatrix} 1 & -j \\ -j & 1 \end{bmatrix}$$

$$\approx \begin{bmatrix} \cos \bar{\kappa}_1 & \sin \bar{\kappa}_1 \\ -\sin \bar{\kappa}_1 & \cos \bar{\kappa}_1 \end{bmatrix},$$

(5.38a)

where $\bar{\kappa}_1 = (1 - \kappa_1)\tau_1 L_1$.

$$\mathbf{T}_2 = \mathbf{O}_1 \tilde{\Lambda} \mathbf{O}_2^{-1} \approx \begin{bmatrix} 1 & -j \\ -j & 1 \end{bmatrix} \begin{bmatrix} e^{j(1-\kappa_2)\tau_2 L_2} & 0 \\ 0 & e^{-j(1-\kappa_2)\tau_2 L_2} \end{bmatrix} \begin{bmatrix} 1 & j \\ j & 1 \end{bmatrix}$$

$$= \begin{bmatrix} \cos \bar{\kappa}_2 & -\sin \bar{\kappa}_2 \\ \sin \bar{\kappa}_2 & \cos \bar{\kappa}_2 \end{bmatrix}$$

(5.38b)

where $\bar{\kappa}_2 = (1 - \kappa_2)\tau_2 L_2$.

As before, an offset of the principal axes of the two fiber sections is described by a rotation matrix $\mathbf{R}(\sigma)$, given by Eq. (5.34). The overall transfer matrix in the local coordinates is then given by

$$\mathbf{T} = \mathbf{T}_2 \mathbf{R}(\sigma) \mathbf{T}_1 = \begin{bmatrix} \cos \bar{\vartheta} & \sin \bar{\vartheta} \\ -\sin \bar{\vartheta} & \cos \bar{\vartheta} \end{bmatrix}$$

(5.39)

$$\bar{\vartheta} = \sigma - j(\bar{\kappa}_2 - \bar{\kappa}_1)$$

where $\bar{\kappa}_1$ and $\bar{\kappa}_2$ are defined above as for Eqs. (5.38a, b).

The corresponding transfer matrix $\hat{\mathbf{T}}$ of the Screw fiber in the fixed coordinates can be derived in a similar way, with $\mathbf{T}_1, \mathbf{T}_2$ replaced by $\hat{\mathbf{T}}_1, \hat{\mathbf{T}}_2$. The derivation is routine but tedious, and so will not be displayed here. The end result is formally the same as Eq. (5.39), except for the changes that $(1 - \kappa_1)\tau_1 \to -\kappa_1\tau_1$ and $(1 - \kappa_2)\tau_2 \to -\kappa_2\tau_2$.

The overall transfer matrix remains a matrix of rotation of the coordinates. Thus, when a circular light propagates through the offset joint of two fibers of opposite senses of spin, only the phase factor will undergo a change, but the SOP will not change. On the other hand, when a linear light propagates through the jointed fiber, the light will keep linear, but its orientation angle will change. Generally, for an elliptical light, the ellipticity of light will be preserved in transmission, but the orientation will change by the same amount.

5.9   EXPERIMENTS ON SCREW FIBER TRANSMISSION CHARACTERISTICS

Experiments on some initial Screw fiber specimen were made in our laboratory, with the data shown by Figures 5.3–5.7. The working wavelength was
0.6328 μm. The pen-speed of the x-y recorder was 2 s/cm. The scale of the
ordinate was not calibrated, but we measured I_{max} and I_{min} (maximum and
minimum light power, respectively) for both the input light and the output
light. For Figures 5.3–5.6 and 5.7a, the value of $10 \, Log_{10}(I_{max}/I_{min})$ was
$\approx 0.05 \, dB$ for the input light, and $\leq 0.2 \, dB$ for the output light. These
measured figures signify quantitatively the circularity of light at input and
output. Based on our preliminary experimental results, the following topics
collectively represent an initial report on the circular polarization-maintaining capability of the Screw fiber.

Existence of Circular Eigenmodes in Screw Fiber

The first experiment was to test whether circular light of either sense is really
an eigenmode of the Screw fiber, as so asserted by Eqs. (5.19) and (5.23). In
this experiment, we excite the tested Screw fiber specimen (some meters
long) with a circular light, and observe if the output light is also circular in
the same sense, and moreover, if this circular SOP of light is invariant in the
transmission direction z. We perform this experiment by cutting the output
end of fiber successively for a relatively large number of times, each time by a
tiny length (1 or 2 mm), and observe whether or not the SOP of the output
light is exactly length-independent.

In Figure 5.3, the upper curve is the circular light that excites the Screw
fiber. The lower curves denote the output SOPs after successive cuttings of
the output end. The numerals $1, 2, \ldots, 21$ refer to the number of the
fiber-end cuttings. More than 20 such cuttings were made during the experiment (only 3 of them are shown in Fig. 5.3). The output SOPs are found to
be length-independent. A change in the sense of the incident circular SOP
yielded a circular light of reversed sense at the output, but its SOP is again
length-independent. This invariance of the output SOP is exactly the experimental evidence that confirms the existence in the Screw fiber of a pair of
eigenmodes of right and left circular polarizations.

Jointing and Splicing of Screw Fiber Sections

From the application standpoint, ease of joining or splicing is probably the
most attractive aspect of a circular hi-bi fiber. We noted in connection with
Eq. (5.37) that a "rotation matrix" will only introduce a phase factor to a
circular light, but will not affect the circular SOP of light. Two circular hi-bi
fiber sections are jointed by aligning the cores only. Since the stress filaments
on both sides of the joint are not aligned, the coordinate axes of the second
fiber section are skewed by an angle (denoted by σ) from the coordinate axes
of the first fiber section. According to the analytic result derived in the

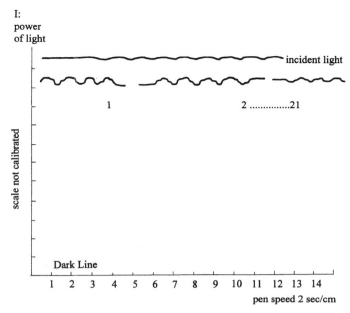

FIGURE 5.3 Circular eigenmodes in Screw fiber independent of fiber length (numerals underneath curve indicate number of fiber-end cuttings); 10 $\text{Log}_{10} (I_{max}/I_{min})$: incident light ≈ 0.05 dB, output light < 0.2dB.

preceding section, this will introduce an extra *constant* phase factor $e^{j\sigma}$ in the final output light, but the circular SOP will still be maintained.

The above analytic result is predictable by intuitive reasoning. This seemingly technological trifle (concerning a joint) is actually a crucial factor that determines whether or not the circular hi-bi Screw fiber is practical in actual application. Clearly, it is nearly impossible in practice (not in principle) to align the whirling stress filaments on both sides at a joint.

The transmission characteristics of Screw fiber involving a joint or splice of two fiber sections was observed in our experiment. The splicing machine available in our laboratory is a homemade setup (simple "spark welder"), and was originally manufactured for multimode fiber splicing. To align the cores of single-mode fibers the machine needs a little more manual adjustment, but this is not very difficult to do. Figure 5.4 shows the experimental results. In the figure, the letter N indicates the SOP curve of the emergent light at the far end of one or the other Screw fiber sections before joining or splicing, with the piece of fiber being laid in a natural course. The letter J indicates the SOP curve when the two fiber sections are cascaded end-to-end, but not yet spliced together by the welder. The letter S indicates the SOP curve of the emergent light after the two fiber sections are spliced together (only the cores being aligned) to form a single fiber piece. The experimental curves in Figure 5.4 show the very desirable result that favors the usefulness of the Screw fiber in a practical fiber-optic circuitry or system. Jointing and splicing

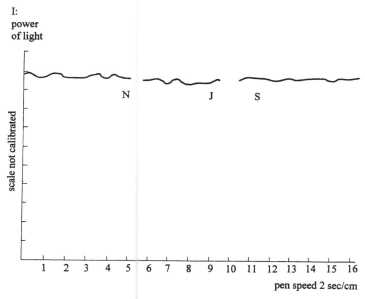

FIGURE 5.4   Jointing and splicing of two Screw fiber segments: same sense of spin, only cores aligned. J, fiber segments butt-jointed; S, fiber segments spliced; N, fiber placed in natural course; $10 \, \mathrm{Log}_{10}(I_{max}/I_{min})$: incident light ≈ 0.05 dB, output light < 0.2 dB.

of two Screw fiber sections are just as simple and easy as in a conventional single-mode fiber. A joint (or splice) in a Screw fiber line made in this way will not spoil the circular SOP of light passing the joint or splice.

In passing, we note here that the measured insertion loss of a splice of the Screw fiber is at a level nearly as low as that of a conventional single-mode fiber splice.

Jointing and Splicing of Screw Fibers of Opposite Senses of Spin

One pertinent aspect of a section of Screw fiber is the sense of its "built-in spin" (clockwise or anticlockwise). This sense of spin can be determined, if necessary, by observing the sense of precession of a linear light traversing the circular hi-bi fiber, according to Eq. (5.26). From the standpoint of application, indeed, it makes no sense to determine the handedness of the built-in spin of a Screw fiber section before jointing it to a fiber-optic circuitry. The experiments for Figure 5.4 employed two Screw fiber sections of the same sense of spin. If the second fiber section has a reversed sense of spin, then the coupled-mode analysis yields a final output light whose circular SOP is still maintained. The only change due to reversal of the sense of spin of the second fiber section is reflected in the phase factor, according to Eq. (5.39).

Our experimental data on jointing and splicing two Screw fibers of opposite senses of spin are shown in Figure 5.5. In the figure, the letter J

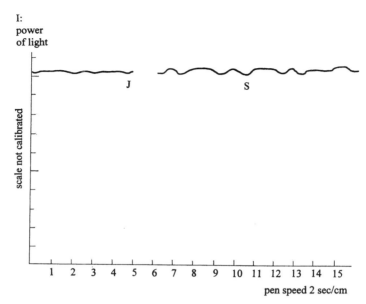

FIGURE 5.5   Jointing and splicing of two Screw fiber segments: opposite senses of spin, only cores aligned. J, fiber segments butt-jointed; S, fiber segments spliced; $10 \log_{10} (I_{max}/I_{min})$: incident light ≈ 0.05 dB, output light < 0.2 dB.

denotes the SOP curve for two fiber sections (opposite in the sense of spin) cascaded end-to-end, and the letter S for the same two fiber sections spliced or welded together. Experiments confirm the analytic result.

Effect of Macroscopic Bending on Screw Fiber

Macroscopic bending of fiber is always unavoidable in a practical fiber-optic system in order to wind the fiber on a form or to negotiate a natural course in fiber line layout. To test the behavior of the Screw fiber involving a circling, we made this experiment on our Screw fiber specimen, with the data shown in Figure 5.6. Somewhere along the length of fiber, a small portion of the fiber is coiled into a circling form of different diameters. The mark $\phi 3$ below the curve denotes a circling of 3 cm diameter, and $\phi 5$, of 5 cm diameter. The rightmost curve in the figure denotes the SOP of light at the output end of the fiber section placed in its natural course without being coiled anywhere. The experimental data show that the circular hi-bi Screw fiber tolerates macroscopic bending of comparatively small radius of curvature.

The "Twisting Experiment"

The foregoing experiments show the behaviors of Screw fiber with emphasis on its feasibility for circular light transmission. To view an enhanced contrast

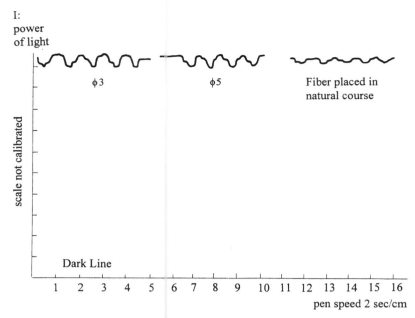

FIGURE 5.6 Macroscopic bending of Screw fiber. Bending diameters: $\phi 3 = 3$ cm, $\phi 5 = 5$cm; $10 \, \mathrm{Log}_{10} \, (I_{max}/I_{min})$: incident light ≈ 0.05 dB, output light < 0.2 dB.

between the behavior of a Screw fiber and that of a conventional fiber, we made the "twisting experiment," which involves twisting the two different fibers only slightly, one at a time, to see what would happen in the respective output SOPs. We understand that the effect of twisting a fiber in the "cool" state is to introduce a circular-bi component whose magnitude is proportional to the twisting rate. In the case of conventional fiber, if the twisting rate is slight such that the twist-induced circular-bi component is comparable to the residual linear-bi component, the resulting SOP of light at the output fluctuates erratically [24]. On the other hand, in the case of Screw fiber, the twist-induced circular-bi will be superposed on the inherent high circular birefringence of this fiber, such that the resulting SOP of light at the output will essentially remain circular. Thus, this simple "twisting experiment" may serve as a criterion that distinguishes a Screw fiber from a conventional fiber.

In Figures 5.7a and b, the numerals (0.5, 1, 1.5) denote the number of turns (revolutions) of twist applied on the respective fiber. Both fibers are several meters long, and excited in succession by the same circular light. The rightmost curve in Figure 5.7a shows the SOP of output light when the Screw fiber specimen is laid in a natural course without intentional twisting. Figures 5.7a and 5.7b unequivocally distinguish the behavior of a Screw fiber from that of a conventional fiber.

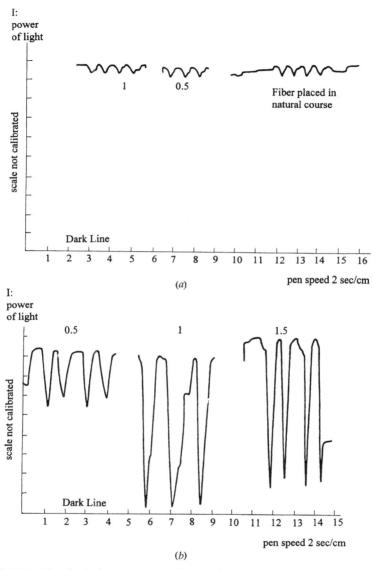

FIGURE 5.7 The "twisting experiment" that detects a circular hi-bi Screw fiber. (*a*) Screw fiber immune to slight twist (numeral under a curve: number of twist turns. (*b*) Conventional fiber strongly affected by a slight twist (numerals, same meaning as above).

Beat-Length Measurement of Screw Fiber

The beat length is a convenient parameter that measures the capability of fiber to tolerate a macroscopic bend, or to resist certain kinds of extrinsic perturbation. We made beat-length measurements on our circular hi-bi Screw fiber specimen. In the previous measurements for Figures 5.3–5.7, the incident light is always a circular light. In the present measurement, however, a circular incident light will not do, because no beating pattern will be produced in the fiber by a single circular eigenmode.

A linear light is *not* an eigenmode of the circular hi-bi Screw fiber. According to Eq. (5.27), an incident linear light will excite both eigenmodes in the fiber, thereby producing a beating pattern in the form of a precessing linear light. The beat length is equal to the distance of a complete precession. Therefore the familiar "cutback" method for beat-length measurement works. On our initial Screw fiber specimen (made about 1994), preliminary measurements gave the beat length in the 2–3-mm range. Realization of still shorter beat lengths should not be a problem.

5.10 APPLICATION EXAMPLE I: FARADAY EFFECT SENSING USING CIRCULAR Hi-Bi FIBER

The effect of Faraday rotation provides an important classical means of measuring a magnetic field, and hence the electric current that generates the magnetic field. Since this effect universally exists in all materials, diamagnetic or paramagnetic, it is only natural to conceive the idea of using optical fiber to make a current meter. While the Faraday effect is fairly weak in a diamagnetic glass fiber, the length of fiber can be long enough so that the effect becomes detectable. Among the many advantages of a fiber-optic current meter, one unique feature is its immunity to electromagnetic interference. This feature is most desirable in hazardous regions covering a distribution of high-voltage electric lines.

The early work on a fiber-optic current meter can be traced back to the beginning of the last decade or earlier. Over the last decade and a half, great efforts have been made worldwide in various laboratories toward this very attractive technological objective. The ideal is not yet realized completely, however. The difficulty is not that the Faraday effect is unobserved in optical fiber, but that the observed Faraday effect is unstable, not easy to maintain through transmission. The issue is that the attempt of making a practical fiber-optic current meter remains a "hard-nut" problem that presents a challenge to the fiber scientists.

According to the relevant literature, the reason that causes the Faraday effect to be unstable in fiber has been mainly attributed to the temperature effect. Remarkable effort has therefore been directed toward devising a variety of sophisticated schemes whose function is to eliminate or reduce the temperature effect. However, it is our belief that, in the final analysis, what causes an unstable Faraday effect in fiber is rooted in the innate property, or the inherent transmission characteristics, of the type of fiber that is used to make the Faraday-effect current meter. The environmental temperature perturbation indeed represents a major annoyance, but we believe that the temperature effect should be of secondary importance when compared to the transmission characteristics of the fiber in which the Faraday rotation occurs. In the following discussion, we shall therefore concentrate on the transmission aspect of the fiber.

In retrospect, it was thought at the initial stage of research that an ultra-lo-bi fiber should be an ideal fiber version that supports the Faraday rotation. Later experience revealed that such fiber, because of its high degree of circular symmetry, does support a Faraday rotation, but the Faraday rotation produced can never be stable just because of this ultra-lo-bi nature of the fiber.

We owe the pioneering scientists for their work more than one and a half decades ago on detecting and measuring the Faraday effect in optical fiber, at a time when the only kind of fiber available was the "conventional" fiber of a nominally circular symmetrical structure (see Refs. [25–28]). Apart from its archival value, certain of the ideas implied in the early work lend perspectives that are of fundamental relevance in the present day. For example, in one early measuring setup, four turns of single mode fiber were wound around a 12-cm-radius former mounted on a standard 14-cm-diameter busbar. As noted in the publication [25], the reason for the use of the former, rather than winding the fiber directly onto the busbar, is that the critical bend radius for the "conventional" fiber then used was $\cong 10$ cm below which bending losses would become severe.

Twisted fiber and helical fiber, among some others, were used in the Faraday effect experiment, until the advent of elliptically birefringent fiber (spun Bow-Tie), which draws much attention in the relevant scientific circle. But the elliptically birefringent fiber is impractical in either the single-eigenmode or dual-eigenmode transmission scheme (see Chapter 6). For the former scheme, the difficulty is in the problem of elliptical light excitation, which is required to match an eigenmode of the fiber both in ellipticity and in the principal axes. For the latter scheme, the difficulty is in the length-sensitive behavior of the beating modes.

It was through a careful review of this technical background that I attempted a new approach that led to the discovery of the Screw fiber [2], whose feasibility in detecting the Faraday rotation is analyzed below.

Coupled-Mode Analysis of the Faraday Effect

According to the previous analysis, the Screw fiber is a circular hi-bi fiber whose eigenmodes are right and left circular light. In the single-eigenmode regime, the fiber is excited with either a right or a left circular light. The circular SOP of light will always be maintained in the fiber, except for a phase shift, as given by Eqs. (5.19) or (5.23).

When a magnetic field is applied along the transmission direction of the fiber, the Faraday effect occurs. We start with the modified coupled-mode equations, which imply the additional coupling terms due to the Faraday effect. Following Ref. [6], the Faraday effect introduces an additional term α_F to the coupling coefficient such that the modified coupling coefficient becomes

$$k_{21} = -k_{12} = \kappa\tau \pm \alpha_F \tag{5.40}$$

$$\alpha_F = VH_z \tag{5.41}$$

where H_z is the longitudinal magnetic field applied to the fiber, and V is the Verdet constant. The choice of the $\pm$ signs in Eq. (5.40) is determined by the relative senses of the Faraday effect term α_F and the characteristic parameter $\kappa\tau$ of the Screw fiber. The Verdet constant of fused silica is 1.61×10^{-2} min A^{-1} at a wavelength of 0.6328 μm. The Verdet constant of the fiber specimen used in an initial experiment [25] was found to be

$$\begin{aligned} V &= 1.56 \times 10^{-2} \text{ min } A^{-1} \\ &= 2.6 \times 10^{-4} \text{ deg } A^{-1} \end{aligned} \tag{5.42}$$

The Faraday effect does not introduce modification terms in the propagation constants [6]. The resulting coupled-mode equations for a magnetized fast-spun Screw fiber in the fixed coordinates take the form:

$$\begin{aligned} \frac{d\hat{A}_x}{dz} &= j\frac{\delta\beta}{2}\hat{A}_x - (\kappa\tau + \alpha_F)\hat{A}_y \\ \frac{d\hat{A}_y}{dz} &= (\kappa\tau + \alpha_F)\hat{A}_x - j\frac{\delta\beta}{2}\hat{A}_y \end{aligned} \tag{5.43}$$

where $\alpha_F = VH_z$. By Eq. (5.43), the composite coupling coefficient $(\kappa\tau + \alpha_F)$ now contains two terms: $\kappa\tau$ and α_F $(= VH_z)$. Before the longitudinal magnetic field H_z is applied to the fiber, the new term $VH_z = 0$, such that the only nonzero coupling term is $\kappa\tau$, and Eq. (5.43) reduces to the coupled-mode equations for an unmagnetized Screw fiber in the fixed coordinates.

The implied physical mechanisms reveal that the term $\kappa\tau$ is descriptive of the waveguiding effect produced by the fiber structure (a single built-in stress filament helically coiling the core), while the term VH_z is descriptive of the Faraday effect due to the longitudinal magnetic field acting on the lightwave propagating in the fiber. Equation (5.43) shows that the two terms $\kappa\tau$ and VH_z are on equal footing in the equation in setting up the pattern of the lightwave propagation in the fiber.

In Eq. (5.43), the $+$ sign in $(\kappa\tau + \alpha_F)$ refers to the case where the Faraday effect and the waveguiding effect are in the same sense. In the case where they are in opposite senses, the $+$ sign should be changed to a $-$ sign, that is, $(\kappa\tau + \alpha_F)$ is replaced by $(\kappa\tau - \alpha_F)$.

Single Eigenmode in Magnetized Fiber

Previous analysis of an unmagnetized Screw fiber has shown that, if the input light matches an eigenmode of the fiber (circular light of either sense), the propagating lightwave will keep this same single eigenmode (SOP) in fiber throughout the entire course of transmission. Let the circular light be right-handed. By Eq. (5.19), the lightwave will only undergo a phase change $e^{-j\kappa\tau z}$ due to the waveguiding effect. When a magnetic field is applied to the fiber, the single-eigenmode solution becomes

$$\hat{\mathbf{A}}(0) = \frac{1}{\sqrt{2}}\begin{bmatrix} 1 \\ j \end{bmatrix}$$

$$\hat{\mathbf{A}}(z) = \frac{1}{\sqrt{2}}\begin{bmatrix} 1 \\ j \end{bmatrix} e^{-j(\kappa\tau + VH_z)z} \qquad (5.44)$$

where VH_z is used for α_F. Thus, in the single-eigenmode process, while the SOP of light remains unchanged, the Faraday effect will occur as an extra phase shift $e^{-jVH_z z}$.

For simplicity, the Faraday effect has been considered to be a constant, invariant in the transmission direction z. In actual application, it is a common practice to have the fiber coiled around the metallic conductor that carries the electric current. For such actual device configuration, the extra phase factor in Eq. (5.44) for a fiber of length L due to the Faraday effect should be replaced by an integral, that is,

$$\alpha_F L \rightarrow V\int_0^L H_z\, dz \qquad (5.45)$$

$$\approx V(AT)$$

where $(AT) = \oint H_l\, dl$ is the ampere-turns that produces the overall magnetic field, and $\int_0^L H_z\, dz$ is the part of $\oint H_l\, dl$ that contributes to the Faraday effect in the magnetized circular hi-bi fiber.

To be exact, H_z, and hence α_F, are not constant, but are functions of z. Equation (5.43) then becomes differential equations of variable coefficients, whose mathematical solution is generally not easy to obtain. Nevertheless, as the circular hi-bi Screw fiber necessarily satisfies the fast-spun condition, the SOP of a circular light remains unchanged. Under the fast-spun condition, solution (5.44) remains valid if we replace only the phase factor in this solution by an exponential function of integral argument

$$e^{-jVH_z L} \rightarrow e^{-jV\int_0^L H_z\, dz} \qquad (5.46)$$

For practical purpose, however, it suffices to employ Eq. (5.44) in an actual device configuration that works on the Faraday effect.

Faraday Rotation Sensing by Dual Eigenmodes in Circular Hi-Bi Fiber

Let the initial condition be a linear light of an arbitrary orientation as given by Eq. (5.27). Before applying the electric current and hence the magnetic field to the fiber, light induced in the fiber is a precessing linear light constituting a right and left circular-eigenmode pair described by Eq. (5.27a). As soon as the electric current is applied so that a longitudinal magnetic field occurs in the fiber, the pattern of light transmission will change. By Eq. (5.43), for a circular hi-bi fiber section of length L, the solution for linear light excitation is given by

$$\hat{A}(0) = \begin{bmatrix} \cos\theta \\ \sin\theta \end{bmatrix} = \frac{e^{j\theta}}{2}\begin{bmatrix} 1 \\ -j \end{bmatrix} + \frac{e^{-j\theta}}{2}\begin{bmatrix} 1 \\ j \end{bmatrix}$$

$$\hat{A}(L) = \frac{e^{j(\theta + \kappa\tau L)}}{2}\begin{bmatrix} 1 \\ -j \end{bmatrix}e^{jVH_zL} + \frac{e^{-j(\theta + \kappa\tau L)}}{2}\begin{bmatrix} 1 \\ j \end{bmatrix}e^{-jVH_zL} \qquad (5.47)$$

$$= \begin{bmatrix} \cos\left[(\theta + \kappa\tau L) + VH_zL\right] \\ \sin\left[(\theta + \kappa\tau L) + VH_zL\right] \end{bmatrix}$$

The result shows that in a magnetized circular hi-bi Screw fiber a linear light will precess by two mechanisms: one is due to the waveguiding effect equal to $\kappa\tau L$, and the other is due to the Faraday effect equal to VH_zL. In the equation, $(\theta + \kappa\tau L)$ is the sum of the input light orientation angle θ and the precessional angle $\kappa\tau L$ due to waveguiding effect of fiber. This sum of angles is *constant* (i.e., H-independent) for a given section of fiber and a given initial condition. On the other hand, the term VH_zL in Eq. (5.47) is proportional to the applied longitudinal magnetic field H_z in fiber. This is exactly the Faraday rotation angle ϑ_F of the precessing linear light due to the applied longitudinal magnetic field:

$$\vartheta_F = VH_zL \qquad (5.48a)$$

$$\vartheta_F = V\int_0^L H_z\,dz \qquad (5.48b)$$

Equation (5.48a) applies to the case wherein the longitudinal magnetic field is uniform over the entire length of the fiber. If the magnetic field component is not a constant in the direction of light transmission, such as the actual case of the coiled configuration of either the fiber or the electric conductor or both, then the equation for Faraday rotation angle assumes the more refined integral form given by Eq. (5.48b).

Generally, a wave field is difficult to manage and to use if it is not an eigenmode of the waveguiding structure. That is why the single eigenmode (here referring to as a circular light) is generally desirable for a stable transmission of the wave field. Linear light is not an eigenmode of the circular hi-bi fiber, being composed of the two eigenmodes. But the eigenmode-like precessing linear light is manageable and does work in the present example of Faraday rotation measurement. This is really a rare special case. Perhaps such a case occurs only in circular hi-bi fiber, and is not to be found in other polarization-maintaining fiber versions.

Experimental evidence is still insufficient to characterize linear light versus circular light in Faraday-effect measurement with a circular hi-bi fiber. By intuition, it is reasonable to expect that linear light in a circular hi-bi fiber is more liable to be disturbed or spoiled by an outside perturbation, in particular when this perturbation causes different effects on the two constituent circular eigenmodes of the linear light. If this is the case, we shall be forced to resort to the single eigenmode (circular light of either sense) for Faraday-effect measurement. Then, what is to be measured is not angular rotation, but the phase. The measuring setup is thus likely to be networked in the form of a fiber interferometer.

The Simple Additive Property in Faraday-Effect Measurement

The coupled-mode solution, Eq. (5.44), for circular light in a circular hi-bi fiber shows that the extra phase-shift term $e^{-jVH_z z}$ due to the Faraday effect is simply additive to the phase-shift term $e^{-j\kappa\tau z}$ due to the waveguiding effect. Correspondingly, the coupled-mode solution, Eq. (5.47), for linear light in a circular hi-bi fiber shows that the two parts of the angular rotation (Faraday rotation and linear light precession in fiber) are likewise simply additive.

The said simple additive property of the phase or of the angle is a particular advantageous feature of a circular hi-bi fiber in view of ease of measurement. Being an inherent attribute of the circular hi-bi fiber structure, this simple additive property does not involve approximation in the sense of perturbation.

It is interesting to note that the desirous "simple additive property" does not exist in a fiber version that is not circularly hi-bi. A slightly twisted fiber or an elliptically birefringent (spun hi-bi) fiber, for example, does not have this simple additive property of the phase or of the angle for either circular or linear light excitation. A simple and straightforward mathematical derivation shows that, if any of these fibers (not circular hi-bi) is excited by a linear or circular light, the polarization pattern of light at the output will become rather intricate (see Chapter 6). The phase and orientation of the output light will be related in a complicated way to the Faraday effect and the waveguiding effect. These two effects become inseparable in the output light, such that no simple additive relation exists either in phase change or in angular rotation.

Nonreciprocal Behavior of the Faraday Effect

In a magnetized circular hi-bi fiber, a linear light continuously precesses in the transmission direction, such that it undergoes an angular rotation angle at the end of the fiber of length L. This angular rotation angle consists of two parts, $\kappa\tau L$ and $VH_z L$, due to the waveguiding effect and the Faraday rotation effect, respectively. While these two parts are simply additive, a fundamental difference exists between them. That is, the waveguiding rotation is "reciprocal," while the Faraday rotation is "*nonreciprocal.*"

Consider the case of reciprocal rotation. The sense of a linear light precession in a circular hi-bi fiber, whether clockwise or anticlockwise, depends on the relative sense of the handedness of the spin of the fiber with respect to the transmission direction of light. Let the linear light take a two-way (forth and back) travel along the circular hi-bi fiber. Because both the handedness of spin of the fiber and the transmission direction of light change their signs on the backward travel, their relative sense remains unchanged. The effect of the circular birefringence on the sense of precession of a linear light is therefore reciprocal inasmuch as it is independent of the direction of transmission. When the linear light travels from one end to the other, and then back, it will make a precessional angle of one sign in the forward direction, but on the backward travel its precessional angle will be opposite in sign, such that the net precessional angle through the two-way travel becomes zero, that is, the linear light will come back again to its original place without a net change.

In contrast to the above-described reciprocal behavior, the Faraday rotation effect in fiber is a *nonreciprocal* process. The sense (handedness) of Faraday rotation depends on the direction of the applied longitudinal magnetic field only, irrespective of this direction relative to the direction of lightwave transmission. For a simple illustration, again consider a linear light making a two-way travel in a magnetized medium. In the forward direction, the linear light makes a Faraday rotation angle $VH_z L$ of one handedness. In the backward direction, this linear light will again make a Faraday rotation angle $VH_z L$ of the same handedness, such that after the two-way "forth and back" travel, the total Faraday rotation will be doubled, equaling $2VH_z L$.

As a result, when a linear light makes a two-way travel in a magnetized circular hi-bi fiber, since the net precessional angle due to the waveguiding effect is zero, only the doubled Faraday rotation angle is left in the total angular rotation.

In the above discussion we considered the case of linear light transmission. The case of circular light is predictable. The angular change in linear light now becomes the phase change in the circular light. Nevertheless, a fundamental difference still exists in that the phase change due to the waveguiding effect is reciprocal, while the phase shift due to the Faraday effect is nonreciprocal. Thus, when a circular light makes a two-way travel in the magnetized circular hi-bi fiber, the part of the phase change due to the reciprocal waveguiding effect on the backward travel will exactly cancel that in the forward travel, but the part of the phase change due to the non-reciprocal Faraday effect will be doubled.

The nonreciprocal behavior of the Faraday rotation effect due to a magnetic field is unique. It is unknown in any natural optic-active material; nor does it exist in any birefringent fiber structure when the structure is not magnetized. Such a peculiar magnetooptic effect was discovered experimentally by Michael Faraday in 1845. The nonreciprocal behavior of this effect is to be regarded as an experimental finding, rather than a theoretical prediction or an analytic deduction. As a matter of fact, the basic formula given in Eq. (5.41) is an empirically determined expression. To our knowledge, a thorough theoretical treatment of the nonreciprocal property of the Faraday rotation is beyond the analytic framework of Maxwell's equations. In this application-oriented book, it will suffice if we have made clear the basic distinction between the reciprocal effect and the nonreciprocal effect.

Initial Observation of Faraday Effect as an Angular Rotation

The Faraday-effect measurement using a circular hi-bi Screw fiber was made for the first time in our laboratory (about the end of 1995 to early in 1996). We did not attempt the task of structuring a kind of complete device or apparatus. Our aim was to explore, by an *experimental* approach, whether or not the new fiber is really applicable to the Faraday-effect measurement.

In our initial experiment, we used a simple regular optical measuring bench with several lenses, waveplates, and an analyzer. The measuring setup was just that simple, without inclusion of any tunable electronics circuitry. The customarily adopted method of power splitting, as well as the familiar formula $(I_1 - I_2)/(I_1 + I_2)$, were not employed in our initial experiment on the Faraday effect measurement. Our endeavor was to get rid of all the side effects, so that we can single out the "Faraday rotation angle" as our sole measuring objective. Linear light was injected into the Screw fiber. On our very simple measuring bench, we took the analyzer's angular reading when the current was applied. This reading minus the initial reading before application of the current then simply and genuinely represents the contribution of the Faraday rotation angle.

To be exact, linear light in the fiber will undergo an angular precession due to the waveguiding effect of the circular birefringent fiber. Nevertheless, this precession angle is a constant characteristic of the circular hi-bi fiber structure, and is the same before and after application of the electric current, such that subtraction of the first (no current) reading from the second (current applied) reading is always the Faraday rotation angle, irrespective of the precession angle of linear light produced by the waveguiding effect.

In our measurement, we used a solenoid of 540 turns that surrounded a section of Screw fiber of about 1.5 m. We read the analyzer readings when d.c. electric current of 10, 20, 30 A passed the solenoid. Within the range of

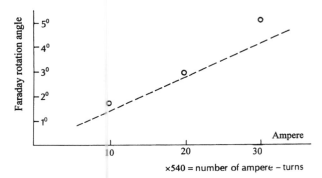

FIGURE 5.8   Experimental data of Faraday rotation in Screw fiber. _ _ _ _ _ , theoretical; o, experimental.

experimental error, the rotation of the analyzer was observed to be essentially linearly proportional to the ampere-turns, as shown in Figure 5.8. The proportionality, that is, the Verdet constant, was estimated to be $\approx 3 \times 10^{-4}$ deg/A. In our experiment, we used a HeNe laser that emitted a visible light at 0.6328 μm. In a figurative sense, we can say that really we did "see" the Faraday rotation phenomenon directly as an angular rotation.

5.11   APPLICATION EXAMPLE II: GYROSCOPIC ARCHITECTURE USING CIRCULAR Hi-Bi FIBER

Central to the fiber-optic architecture of a gyroscope is the Sagnac loop onto which two beams of light are launched, separately and in counterpropagating directions. A gyroscope falls into the category of sensor technology, being employed to sense the rotational rate of a system. In its stationary state, the two counterpropagating beams in the fiber-optic circuitry of the gyroscope overlap. In the rotational state, however, the rotation effectively shortens the path of one beam in comparison to the path of the other beam. When the two beams recombine and interfere, the result is a fringe shift that is proportional to the angular speed Ω of the rotation.

By intuition, it is not difficult to accept the established working principle of a Sagnac loop, that is, a rotation effectively shortens the path of one beam in comparison to that of the other. However, a classical treatment of the relevant problem is lacking, and does not appear possible within the theoretical framework of Maxwell's equations. The rules of the theory of relativity may prevail in the system concerned. From the application standpoint, suffice it to say that the Sagnac effect employed to measure a rotational speed of a system is based on the experimental evidence initially discovered by Sagnac in 1911.

Before a practical circular hi-bi fiber is in existence, the choice of linear hi-bi fiber to structure the Sagnac loop poses a difficulty in splicing the terminals of the loop to the rest of the fiber-optic circuitry. The difficulty is in the technique of aligning the stress filaments. In the simple case of splicing two linear hi-bi fibers, alignment of the stress filaments, in addition to the alignment of the cores, on both sides of the spliced joint can be helped by power monitoring. That is, the principal-axes alignment of the two fiber segments is regarded as acceptable when maximum power is analyzed at the output of the two-piece fiber with a splice. In the case of a Sagnac fiber loop, however, alignment of the principal axes on both sides of the last joint cannot be aided by power monitoring, because two beams are propagating through this joint in opposite directions.

Using circular hi-bi fiber to structure a Sagnac loop can remove this technological difficulty inherent to a linear hi-bi fiber. What is now required in splicing the last joint of the loop is simply to align the cores on both sides of the joint. The task is just as simple as what is required in splicing two pieces of conventional fiber.

The fiber-optic circuitry shown in Figure 5.9 is essentially self-explanatory with reference to the rotating Sagnac interferometer in classical bulk optics. In the figure, C_1 and C_2 are two directional couplers, both preferably of 3 dB. The use of two couplers in the circuitry ensures the entire identity of the paths of the two counterpropagating beams when $\Omega = 0$. Each L is a matching load; D is a detector, M, an on-line phase modulator, and S is a light source delivering a linear light which is readily available in common practice. PPT is the abbreviation for passive polarization transformer which is capable of transforming a linear light automatically to a circular light without having to be adjusted or tuned [29, 30] (see Chapter 7).

It is of much practical interest to see whether the fiber-optic architecture employing circular hi-bi fiber throughout also works on a linear light. As previously analyzed, a linear light in a circular hi-bi fiber actually consists of two circular eigenmodes of opposite senses, and the beating of these two modes forms a precessing linear light. Assuming that the external disturbances affect the two constituent circular eigenmodes equally, the linear light (dual-eigenmode regime) works equally well in the Sagnac loop shown in

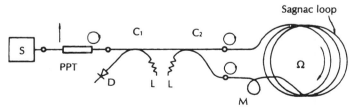

FIGURE 5.9 Proposed gyroscopic architecture using circular hi-bi fiber. The 4 minicircles are splices; the vertical arrow: linear light; the circle with arrow: circular light.

Figure 5.9. Because the linear light precession in the circular-bi fiber is a reciprocal process, the two counterpropagating beams will make the same total precession angle at the two ends of the loop, and in the same sense, such that they also overlap as long as the loop is stationary. When the loop rotates, the Sagnac effect will emerge. More likely, the external disturbances affect the two constituent circular eigenmodes of the circular hi-bi fiber differently, such that the linear light scheme may not work adequately. Then, we can only resort to the single eigenmode regime by using circular light in the fiber circuitry.

With reference to Figure 5.9, the proposed gyroscopic fiber circuitry works entirely on circular light. The linear light of the source is transformed by the PPT into circular light, which feeds the fiber circuitry. Such simple architecture is possible only after the invention of the circular hi-bi Screw fiber [2]. It is noted here that in an earlier (1993) publication [31] of mine, a gyroscopic architecture was proposed that looks similar to that in Figure 5.9, but with the fundamental difference that the Sagnac loop therein worked not on circular light, but on principal-axis-aligned linear light. The advantage of the earlier proposal is that it is easily adapted to linear light technology. The disadvantage in this earlier proposal is that it required two more PPTs, and hence its structure was more complex than that in Figure 5.9.

5.12 APPLICATION EXAMPLE III: COHERENT TRANSMISSION USING CIRCULAR Hi-Bi FIBER

The idea of coherent optical fiber communication comes naturally from the old and established technique of the superheterodyne radio. One advantage of coherent optical communication is the possible improvement of the sensitivity of the receiver by 15–20 dB as compared with the direct-detection system. But this feature has lost much of its charm since the successful development of the optical fiber amplifier. The second advantage of the coherent system is the significant improvement of the selectivity of the transmission channels. This frequency-selective feature remains a major impetus for the continuing effort in this research area.

Two approaches have been adopted to achieve the condition of SOP matching of the incoming signal light and the local light in a coherent receiver. One is to use the method of active control, and the other is to use hi-bi fiber to preserve the SOP of light in transmission. For the latter, because linear hi-bi fibers pose a severe problem in principal-axes aligning at fiber joints, circular hi-bi fiber was originally proposed for use in polarization-maintaining single-mode fiber cable design [3]. Nevertheless, the only circular-bi fiber that could be considered until now was the twisted fiber. Meanwhile, in regard to transforming a linear light of the source to a circular light for transmission, the only means that could be considered until now was

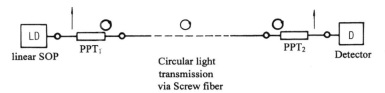

FIGURE 5.10 Proposed coherent optical fiber transmission using circular hi-bi (Screw) fiber. The 4 minicircles are splices; the vertical arrow: linear light; the circle with arrow: circular light.

the bulk-optic quarter-wave plate. Such early proposed arrangement, consisting of twisted fiber and bulk-optic quarter-wave plates did not prove practical.

The two inventions (1995, 1990) [1, 29] of mine are just right for structuring a simple and practical *all-fiber* coherent optical fiber transmission architecture, as shown in Figure 5.10. A detailed description of the PPT in Figure 5.10 can be found in the relevant U.S. patents [29–31]. Suffice it to say here that the PPT (practical polarization transformer) is simply the fiber-optic analog of a bulk-optic quarter-wave plate. Figure 5.10 is all self-explanatory. Note that the PPT on the input side can be spliced (in the form of a tail fiber) with the LD source to make an integrated module. This PPT also serves as a fiber-optic isolator, besides being an SOP transformer.

Clearly, in a coherent optical fiber system employing the circular hi-bi Screw fiber, only single-eigenmode (circular light) transmission works. Matching incoming signal light and the local light is always achieved automatically when both signal light and local light are circular waves. The dual-eigenmode (linear light) transmission scheme of the circular hi-bi fiber will not do for SOP matching in a coherent optical receiver.

5.13 NOTES

The analytic part of the Screw fiber treated in this chapter is essentially analogous to that of the *strongly* twisted fiber treated in Chapter 4. Many of the formulas are the same, with the recognition that the characteristic figure ς of the strongly twisted fiber corresponds to the characteristic figure κ of the Screw fiber. The practical parts of these two fibers, however, are drastically different. Technologically, the difficulties inherent to the strongly twisted fiber (mechanical strength limitation and fiber handling, in particular) are avoided in the Screw fiber.

Analysis of Error in the Asymptotic Approximation

The transverse structure of the Screw fiber is asymmetrical, for which complex mathematics is required for a thorough analytic treatment. In view of simplicity and practicability, we consistently use the method of diagonalization, as well as the asymptotic approximations that yield analytic solutions in closed form. Before ending this chapter, it is of interest to examine whether these asymptotic solutions are sufficiently accurate for practical purposes.

Using the method of diagonalization, it is straightforward to derive the transfer matrix of a circular hi-bi fiber:

$$\mathbf{T} = \mathbf{O}\tilde{\Lambda}\mathbf{O}^{-1} \tag{5.49}$$

where the matrices from left to right are

$$\mathbf{O}^{-1} = \begin{bmatrix} \cos\phi & -j\sin\phi \\ -j\sin\phi & \cos\phi \end{bmatrix}$$

$$\tilde{\Lambda} = \begin{bmatrix} e^{jgz} & 0 \\ 0 & e^{-jgz} \end{bmatrix} \tag{5.50}$$

$$\mathbf{O} = \begin{bmatrix} \cos\phi & j\sin\phi \\ j\sin\phi & \cos\phi \end{bmatrix}$$

Substitution of Eq. (5.50) into Eq. (5.49) yields

$$\mathbf{T} = \begin{bmatrix} \cos(gz) & \sin(gz) \\ -\sin(gz) & \cos(gz) \end{bmatrix} + \sin(gz)\begin{bmatrix} j\cos(2\phi) & \sin(2\phi)-1 \\ 1-\sin(2\phi) & -j\cos(2\phi) \end{bmatrix} \tag{5.51}$$

We see that the first matrix on the right is the asymptotic expression of $\mathbf{T}$ with the approximation of $\tan(2\phi) = \tau/(\delta\beta/2) \to \infty$, or $\phi \to 45°$, for which the second matrix on the right becomes zero. In the actual case, $\tau/(\delta\beta/2)$ is finite, such that ϕ is not exactly equal to $45°$, but only close to $45°$. We therefore write

$$\phi = 45° - \bar{\varepsilon} \tag{5.52}$$

where $\bar{\varepsilon} \ll 45°$ is a small deviation of ϕ in the neighborhood of $45°$. The deviation (or error) of an actual circular hi-bi fiber from a perfect circular hi-bi fiber is therefore given by the second matrix of Eq. (5.51), which, by approximation of first order, reduces to

$$\delta\mathbf{T} \approx j\sin(2\bar{\varepsilon})\sin(gz)\begin{bmatrix} 1 & 0 \\ 0 & -1 \end{bmatrix}$$
$$\approx j(2Q)^{-1}\sin(gz)\begin{bmatrix} 1 & 0 \\ 0 & -1 \end{bmatrix} \tag{5.53}$$

where $2\bar{\varepsilon} = 90° - 2\phi$ is a small angle, such that $\sin(2\bar{\varepsilon}) = (\delta\beta/2) \approx (2Q)^{-1}$.

Consider the case of precessing linear light (dual-eigenmode) transmission in a circular hi-bi fiber. Since $|\delta T| \le (\delta\beta/2)/g$ by Eq. (5.53), the estimated upper bound of the maximum to minimum power ratio is

$$\frac{P_{max}}{P_{min}} \approx \frac{1}{[(\delta\beta/2)/g]^2} \approx \tan^2(2\phi) = 4Q^2 \qquad (5.54)$$

For circular light (single-eigenmode) transmission in a circular hi-bi fiber, the upper bound of the ratio of maximum power to minimum power is estimated to be of the order

$$\frac{P_{max}}{P_{min}} \approx \frac{[1 + ((\delta\beta/2)/g)]^2}{[1 - ((\delta\beta/2)/g)]^2} \approx \frac{1 + (\delta\beta/g)}{1 - (\delta\beta/g)}$$

$$\approx 1 + \frac{2\delta\beta}{g} \approx 1 + \frac{4}{\tan(2\phi)} = 1 + 2Q^{-1} \qquad (5.55)$$

Circular light maintenance therefore requires a fairly high value of Q. Thus, besides requiring a high spin-rate, the residual linear birefringence $\delta\beta$ should be kept as small as possible.

The Fast-Spun Condition

The fast-spun condition $|(1 - \kappa)\tau/(\delta\beta)| \gg 1$ given by Eq. (5.6) is defined on the basis of Eq. (5.1), and hence refers to the local coordinates. From the regular mathematical procedures for transformation of coordinates, it should be obvious that the fast-spun condition referring to the fixed coordinates is given by $|\kappa\tau/\delta\beta| \gg 1$. This expression, which is likely to be more useful in fiber fabrication practice, demands a higher spin-rate in order to fulfill the fast-spun condition in the fixed coordinates.

REFERENCES

[1] H. C. Huang, U.S. Patent No. 5,452,394 (September 1995).

[2] H. C. Huang, "Practical circular polarization maintaining optical fiber," *Appl. Opt.*, vol. 36, pp. 6968–6975 (1997).

[3] L. Jeunhomme and M. Monerie, "Polarization-maintaining single-mode fibre cable design," *Electron. Lett.*, vol. 16, pp. 921–922 (1980).

[4] S. Machida, J. Sakai, and T. Kimura, "Polarization preservation in long-length twisted single-mode optical fibers," *Trans. IECE of Japan*, vol. E65, No. 11, pp. 642–647 (1982).

[5] J. Sakai, S. Machida, and T. Kimura, "Existence of eigen polarization modes in anisotropic single-mode optical fibers," *Opt. Lett.*, vol. 6, pp. 496–498 (1981).

[6] R. Ulrich and A. Simon, "Polarization optics of twisted single-mode fibers," *Appl. Opt.*, vol. 18, pp. 2241–2251 (1979).

[7] A. J. Barlow and D. N. Payne, "Polarization maintenance in circularly birefringent fibers," *Electron. Lett.*, vol. 17, pp. 388–389 (1981).

[8] F. Gauthier, J. Dubos, S. Blaison, Ph. Graindorge, and H. J. Additty, "Attempt to draw a circular polarization conserving fiber," *Proc. Int. Congr. Fiberoptic Rotation Sensors*, MIT, Cambridge, pp. 196–200 (Nov. 1981).

[9] M. P. Varnham, R. D. Birch, and D. N. Payne, "Helical-core circularly birefringent fibres," *Proc. Digest, IOOC-ECOC*, Venice, pp. 135–138 (1985).

[10] M. P. Varnham, R. D. Birch, D. N. Payne, and J. D. Love, "Design of helical core circularly birefringent fibres," *Proc. OFC*, p. 68 (1986).

[11] R. D. Birch, "Fabrication and characterization of circularly birefringent helical fibres," *Electron. Lett.*, vol. 23, pp. 50–51 (1987).

[12] X. S. Fang and Z. Q. Lin, "Field in single-mode helically wound optical fibers," *IEEE Trans. Microwave Theory Tech.*, vol. MTT-33, pp. 1150–1154 (1985).

[13] F. Mystre and A. Bertholds, "Magneto-optical current sensor using a helical-fiber Fabry-Perot resonator," *Opt. Lett.*, vol. 14, pp. 587–589 (1989).

[14] A. J. Rogers, J. Xu, and J. Yao, "Vibration immunity for optical-fiber current measurement," *IEEE J. Lightwave Tech.*, vol. LT-13, pp. 1371–1377 (1995).

[15] R. Dandliker, "Rotational effects of polarization in optical fibers," in *Anisotropic and Nonlinear Optical Waveguides*, C. G. Someda and G. Stegeman, eds., Book 2 of *Optical Wave Sciences and Technology* (series editor: H. C. Huang), Elsevier Amsterdam, pp. 39–76 (1992).

[16] C. D. Hussey, R. D. Birch, and Y. Fujii, "Circularly birefringent single-mode optical fibres," *Electron. Lett.*, vol. 22, pp. 129–130 (1986).

[17] Y. Fujii and C. D. Hussey, "Design considerations for circularly form-birefringent optical fibres," *Proc. IEE J.*, vol. 133, pp. 249–255 (1986).

[18] C. G. Someda, Italian Patent No. 41584A/85 (July 1985).

[19] C. G. Someda, Italian Patent No. 41638A/86 (Dec. 1986).

[20] R. Castelli, F. Irrera, and C. G. Someda, "Circularly birefringent optical fibres: New proposals. Pt. I Field analysis," *Opt. Quantum. Electron.*, vol. 21, pp. 35–46 (1989).

[21] C. G. Someda, "Circularly birefringent optical fibres: new proposals Pt. II Birefringence and coupling loss," *Opt. Quantum. Electron.*, vol. 23, pp. 713–725 (1991).

[22] V. Ramaswamy, W. G. French, and R. D. Standley, "Polarization characteristics of noncircular core single-mode fibers," *Appl. Opt.*, vol. 17, pp. 3014–3017 (1978).

[23] H. C. Huang, "Generalized theory of coupled local normal modes in multi-wave guides," *Sci. Sin.*, vol. 9, pp. 142–154 (1960).

[24] H. C. Huang and Y. C. He, "Polarization behavior of spun fiber versus conventional fiber under strong and slight twisting," *Microwave Opt. Tech. Lett.*, vol. 9, pp. 37–41 (1995).

[25] A. M. Smith, "Polarization and magnetooptic properties of single-mode optical fiber," *Appl. Opt.*, vol. 17, pp. 52–56 (1978).

[26] A. Papp and H. Harms, "Magnetooptical current transformer. 1: Principles," *Appl. Opt.*, vol. 19, pp. 3729–3734 (1980).

[27] H. Aulich, W. Beck, N. Douklias, H. Harms, A. Papp, and H. Schneider, "Magnetooptical current transformer. 2: Components," *Appl. Opt.*, vol. 19, pp. 3735–3740 (1980).

[28] H. Harms and A. Papp, "Magnetooptical current transformer. 3: Measurements," *Appl. Opt.*, vol. 3741–3745 (1980).

[29] H. C. Huang, U.S. Patent No. 4,943,132 (July 1990).

[30] H. C. Huang, U.S. Patent No. 5,096,312 (March 1992).

[31] H. C. Huang, "Passive polarization-controlled all-fiber gyroscope and other interferometic architectures," *Fiber and Integrated Optics*, vol. 12, pp. 21–29 (1993).

BIBLIOGRAPHY

I. M. Bassett, "Design principles for a circularly birefringent optical fiber," *Opt. Lett.*, vol. 13, pp. 844–846 (1988).

J. L. Cruz, M. V. Andres, and M. A. Hernandez, "Faraday effect in standard optical fibers: dispersion of the effective Verdet constant," *Appl. Opt.*, vol. 35, pp. 922–927 (1996).

J. R. Qian and C. D. Hussey, "Circular birefringence in helical-core fibre," *Electron. Lett.*, vol. 22, pp. 515–517 (1986).

CHAPTER SIX

Elliptically Birefringent Fiber Transmission Characteristics

The state of elliptical birefringence, comprising both linear birefringence and circular birefringence, is actually a common occurrence in fiber transmission practice. To be exact, a fiber is never purely linearly birefringent or purely circularly birefringent. If, through the introduction of an anisotropic stress, for example, the linear birefringence is intentionally enhanced to a predominant proportion, the fiber then behaves practically linearly birefringent. On the other hand, if the circular birefringence becomes predominant (by way of a strong twist, for example), then the fiber is practically circularly birefringent. Thus, in a sense, linear and circular birefringent fibers can be viewed as two important special cases of the general class of elliptically birefringent fibers.

In Section 4.8 we touched on the problem of elliptical-bi in a slightly twisted conventional fiber, wherein the twist-induced circular-bi and the residual linear-bi due to tiny deformation of the core are both small, thereby causing the fiber to behave elliptically birefringent. A fiber under a constrained state of slight twisting, however, is not exactly qualified to be a kind of special fiber from the practical viewpoint. Naturally, the analysis in Section 4.8 on the subject of elliptical birefringence in fiber could only be very brief.

The subject of this chapter concerns a particular fiber version featured by special design parameters such that the fiber behaves inherently elliptically birefringent. This fiber is made by spinning in the hot state a linear highly birefringent (hi-bi) fiber during the linear draw. Any kind of linear hi-bi fiber will serve the purpose. The fibers thus made are called the "spun hi-bi" fibers, of which the spun Bow-Tie fiber and the spun elliptical-cladding fiber are best known. In a spun hi-bi fiber, since the inherent linear birefringence is intentionally chosen to be high (not just a small residual), and the spinning rate can be relatively fast in the hot state (unlike postdraw cool twisting), the inherent birefringence of the fiber is necessarily elliptical, with the linear birefringence component comparable in magnitude with the circular birefringence component.

As far as the transmission characteristic is concerned, the spun hi-bi fiber and the slightly twisted lo-bi fiber behave similarly. In either kind of fibers the eigenmodes are (right and left) elliptical light. Since it is the Q-factor (ratio of circular-bi to linear-bi) that determines the state of polarization (SOP) of light, the SOP pattern will be similar, irrespective of whether the circular-bi and the linear-bi are both large (for spun hi-bi fiber), or they are both small (for slightly twisted hi-bi fiber).

On the other hand, the transmission characteristics of spun hi-bi fiber are entirely different from those of the conventional spun fiber, which were covered in Chapter 2, though both fiber versions are fabricated using the process of spinning in the hot state. While the spun hi-bi fiber is elliptically birefringent in nature, the conventional spun fiber is lo-bi for any state of polarization (SOP) of light.

In view of the mathematical theory, elliptical birefringence is a topic of more generality, including linear birefringence and circular birefringence as two limiting special cases. A separate chapter wholly devoted to the subject of elliptically birefringent fiber will therefore be of help, not only in understanding the spun hi-bi fiber in a narrow sense, but also for a more in-depth comparative study of the other kinds of special fibers.

6.1   TECHNICAL BACKGROUND

The major objective that prompted the idea of making the kind of spun Bow-Tie fiber was to fabricate an all-fiber current meter, which appears most attractive for its immunity in an electromagnetic hazardous environment. In this important application area, the linear hi-bi fiber is unsuitable by nature, inasmuch as the Faraday effect in glass fiber is small per unit length, such that it is easily quenched by a linear birefringence in fiber. In the research and development of the Faraday-rotation device, it was originally thought that the ultra-lo-bi spun (conventional) fiber would be an ideal medium. This thought is correct, if the Faraday-rotation effect alone is concerned, because an idealized ultra-lo-bi fiber would efficiently reflect the Faraday effect, without suffering a distributed quenching due to the intrinsic linear birefringence in fiber. In a practical system, however, the ultra-lo-bi fiber is oversensitive to extrinsic perturbations that likewise may quench the Faraday effect.

The advantages of circular light transmission were recognized early by scientists engaged in the art of fiber optics. Unfortunately, as compared with the case of linear hi-bi fibers, advancement in the search for practical circular polarization-maintaining fiber has been far less fruitful during the past period of more than one and half decades. One attempt was directed toward the helical-core fiber. Another attempt was directed toward a variety of form-birefringent fibers. Despite all these attempts, the only useful means yet available to make a practical circular polarization-maintaining fiber is still the early inefficient method of fiber-twisting, that is, to twist a conventional fiber in its "cool" state. In early Faraday rotation experiments, "twisted" fiber was even used in an attempt to suppress the distributed linear birefringence in the fiber circuitry. But the method of twisting is not practical in the final analysis, as described in Chapter 4.

The advent of elliptically birefringent fiber around mid-1980s reflects exactly the technical background at that time. There was a dilemma in the choice of fiber for magnetic field or electric current measurement. The lo-bi fiber failed because of lack of capability to hold the sensed Faraday rotation; the linear hi-bi fibers are not suitable in this application; twisting of fiber is impractical, and so on. Under these circumstances, it was only natural for fiber-optic scientists to try other possible approaches. The spun Bow-Tie fiber represents one major attempt toward the said objective. Since its advent at Southampton in 1986 [1], the spun Bow-Tie fiber has received worldwide attention. After the United Kingdom, the technique was also remarkably developed in Australia, where the spun elliptical-cladding fiber was used relatively extensively in trial fiber systems.

The intuitive thought in the spun hi-bi fiber approach is to compromise between the response to the Faraday effect and the capability of resisting extrinsic perturbations. Such a compromise was thought reasonable on the ground that, on the one hand, the spun hi-bi fiber, because of its spinning aspect, may be able to respond to the Faraday-rotation effect, while on the other hand, the same fiber, because of its original (pre-spin) linear hi-bi characteristics, may still retain to some extent the birefringent property for perturbation resistance. Whether or not this presumed compromise is actually possible is a question that needs to be answered on a rigorous theoretical ground.

6.2   INITIAL-VALUE PROBLEM APPROACH TO SPUN Hi-Bi FIBER

We consistently adopt the coupled-mode theoretical approach in our analytic study of the spun hi-bi fiber. At the start of this analysis, it is noted that the mathematical manipulation for elliptically birefringent fiber is actually always more laborious than it is for either linearly birefringent fiber or circularly birefringent fiber. The reason is obvious. In the case of linear or circular birefringent fiber, since one kind of birefringence far exceeds the other kind of birefringence, it is more likely that certain mathematical simplification can be allowed during the analysis. For elliptical birefringent fiber, however, since the magnitudes of the linear birefringence component and of the circular birefringence component are comparable, mathematical simplification is scarcely permissible.

As afore-discussed, an elliptically birefringent fiber requires that the two birefringence components (linear-bi and circular-bi), that is, $\delta\beta$ and τ, be comparable by order of magnitude. Because of this constraint, the choice of either $\delta\beta$ or τ becomes unusually inflexible. For the parameter $\delta\beta$ (or $L_b = 2\pi/\delta\beta$, the unspun beat length), its magnitude cannot be too large or too small. If $\delta\beta$ is too large, it will quench the Faraday rotation effect. On the other hand, if $\delta\beta$ is too small, it will mean a sacrifice of the eventual perturbation-resistive capability of the spun hi-bi fiber. Meanwhile, the parameter τ (or $L_s = 2\pi/\tau$, the spin pitch) also cannot be too large or too small. Concerning the response to the Faraday rotation, it is obvious that the choice of a relatively large value of τ is advantageous. Nevertheless, an additional restrictive factor exists that needs careful consideration. As was shown in Chapter 2, should the spin rate be raised high enough, the "averaging" effect would become effective such that any originally hi-bi fiber would become a lo-bi fiber. In view of all such rather conflicting considerations, the only way to choose the two birefringence components $\delta\beta$ and τ is to let them be comparable by order of magnitude. Bearing this fundamental specialty of the spun hi-bi fiber in mind, we proceed to the coupled-mode analysis, aiming at a rigorous description of this particular kind of fiber.

Coupled-Mode Equations for Spun Hi-Bi Fiber in Local Coordinates

With respect to the mathematical form, the coupled-mode equations derived previously for the spun lo-bi fiber is equally valid for the spun hi-bi fiber, which is our present concern. We can therefore directly borrow Eq. (2.12), or equivalently Eqs. (2.13) and (2.14), for the present study of spun hi-bi (elliptical-bi) fiber. The "method of diagonalization" applies to the present case as well. However, we cannot go further to make use of the mathematical step of "asymptotic approximation" given by Eqs. (2.34)–(2.37). These equations did substantially simplify the analytic study of the spun lo-bi fiber, but now they no longer apply to the spun hi-bi fiber. In an elliptically birefringent fiber, $\delta\beta \approx \tau$ by order of magnitude, such that the said mathematical simplification is no longer allowable.

Like the earlier cases, mathematical formulation of the coupled-mode equations for spun hi-bi fiber depends on the choice of the reference coordinates, whether local or fixed. Particularly in the case of spun hi-bi fiber, it is feasible to employ the local coordinates for coupled-mode solutions. The corresponding solutions in the fixed coordinates can be derived afterwards by coordinate transformation.

Following the same way of derivation leading to Eqs. (2.13) and (2.14), the coupled-mode equations for spun hi-bi fiber in the local coordinates are written as:

$$\frac{d\mathbf{A}}{dz} = \mathbf{K}\mathbf{A} \tag{6.1}$$

$$\mathbf{A} = \begin{bmatrix} A_x \\ A_y \end{bmatrix}$$

$$\mathbf{K} = \begin{bmatrix} j(\delta\beta/2) & \pm\tau \\ \mp\tau & -j(\delta\beta/2) \end{bmatrix} = \begin{bmatrix} j\pi/L_b & \pm 2\pi/L_s \\ \mp 2\pi/L_s & -j\pi/L_b \end{bmatrix} \tag{6.2}$$

where $L_b(=2\pi/\delta\beta)$ is the local beat length of the fiber in its unspun state; $L_s(=2\pi/\tau)$ is the spin pitch; the upper and lower $\pm$ signs refer to the senses of spin with respect to the transmission direction z; and $\delta\beta$ and τ are the inherent linear birefringence and the spin-rate, respectively. Equations (6.1) and (6.2) for spun hi-bi fiber are the same as Eqs. (2.13) and (2.14) for spun lo-bi fiber, except that the elements of $\mathbf{K}$ are now expressed both in terms of $(\delta\beta, \tau)$ and of the readily measurable quantities (L_b, L_s).

Nevertheless, the physical implication of Eqs. (6.1) and (6.2) for spun hi-bi fiber are drastically different from those of Eqs. (2.13) and (2.14) for spun lo-bi fiber. In the case of spun lo-bi fiber, the fiber parameters are required to satisfy the fast-spun condition: $\tau/\delta\beta = L_b/L_s \gg 1$. This condition was easily satisfied in the previous case of spun lo-bi fiber where $\delta\beta$ represents the small residual linear birefringence. In the present case of spun hi-bi fiber, however, what is required is that τ and L_b are of the same order of magnitude, that is

$$\tau \approx \delta\beta \tag{6.3}$$

$$L_s \approx L_b \tag{6.3a}$$

That is to say, $(\tau/\delta\beta) = (L_b/L_s) \approx 10^0$ by order. The value of $\delta\beta$ is necessarily high, because the fiber at its unspun state is a linear hi-bi fiber. The spinning rate cannot be too high, in order to avoid the "averaging effect" that would reduce the perturbation resistivity. Also it cannot be too low, in order to favor the effect of Faraday rotation. Thus, Eq. (6.3) or Eq. (6.3a) manifests the special feature inherent in a spun hi-bi fiber that makes the fiber elliptically birefringent. Under the condition of Eq. (6.3) or Eq. (6.3a), the analysis of a spun hi-bi fiber by the coupled-mode theoretical approach becomes more laborious than the analysis of other fiber versions dealt with previously. The reason is simply that, in the present case it is not legitimate to adopt a mathematical approximation by neglecting one or another term, or to take advantage of some limiting process in the derivation.

The Method of Diagonalization and General Solutions

Equations (2.13)–(2.32) remain useful in this analytic study of spun hi-bi fiber. In the following, we outline the pertinent equations, with only slight alterations in the form of these equations in order to suit the needs of the present analysis. To begin with, we introduce the transformation:

$$\mathbf{A}(z) = \mathbf{OW}(z) \tag{6.4}$$

$$\mathbf{O} = \begin{bmatrix} \cos\phi & \pm j\sin\phi \\ \pm j\sin\phi & \cos\phi \end{bmatrix}$$

$$\phi = \frac{1}{2}\arctan\left(\frac{2\tau}{\delta\beta}\right) \tag{6.4a}$$

$$= \frac{1}{2}\arctan\left(\frac{2L_b}{L_s}\right)$$

where the $(+)$ and $(-)$ of the $\pm$ signs before the off-diagonal elements of $\mathbf{O}$ (the diagonalizing matrix) refer to the positive and negative senses respectively, of spin of fiber. Substituting Eq. (6.4) into Eq. (6.1) yields

$$\frac{d}{dz}\mathbf{W}(z) = \Lambda\mathbf{W}(z)$$

$$\Lambda = \mathbf{O}^{-1}\mathbf{KO} = \begin{bmatrix} \lambda & 0 \\ 0 & -\lambda \end{bmatrix} \tag{6.5}$$

$$\lambda = jg$$

$$g = \frac{\pi}{L_b}\left[1 + 4\left(\frac{L_b}{L_s}\right)^2\right]^{1/2} \tag{6.5a}$$

Equation (6.5) yields two independent exponential solutions in the form:

$$W_1(z) = e^{jgz}W_1(0)$$
$$W_2(z) = e^{-jgz}W_2(0) \tag{6.6}$$

such that

$$\mathbf{W}(z) = \tilde{\Lambda}\mathbf{W}(0)$$

$$\tilde{\Lambda} = \begin{bmatrix} e^{jgz} & 0 \\ 0 & e^{-jgz} \end{bmatrix} \tag{6.6a}$$

As known, $\lambda = jg$ and $-\lambda = -jg$ in Eqs. (6.5) and (6.5a) are the eigenvalues. The two exponential functions in Eqs. (6.6) and (6.6a) are the eigenfunctions. The diagonal matrix $\tilde{\Lambda}$ in Eq. (6.6a) is the integration of the diagonal matrix Λ in Eq. (6.5).

The Local Transfer Matrix of Spun Hi-Bi Fiber

By Eqs. (6.4) and (6.6a), $\mathbf{A}(z) = \mathbf{O}\tilde{\Lambda}\mathbf{W}(0)$, where $\mathbf{W}(0) = \mathbf{O}^{-1}\mathbf{A}(0)$, and by the inverse of Eq. (6.4), the solution of Eq. (6.1) can be written in the following simple form:

$$\mathbf{A}(z) = \mathbf{T}_l\mathbf{A}(0) \tag{6.7}$$

$$\mathbf{T}_l = \mathbf{O}\tilde{\Lambda}\mathbf{O}^{-1} \tag{6.7a}$$

where $\mathbf{T}_l$ is the "transfer matrix" of the spun hi-bi fiber with reference to the local coordinates.

6.3 LOCAL ELLIPTICAL EIGENMODES IN SPUN Hi-Bi FIBER

In mathematical physics, independent exponential solutions of Eq. (6.6) are called the "normal modes," that is, the eigenmodes in the normal coordinates. A more detailed description of this definition is relegated to Appendix B. If the common wave factor $e^{j(\omega t - \beta z)}$ is written out, it can be seen that the exponential functions with positive and negative exponents represent the fast and slow eigenmodes, respectively. Thus, we have

$$\mathbf{W}^f(z) = \begin{bmatrix} 1 \\ 0 \end{bmatrix} e^{jgz} \tag{6.8a}$$

$$\mathbf{W}^s(z) = \begin{bmatrix} 0 \\ 1 \end{bmatrix} e^{-jgz} \tag{6.8b}$$

where the superscripts f and s designate that the normal modes are fast mode and slow mode, respectively. The local modes referring to the local coordinates can be derived from these normal modes by the transformation, Eq. (6.4). Thus, for the fast normal mode and the slow normal mode, respectively, Eq. (6.4) yields

$$\begin{aligned}
\mathbf{A}^f(z) &= \mathbf{O}\mathbf{W}^f(z) \\
&= \begin{bmatrix} \cos\phi & j\sin\phi \\ j\sin\phi & \cos\phi \end{bmatrix} \begin{bmatrix} 1 \\ 0 \end{bmatrix} e^{jgz} \\
&= \begin{bmatrix} \cos\phi \\ j\sin\phi \end{bmatrix} e^{jgz}
\end{aligned} \tag{6.9}$$

and

$$\mathbf{A}^s(z) = \mathbf{OW}^s(z)$$

$$= \begin{bmatrix} \sin \phi \\ -j \cos \phi \end{bmatrix} e^{-j(gz - \pi/2)} \qquad (6.10)$$

The local eigenmodes (short for eigenmodes referring to local coordinates) of a spun hi-bi fiber are thus found to be a right-fast elliptical light and a left-slow elliptical light. In geometrical representation, the electric-field vector of the fast mode traces an ellipse in the right-handed (clockwise) sense at a faster angular rate, while the electric-field vector of the slow mode traces an ellipse in the left-handed (anticlockwise) sense at a slower angular rate. The two ellipses are orthogonal and aligned with the local coordinates, with the major axis of one ellipse coinciding with the minor axis of the other ellipse, and vice versa.

The two elliptical eigenmodes have the same ellipticity (ratio of minor axis to major axis) equal to

$$\varepsilon = \tan \phi \qquad (6.11)$$

This expression can be seen immediately from Eqs. (6.9) and (6.10).

6.4   SINGLE-EIGENMODE TRANSMISSION IN SPUN Hi-Bi FIBER

Knowledge of the eigenmodes is necessary for an understanding of the waveguiding property of the fiber structure concerned. But this knowledge alone is not sufficient to show what kind of wave field is actually propagating in the fiber. What we also need to know is the *excitation* condition at the input. For the same fiber, the lightwave transmission pattern may change completely if the excitation condition is changed.

Matching Condition at Input of Spun Hi-Bi Fiber

It should be easy to understand now that, for the same fiber, the possible transmission patterns are as numerous as the possible excitation conditions. For convenience, we can classify the multitude of possible transmission behaviors into two kinds, according to the involvement of a single eigenmode and the involvement of both eigenmodes, respectively, during the course of transmission [2]. The first kind is realized if and only if the SOP of the input light *matches* the SOP of one or the other eigenmode at the initial local point (i.e., at the input $z = 0$). The second kind then embodies all the other cases that do not satisfy the matching condition at the input of the fiber.

According to Eqs. (6.9) and (6.10), the eigenmodes of spun hi-bi fiber at the input ($z = 0$) are given by:

$$\mathbf{A}^R(0) = \begin{bmatrix} \cos \phi \\ j \sin \phi \end{bmatrix} \tag{6.12}$$

$$\mathbf{A}^L(0) = \begin{bmatrix} j \sin \phi \\ \cos \phi \end{bmatrix} \tag{6.13}$$

where the superscripts indicate that the first and the second matrices are right elliptical light and left elliptical light, respectively. Let the fiber be excited by an incident light that matches one eigenmode of the spun hi-bi fiber (say, the right elliptical light). By the regular procedure of the method of diagonalization, we have

$$\mathbf{A}(0) = \mathbf{A}^R(0) = \begin{bmatrix} \cos \phi \\ j \sin \phi \end{bmatrix}$$

$$\mathbf{W}(0) = \mathbf{O}^{-1}\mathbf{A}(0) = \begin{bmatrix} 1 \\ 0 \end{bmatrix}$$

$$\mathbf{W}(z) = \Lambda\mathbf{W}(0) = \begin{bmatrix} 1 \\ 0 \end{bmatrix} e^{jgz} \tag{6.14}$$

$$\mathbf{A}(z) = \mathbf{O}\mathbf{W}(z) = \begin{bmatrix} \cos \phi \\ j \sin \phi \end{bmatrix} e^{jgz}$$

Similarly, if the incident light matches the other eigenmode of the spun hi-bi fiber (the left elliptical light), we have:

$$\mathbf{A}(0) = \mathbf{A}^L(0) = \begin{bmatrix} j \sin \phi \\ \cos \phi \end{bmatrix} \tag{6.15}$$

$$\mathbf{A}(z) = \begin{bmatrix} j \sin \phi \\ \cos \phi \end{bmatrix} e^{-jgz} \tag{6.16}$$

The above results thus verify the fundamental concept that, if the matching condition is satisfied at the input, then lightwave transmission in the fiber will be in single eigenmode in the entire course.

Discussion on the Excitation Condition

According to waveguide theory, a regular method for ensuring a stable SOP of light all the way from the input end to the output end is to keep supporting a *single* eigenmode in the entire course of light transmission along the fiber line. Recall that the prefix "eigen" to the word "mode" actually implies that the transverse field pattern of the mode is maintained everywhere along the fiber during transmission. After traversing a length of fiber an eigenmode undergoes a change only in its phase, with the transverse-field pattern (or SOP of light) always z-invariant. In the case of spun hi-bi fiber, therefore, a stable transmission of light can be ensured only if we employ either a fast right elliptical light, whose transverse revolving ellipse is given by Eq. (6.9), or a slow left elliptical light, whose transverse revolving ellipse is given by Eq. (6.10). In the single-eigenmode regime, the transmission of a polarization mode, or of an SOP of light, is z-invariant, and hence length-insensitive.

Most of the existing literature on spun hi-bi fiber puts sufficient emphasis on the existence of elliptical eigenmodes in such fiber. Nevertheless, the topic of single-eigenmode transmission and the related topic of the excitation condition have scarcely received adequate attention in the relevant publications.

It is therefore worth emphasizing that a stable and z-invariant SOP of light in a spun hi-bi fiber can only be secured when a single eigenmode is excited. In other words, the incident light at $z = 0$ cannot be an arbitrary elliptically polarized light, but should be the one whose principal axes exactly match the local coordinate axes of the fiber at the input end, and moreover, the ellipticity of the incident light also cannot be arbitrary, but should be exactly equal to tan ϕ, by Eq. (6.11), in conformity with the ellipticity of either eigenmode of the spun hi-bi fiber. In practice, such strict technological requirement is extremely difficult, if not impossible, to fulfill. For an arbitrary elliptically polarized incident light, the SOP of light at the output end $z = L$ of a spun hi-bi fiber will be highly erratic and practically unpredictable. In attempting the Faraday effect measurement, such an output light will still respond to an applied current, and the response is likely to be stronger for a stronger current, but ultimately, the response will be highly irregular.

At first thought, we might wonder if the required excitation condition is really so strict. To clarify the point, consider the case of a linear polarization-maintaining fiber of any kind. One who has some experience in a fiber-optics laboratory must be impressed by the relative severity of the excitation condition of a linear light in such fiber. The linear light excitation at the input end $z = 0$ must be accurately oriented along either of the principal axes of the linear polarization-maintaining fiber in order for the light transmission to be stable in linear SOP. A small deviation in the linear incident light orientation angle, say by only a few degrees, will cause an erratic change in the SOP picture along the fiber line, with the result that the output light no longer stays linear, but can assume any SOP. By analogy with the excitation condition of a linear polarization-maintaining fiber, we naturally expect that the excitation condition of an elliptically birefringent fiber should be more complex, inasmuch as an elliptical light is specified not only by its orientation, but also by its ellipticity.

From the analytical viewpoint, a spun Bow-Tie or other spun hi-bi fiber is impractical in view of the overstrict excitation requirement, as described above. Simultaneous matching the ellipticity and the principal axes between the incident light and a local eigenmode at $z = 0$ of the spun hi-bi fiber is scarcely achievable in fiber practice.

At this point we recall an early paper by Sakai et al. [3] in which twisted anisotropic fibers were studied under fairly general excitation conditions. The eigenmodes of moderately twisted fiber are elliptically polarized waves of right- and left-handedness. In Ref. [3, figs. 1-3], experimental curves are shown both for non-eigen polarization incidence (linear and circular), and for *eigen polarization* incidence (specific elliptical). With reference to this early work, one therefore might wonder why excitation of a spun Bow-Tie or other spun hi-bi fiber could not have been achieved in a similar way.

To examine this point further, we should know that Ref. [3] actually concerns a rather specific experiment on twisted elliptical-core fiber. According-ing to the relevant description given in Ref. [3], the tested fiber is 3.888 meters long, "held straight in a V groove over the entire length." It is understood that in a twisting experiment, the tested fiber is initially in an untwisted state, such that its principal axes can be determined by a routine measuring procedure using location of the "minimum of minimum" point of power output. Since the fiber is held straight in the V groove over the entire length, the principal axes can always be determined when the fiber is later twisted at a known twist rate, because the twist angle per unit length can be calculated from the total twist applied over the entire length of fiber. With the principal axes known at any twist rate, the only restrictive condition necessary for eigen polarization incidence can be satisfied if the incident light exciting the twisted fiber has the proper ellipticity. In the experiment, therefore, one and only one degree of freedom (the ellipticity of light), which needs adjustment and tuning, is involved.

The technique of exciting a spun Bow-Tie or other spun hi-bi fibers is another matter. Here, the progressive rotation, along the fiber length, of the principal axes is not known a priori, such that both the condition of matching the principal axes and the condition of matching the ellipticity will be satisfied simultaneously. Such experiment therefore implies two degrees of freedom (ellipticity and orientation), so that adjustment and tuning become a very difficult, if not intractable, procedure.

6.5   DUAL-EIGENMODE TRANSMISSION IN SPUN Hi-Bi FIBER

In the preceding section we examined the overstrict excitation condition of a spun hi-bi fiber, and arrived at the conclusion that stable lightwave transmis-sion in single eigenmode is hardly achievable in this kind of fiber, inasmuch as simultaneous matching of ellipticity and orientation between the incident light and the fiber poses severe technological difficulty.

If the input light does not match either local eigenmode of the spun hi-bi fiber at $z = 0$, as is commonly the case, then lightwave transmission in single eigenmode will not be possible. The result is that this input light excites both elliptical eigenmodes in the spun hi-bi fiber line. The type of fiber transmission in which both eigenmodes are involved is then referred to as the dual-eigenmode transmission. The coexistence of two eigenmodes in the fiber results in beating of the eigenmodes, such that the field pattern becomes very complicated and length-sensitive.

The Transfer Matrix of Spun Hi-Bi Fiber

In Section 6.2, we adopted the method of diagonalization and derived the transfer matrix of the spun hi-bi fiber in the local coordinates. In this section we examine the dual-eigenmode transmission characteristics by this transfer-matrix approach. According to Eq. (6.7), we have

$$\mathbf{A}(z) = \mathbf{O}\tilde{\Lambda}\mathbf{O}^{-1}\mathbf{A}(0) \equiv \mathbf{T}_l(z)\mathbf{A}(0) \tag{6.17}$$

Substituting Eqs. (6.6a) and (6.4a) into this equation, we have

$$\mathbf{T}_l(z) \equiv \mathbf{O}\tilde{\Lambda}\mathbf{O}^{-1} = \begin{bmatrix} \cos^2\phi e^{jgz} + \sin^2\phi e^{-jgz} & \sin 2\phi \sin gz \\ -\sin 2\phi \sin gz & \sin^2\phi e^{jgz} + \cos^2\phi e^{-jgz} \end{bmatrix} \tag{6.18}$$

$$g = \left[\left(\frac{\delta\beta}{2}\right)^2 + \tau^2\right]^{1/2} \tag{6.19a}$$

$$\phi = \frac{1}{2}\arctan\left(\frac{2\tau}{\delta\beta}\right) \tag{6.19b}$$

where $\mathbf{T}_l(z)$ is the transfer matrix of the spun hi-bi fiber referring to the *local* coordinates. By simple trigonometric-algebra, Eq. (6.18) can be written as Eq. (2.58). Here we retain the exponential terms so as to show the involvement of two eigenmodes in the general case. We observe that this transfer matrix is a function of the structural parameters g and ϕ, and hence of $\delta\beta$ and τ. Therefore, $\mathbf{T}_l$ describes the property of the spun hi-bi fiber itself, without regard to the input light or initial condition.

Description of the Initial Condition

According to Eqs. (2.44) and (2.45), an arbitrary elliptical light can be formulated by either the (θ, δ)-representation, or the (ε, ψ)-representation. The (θ, δ)-representation is shorter and simpler in form but, in the general case, it does not show explicitly the two features (orientation and ellipticity) of an arbitrary elliptical light. On the other hand, the (ε, ψ)-representation displays these two features explicitly, but looks relatively more expansive. These two matrix expressions are equivalent to each other, with the parameters (θ, δ) of Eq. (2.44) related to the parameters (ε, Ψ) of Eq. (2.45) by Eq. (2.46). Mathematically, both equations are useful in defining the excitation condition in a general way.

In the following derivation, we use the (θ, δ)-representation to describe an elliptical light for the sake of relative simplicity. Clearly, if the (ε, ψ)-representation is used, the mathematical formulation of the end results will be overexpansive, while not yielding much more informative data.

Let an arbitrary elliptical light exciting the spun hi-bi fiber at its input be defined by

$$\mathbf{A}(0) = \begin{bmatrix} \cos\theta \\ \sin\theta e^{j\delta} \end{bmatrix} \tag{6.20}$$

This expression reduces to linear light of orientation θ if $\delta = 0$, and to right or left circular light if $\theta = \pi/4$, and $\delta = \pi/2$ or $-\pi/2$. Excepting for these two special cases, Eq. (6.20) represents an elliptical light. As previously noted, in the general case of an elliptical light, θ does not represent the orientation of either of the principal axes of the polarization ellipse. This orientation is given by $\psi = (1/2)\arctan[\tan(2\theta)\cos\delta]$.

Arbitrary Elliptical Light Exciting a Spun Hi-Bi Fiber

Let an arbitrary elliptical light given by Eq. (6.20) be incident at the input of a spun hi-bi fiber. Putting Eq. (6.20) in Eq. (6.18) yields $\mathbf{A}(z)$. Thus

$$\mathbf{A}(0) = \begin{bmatrix} \cos\theta \\ \sin\theta e^{j\delta} \end{bmatrix}$$

$$\mathbf{A}(z) = \begin{bmatrix} \cos\theta(\cos^2\phi e^{jgz} + \sin\phi e^{-jgz}) + \sin\theta e^{j\delta}\sin 2\phi\sin gz \\ \sin\theta e^{j\delta}(\sin^2\phi e^{jgz} + \cos^2\phi e^{-jgz}) - \cos\theta\sin 2\phi\sin gz \end{bmatrix} \tag{6.21}$$

where $g = [(\delta\beta/2)^2 + \tau^2]$, $\phi = (1/2)\arctan(2\tau/\delta\beta)$.

Note that Eq. (6.21) involves two pairs of parameters: (g, ϕ) and (θ, δ). The former describes the fiber structure, while the latter describes the input light that is externally applied (as a source) to excite the fiber. Equation (6.21) contains both eigenmodes so that they beat. Note that either $\cos gz$ or $\sin gz$ in the elements of Eq. (6.21) is expressible in terms of the two local eigenmodes e^{jgz} and e^{-jgz}. The construct of this dual-eigenmode matrix expression is fairly complex, so that the beating pattern is not easily interpretable.

Circular Light Excitation and Linear Light Excitation

It is interesting to see if the solution is less complex for the two important special cases, that is, the circular light excitation and the linear light excitation. Thus, putting $\theta = \pi/4$ and $\delta = \pm \pi/2$ for circular light in Eq. (6.21) yields

$$
\mathbf{A}(z) = \frac{1}{\sqrt{2}} \left[\begin{array}{c} \cos^2\phi e^{jgz} + \sin^2\phi e^{-jgz} + j\sin 2\phi \sin gz \\ \pm j(\sin^2\phi e^{jgz} + \cos^2\phi e^{-jgz}) - \sin 2\phi \sin gz \end{array} \right] \quad (6.22)
$$

Then, putting $\delta = 0$ for linear light in Eq. (6.21) yields

$$
\mathbf{A}(z) = \left[\begin{array}{c} \cos\theta(\cos^2\phi e^{jgz} + \sin^2\phi e^{-jgz}) + \sin\theta\sin 2\phi \sin gz \\ \sin\theta(\sin^2\phi e^{jgz} + \cos^2\phi e^{-jgz}) - \cos\theta\sin 2\phi \sin gz \end{array} \right] \quad (6.23)
$$

where in both Eqs. (6.22) and (6.23) $g = [(\delta\beta/2)^2 + \tau^2]$, $\phi = (1/2)\arctan(2\tau/\delta\beta)$.

Both equations do not differ much in mathematical complexity from the general Eq. (6.21). Thus, except for the specific case that the matching condition is fulfilled at the input, a spun hi-bi fiber always carries a fairly complex field of lightwave that varies erratically along the fiber, such that the SOP of light at the far end of fiber becomes practically unpredictable.

Beat Length in Local Coordinates

Currently available sources usually deliver light power of linear SOP. But, according to Eq. (6.23) even a simple linear light will excite a wave-field pattern in spun hi-bi fiber that varies in the transmission direction z in a rather complex manner. The field pattern therefore becomes *length-sensitive*. This result is not surprising. Leaving aside the single-eigenmode scheme for which the matching condition is overstrict for practical use, linear light or an arbitrary incident light will excite both eigenmodes in the spun hi-bi fiber so that they beat to form a length-sensitive pattern.

The beat length of a spun hi-bi fiber in the local coordinates can be readily determined by Eq. (6.21), or by Eqs. (6.22) and (6.23). Note that the beat length is a characteristic figure descriptive of the fiber structure only, regardless of what the input light is. In fact, we can also immediately determine this characteristic figure from the transfer-matrix given by Eq. (6.18). Either way, we observe that e^{jgz} and e^{-jgz} are periodic functions of z, such that the wave field or SOP of light reproduces itself at a distance $z = L$ along the fiber where gL becomes equal to an integer multiple of π. That is, the condition for the SOP of light to reproduce itself in the local coordinates is given by

$$
gL = \pi(1 + m), \qquad m = 0, 1, 2 \ldots \quad (6.24)
$$

such that $e^{jgz} = e^{-jgz} = \mp 1$. Under this condition, Eq. (6.21) is greatly simplified to become

$$\mathbf{A}(L) = \mp \begin{bmatrix} \cos \theta \\ \sin \theta e^{j\delta} \end{bmatrix} \qquad (6.25)$$

The shortest length at which the SOP of light can reproduce itself is the beat length. Thus, letting $m = 0$ in Eq. (6.24), the local beat length of a spun hi-bi fiber is given by

$$L_b^l = \frac{\pi}{g} = \frac{L_b L_s}{\sqrt{L_s^2 + 4L_b^2}} \qquad (6.26)$$

This local beat length, defined as the minimum length of fiber at which an arbitrary field pattern reproduces itself, is in principle a measurable quantity with respect to the local coordinate axes. We recall that the parameter "beat length" refers to the fiber property only, regardless of the input light. Thus, we can also derive the same (local) beating condition and the same (local) beat-length formula from Eq. (6.22) or Eq. (6.23), respectively, for circular light excitation or linear light excitation.

Linear light excitation is of particular practical interest. For linear input light, $\delta = 0$ in Eq. (6.25). Under the condition of local beating, $gL_b^l = \pi(1 + m)$, it can be seen from Eq. (6.25) that a linear light with an orientation angle θ will exactly reproduce itself at $z = L_b^l$. It might therefore be thought that this beating property can be utilized for spun hi-bi fiber to work in system application (a current monitor, for example). Nevertheless, such a scheme cannot be practical because the beating pattern along the fiber is highly length-sensitive. Note that when the condition of local beating given by Eq. (6.24) is applied to a length of spun hi-bi fiber of substantial length, the integer m in this equation is very large such that the required accuracy of the fiber length L is severely critical. A slight deviation of the fiber length L from the required critical value may completely spoil the otherwise SOP reproducible scheme. Restoration of the desired SOP with the help of optoelectronics is possible in principle, but is by no means practical.

Questions about Beat Length of Spun Hi-Bi Fiber in Fixed Coordinates

The local beat-length formula given by Eq. (6.26) shows that an arbitrary field pattern in the spun hi-bi fiber reproduces itself in the local coordinates at an integer multiple of this length. On the other hand, this formula gives a reduced beat length in the local coordinates, and therefore seems to predict an enhanced birefringence, as compared with the corresponding value of the hi-bi fiber in its unspun state. This prediction does not appear to conform with the physical concept that spinning a fiber tends to lower its birefringence due to the "averaging effect." Because of this theoretical difficulty, we are tempted to transform the local coordinate axes to fixed coordinate axes to see what the consequences will be.

For a transformation from local coordinate axes to fixed coordinate axes (Fig. 6.1), we employ the rotation matrix that performs a counterrotation of the coordinate axes by an angle $(-\tau z)$:

$$\mathbf{R}(-\tau z) = \begin{bmatrix} \cos(\tau z) & -\sin(\tau z) \\ \sin(\tau z) & \cos(\tau z) \end{bmatrix} \qquad (6.27)$$

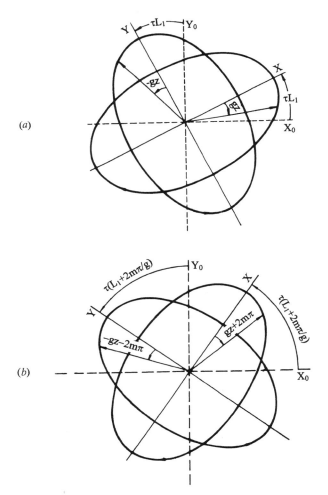

FIGURE 6.1 (*a*) Elliptical eigenmodes in local coordinates. (*b*) Fields in nonreproducible fixed coordinates.

By Eqs. (6.27) and (6.18):

$$\hat{T}(z) = R(-\tau z)T(z) \equiv RO\tilde{A}O^{-1}$$

$$= \begin{bmatrix} \cos(\tau z) & -\sin(\tau z) \\ \sin(\tau z) & \cos(\tau z) \end{bmatrix} \qquad (6.28)$$

$$\times \begin{bmatrix} \cos^2\phi e^{jgz} + \sin^2\phi e^{-jgz} & \sin 2\phi \sin gz \\ -\sin 2\phi \sin gz & \sin^2\phi e^{jgz} + \cos^2\phi e^{-jgz} \end{bmatrix}$$

where $g = [(\delta\beta/2)^2 + \tau^2]$, $\phi = (1/2)\arctan(2\tau/\delta\beta)$.

The matrix-product formulation on the right side of Eq. (6.28) reveals that, in the *fixed* coordinates, the wave field or the SOP of light is no longer repetitive (or periodic) for any fiber length L. Clearly enough, of the two matrices in the above matrix-product, the right matrix has a period $2\pi/g$, while the left matrix has a different period equal to $2\pi/\tau$. Disregarding the idealistic mathematical supposition that the ratio of $2\pi/\tau$ to $2\pi/g$ happens to be a ratio m/n of integers m and n, Eq. (6.28) is never descriptive of a periodic wave field. In view of this result, the beat length of spun hi-bi fiber is undefinable in the fixed coordinates.

6.6 PRACTICAL PROBLEMS RELATING TO SPUN Hi-Bi FIBER

Connecting or splicing of fiber sections is always required in any practical application. From the application standpoint, ease of connection or splicing is one of the major criteria that determines whether or not a kind of fiber is practical.

Offset Joint of Spun Hi-Bi Fiber Sections

Jointing of two spun hi-bi fiber sections poses a far more difficult technical problem than does the case of two linear hi-bi fiber sections. To simplify the problem, let the two spun hi-bi fiber sections be identical in all aspects, except that at the joint the local principal axes of the second fiber section are skewed by an angle σ from the local principal axes of the first fiber section. Light of a single eigenmode (say, the right-fast elliptical) emergent from the first fiber section will then enter the second fiber section with a mismatch of the local principal axes. By Eqs. (6.14) and (6.15), the incident light that excites the second fiber section can be written as

$$A(0) = \begin{bmatrix} \cos\sigma & \sin\sigma \\ -\sin\sigma & \cos\sigma \end{bmatrix} \begin{bmatrix} \cos\phi \\ j\sin\phi \end{bmatrix} \qquad (6.29)$$

where a relative phase shift descriptive of the waveguiding effect of the first fiber section is not written out for simplicity. By Eq. (6.29), the output light from the second fiber section of length L_2 is given by

$$\mathbf{A}(L_2) = \mathbf{T}(L_2)\mathbf{A}(0)$$

$$= \begin{bmatrix} \cos^2\!\phi\, e^{jgL_2} + \sin^2\!\phi\, e^{-jgL_2} & \sin 2\phi \sin gL_2 \\ -\sin 2\phi \sin gL_2 & \sin^2\!\phi\, e^{jgL_2} + \cos^2\!\phi\, e^{-jgL_2} \end{bmatrix} \quad (6.30)$$

$$\times \begin{bmatrix} \cos\sigma\cos\phi + j\sin\sigma\sin\phi \\ -\sin\sigma\cos\phi + j\cos\sigma\sin\phi \end{bmatrix}$$

where $g = [(\delta\beta/2)^2 + \tau^2]$, $\phi = (1/2)\arctan(2\tau/\delta\beta)$. $\mathbf{T}(L_2)$ is the transfer matrix (in local coordinates of the second fiber section), and $\mathbf{A}(0)$ is given by Eq. (6.29). It is thus seen that, for two spun hi-bi fiber to be jointed, even a slight offset of the local principal axes of one fiber section with respect to the other fiber section will dramatically spoil the otherwise simple SOP of the elliptical light in the fiber. The output light from the two-section spun hi-bi fiber piece will become so erratically irregular that it will be scarcely analyzable.

Jointing of Spun Hi-Bi Fibers of Opposite Senses of Spin

Naturally, because the fiber is a spun structure, it is essential to see whether the sense of spinning of fiber is relevant to connecting or splicing of two fiber sections. In actual application, it obviously makes no sense if determination of the sense of spinning of each fiber section is required before connecting one section to the other. We therefore want to see what the consequences will be if two oppositely spun hi-bi fibers are connected together.

Recall that in deriving the preceding equations it was tacitly assumed, for the sake of definiteness, that the spinning of fiber is in the positive sense. With this assumption, the upper $(+)$ sign of $(\pm)$ in Eq. (6.2) was chosen in the analysis. The derivation then yielded a pair of orthogonal elliptical eigenmodes given by Eqs. (6.9) and (6.10). By the same mathematical procedure, except for choosing the lower $(-)$ sign of the $(\pm)$ in Eq. (6.2), we can derive another pair of orthogonal elliptical eigenmodes in a spun hi-bi fiber of negative sense in the following forms:

$$\mathbf{A}^{R^-}(z) = \begin{bmatrix} \sin\phi \\ j\cos\phi \end{bmatrix} e^{-j(gz+\pi/2)} \quad (6.31)$$

$$\mathbf{A}^{L^-}(z) = \begin{bmatrix} \cos\phi \\ -j\sin\phi \end{bmatrix} e^{jgz} \quad (6.32)$$

In contrast to the right-fast and left-slow elliptical eigenmodes given by Eqs. (6.9) and (6.10) for the $(+)$ spun hi-bi fiber, the preceding two expressions for the $(-)$ spun hi-bi fiber represent a right-slow elliptical eigenmode and a left-fast elliptical eigenmode, respectively.

Let an elliptical light conforming to one eigenmode of a $(+)$ spun hi-bi fiber (say, the right-fast elliptical mode) be incident onto a $(-)$ spun hi-bi fiber, such that at $z = 0$ (the input end) of the second fiber section, we have the following initial-condition according to Eq. (6.9):

$$\mathbf{A}^R(0) = \begin{bmatrix} \cos \phi \\ j \sin \phi \end{bmatrix} \tag{6.33}$$

Keeping the choice of $(-)$ sign of the $(\pm)$ in Eq. (6.2), it is straightforward to carry out the derivation and find the SOP evolution in a $(-)$ or anticlockwise spun hi-bi fiber in the following form:

$$\mathbf{A}(z) = \begin{bmatrix} \cos \phi \cos 2\phi e^{jgz} + \sin \phi \sin 2\phi e^{-jgz} \\ -j\sin \phi \cos 2\phi e^{jgz} + j\cos \phi \sin 2\phi e^{-jgz} \end{bmatrix} \tag{6.34}$$

where $g = [(\delta\beta/2)^2 + \tau^2]$, $\phi = (1/2)\arctan(2\tau/\delta\beta)$.

Alternatively, if a left-slow elliptical mode given by Eq. (6.10) is incident onto the $(-)$ spun hi-bi fiber, the derivation will lead to a result of similar complexity. The result shows that the SOP of an eigenmode of a $(+)$ spun hi-bi fiber is not maintained in a $(-)$ spun hi-bi fiber, and vice versa, even though both fiber sections are otherwise completely identical.

Offsets at joint and senses of spinning are among the features that are relevant to the problem of connecting or splicing of spun hi-bi fiber. There are other features that pose additional difficulties in this regard. By intuition, it is obvious that connecting or splicing of two pieces of spun hi-bi fiber requires that all relevant features match. Not only the sense of spinning, but also the orientation, as well as the spinning rate, all need to be matched on both sides of the joint. This very complex circumstance for spun hi-bi fiber can actually be anticipated on the analogy of the circumstance for linear hi-bi fiber. While connecting or splicing of two segments of linear hi-bi fiber requires accurate matching of the principal axes at the joint, connecting or splicing of two segments of elliptical-bi fiber naturally requires matching of the fiber characteristics at the joint in a more complex way.

Faraday Effect in Spun Hi-Bi Fiber

A problem of immediate concern is how spun hi-bi fiber behaves in the presence of the Faraday effect due to an applied electric current or magnetic field. In the existing literature, extensive experimentation has been performed on spun Bow-Tie and spun elliptical-cladding fibers. There has been difficulty in achieving a stable Faraday-effect response in these elliptically birefringent fibers. In most of the relevant publications, this difficulty has been attributed primarily to the temperature effect.

Indeed, the temperature effect is a major source of error in any birefringent fiber whose birefringence is produced by an anisotropic thermal stress. This temperature effect is expected to be more pronounced when the fiber works on the dual-eigenmode scheme such that a temperature perturbation is likely to affect the two eigenmodes differently. In the final analysis, however, we believe that the temperature effect can only be of secondary importance when compared with the effect of the basic transmission characteristics of the fiber. Clearly, any action taken to eliminate or reduce the temperature effect could be really useful only when the fiber behaves properly in the light of waveguide theory. Without ensuring a proper transmission characteristic as a prerequisite, attempt to eliminate or reduce the temperature effect naturally cannot solve the problem of unstable transmission in spun hi-bi fiber, even without the presence of a magnetic field. Applying an electric current, and hence a magnetic field, will only add more complexity to the already intractable transmission pattern of the spun hi-bi fiber. The problem is not solely due to the temperature effect. The primary cause can be traced back to the transmission mechanism inherent in the elliptically birefringent waveguiding medium.

We now make a rigorous examination of the problem of current measurement that uses the spun hi-bi fiber as a working medium. We consistently use the coupled-mode theoretical approach to the analytic study of the Faraday effect in spun hi-bi fiber. When a magnetic field is applied along the z axis of the spun hi-bi fiber, the coupled-mode equations referring to the local coordinates take the form:

$$\frac{dA_x}{dz} = j\frac{\delta\beta}{2}A_x + (\tau - \alpha_F)A_y$$

$$\frac{dA_y}{dz} = -(\tau - \alpha_F)A_x - j\frac{\delta\beta}{2}A_y \qquad (6.35)$$

$$\alpha_F = VH_z$$

where α_F is the Faraday rotation per unit length; H_z is the longitudinal magnetic field; and V is the Verdet constant. According to our specification for the sense of spin with respect to the transmission direction z, the negative sign $(-)$ is used before the coefficient α_F for a longitudinal magnetic field H_z in the $+z$ direction. If the direction of H_z is reversed, the sign before α_F must be changed from $(-)$ to $(+)$.

Here, we note particularly that the coupled-mode equations, Eqs. (6.35), involving the coefficient α_F refer to the *local* coordinates. This is different from the coupled-mode formulation Eq. (5.43) for the circular hi-bi Screw fiber employed in the Faraday rotation measurement. There, the coupled mode formulation was made (simply and directly) in the fixed coordinates, subject to the fast-spun condition for the circular hi-bi Screw fiber. Obviously, the fast-spun condition no longer applies to the present case for the elliptically birefringent spun Bow-Tie fiber for which $\tau \approx \delta\beta$. Consequently, here we can only formulate the coupled-mode equations for spun hi-bi fiber in the local coordinates in the form of Eq. (6.35).

From a *purely mathematical* point of view, the formulation of Eq. (6.35) is similar to the formulation of Eq. (4.3), or its equivalent, Eq. (4.8), referring to the local coordinates. A comparison of these equations shows that the Faraday-effect coefficient α_F in Eq. (6.35) corresponds to the twist-induced elasto-optic coefficient $\alpha(=\varsigma\tau)$ in Eq. (4.3) or Eq. (4.8). This comparison also helps to explain the specification for the sign in front of α_F in Eq. (6.35).

Now, Eq. (6.35) can be readily solved by the method of diagonalization. In a short cut, we simply borrow the earlier solution of Eqs. (6.18) and (6.19), wherein τ is replaced by $(\tau - \alpha_F)$. Consider the case of linear light at the input. The solution follows from Eq. (6.18) in the following form:

$$\mathbf{A}(z) = \begin{bmatrix} \cos^2\phi e^{jgz} + \sin^2\phi e^{-jgz} & \sin 2\phi \sin gz \\ -\sin 2\phi \sin gz & \sin^2\phi e^{jgz} + \cos^2\phi e^{-jgz} \end{bmatrix} \begin{bmatrix} \cos\theta \\ \sin\theta \end{bmatrix} \quad (6.36)$$

$$g = \left[(\beta/2)^2 + (\tau - \alpha_F)^2 \right]^{1/2}$$
$$\phi = (1/2)\arctan\left[2(\tau - \alpha_F)/\delta\beta \right] \quad (6.36a)$$

We see from these equations that the Faraday effect term α_F appears in a complex way not only in the exponential phase terms of exponents $\pm g$, but also in all the amplitude terms involving the parameter ϕ. The Faraday-effect term is not simply additive to the phase or to the orientation angle of the output light. Consequently, an output light in the form of Eqs. (6.36) and (6.36a) will be extremely difficult to interpret and analyze, although it contains the Faraday effect term.

To better understand the preceding analysis, the reader should compare the very complex expression of Eqs. (6.36) and (6.36a) for elliptically birefringent fiber with the corresponding equations for circularly hi-bi fiber (Section 5.10). In the latter, the Faraday effect becomes a simple additive term either in the phase of the output light (for circular incident light), or in the orientation angle of the output light (for linear incident light).

6.7   NOTES

More Discussion on Beat Length and Summary

Recall that the meaning of beat length is related to the beating pattern formed by the superposition of two propagating waves, or modes, of different phase velocities. The beat length can be defined with reference to either the fixed coordinates, or the local coordinates, as long as the modes that beat can be found in the respective coordinates (fixed or local), and as long as the superposed field of the two modes is periodic. The period of this superposed field is then the beat length referring to the respective coordinates (fixed or local).

In view of lightwave transmission in fiber, two aspects relating to the beat length are of practical interest. One is the field-reproducing characteristic of the beating pattern, the other, perhaps more important from the viewpoint of practical application, concerns the perturbation resistivity of the fiber. In the preceding two chapters, we derived the beat-length formulas for strongly twisted fiber and the circularly hi-bi Screw fiber [see Eqs. (4.22) and (5.28)]. Both of these formulas refer to the fixed coordinates, and are indicative of the perturbation resistivity of the respective fibers. In connection with these formulas, we also stressed that, for specification of the capability of holding the propagating mode against perturbation, only the beat length referring to the fixed coordinates proves useful. The beat length referring to the local coordinates will not serve this purpose.

In the present case of spun hi-bi fibers, we have derived the beat-length formula, Eq. (6.26), that refers to the local coordinates. This local beat length describes the field-reproducing characteristics of spun hi-bi fiber in the local coordinates. In the fixed coordinates, since the field-reproducing property of the fiber is lost, the respective beat-length is undefinable. The consequence is that the perturbation resistivity of spun hi-bi fiber cannot be analytically specified by a single parameter like the beat length.

The beat length, if definable on the basis of field-reproducing property, is a measurable quantity. In the measurement, it is convenient to make use of the dual-eigenmode transmission scheme. Preferably, the two eigenmodes excited in the fiber are of comparable amplitudes, such that they beat effectively to form a clear periodic beating pattern.

Nevertheless, we note here that the concept of beat length is not exclusively relevant to the dual-eigenmode transmission involving the propagation of two modes. In fact, the beat length is equally important and useful in describing the single-eigenmode transmission scheme in a practical environment. Thus, in single-eigenmode transmission, perturbations of various sorts are almost always inevitable such that, besides the initial propagating mode (the desired mode), the undesired or spurious mode will also be excited by the perturbations. In the fixed coordinates considered, therefore, the desired mode and the spurious mode will beat. The relevant beat length referring to the fixed coordinates is then indicative of the perturbation resistivity of the desired mode against the spurious mode.

Single-Eigenmode Transmission Derived by Transfer Matrix

With the aid of the transfer matrix $\mathbf{T}_l = \mathbf{O}\tilde{\mathbf{\Lambda}}\mathbf{O}^{-1}$, we have derived Eq. (6.18) for the local mode solution of a spun hi-bi fiber. This equation in its general form has been used to illustrate the dual-eigenmode transmission of a spun hi-bi fiber. But this equation is actually a general solution, valid not only for transmission involving both eigenmodes, but also for transmission involving only a single eigenmode. Here let us see how this is the case.

According to Eqs. (6.9) and (6.10), the local eigenmode fields at the input ($z = 0$) of a spun hi-bi fiber are a pair of elliptical light of opposite senses:

$$\mathbf{A}^R(0) = \begin{bmatrix} \cos \phi \\ j \sin \phi \end{bmatrix} \tag{6.37}$$

$$\mathbf{A}^L(0) = \begin{bmatrix} j \sin \phi \\ \cos \phi \end{bmatrix} \tag{6.38}$$

Compare these equations with the expression of the initial condition Eq. (6.20). We see that, if letting $\theta = \phi$, $\delta = \pi/2$ in Eq. (6.20), the result is the right-fast elliptical light given by Eq. (6.37). On the other hand, if letting $\theta = \pi/2 - \phi$, $\delta = -\pi/2$ in Eq. (6.20), the result is the left-slow elliptical light given by Eq. (6.38).

In principle, if the incident light is an elliptical light exactly specified by either Eq. (6.37) or Eq. (6.38), that is, the matching condition is satisfied at the input of the spun hi-bi fiber, then, according to the analysis in Section 6.4, single-eigenmode transmission will occur.

Consider the case of right-fast elliptical light incidence. We see that, if we substitute $\delta = \pi/2$, $\theta = \phi$ into Eq. (6.21), this complex and rather lengthy matrix equation will reduce to the following strikingly simple form:

$$\mathbf{A}(z) = \begin{bmatrix} \cos \phi \\ j \sin \phi \end{bmatrix} e^{jgz} \tag{6.39}$$

This simple result is exactly what is expected. Comparing this equation with Eq. (6.14) shows that this is the equation of single-eigenmode transmission. It is therefore seen that, if the input light is exactly specified by $\delta = \pi/2$, and $\theta = \phi$, then light transmission in the spun hi-bi fiber will be in single eigenmode. Note that θ is a parameter descriptive of the input light, while ϕ is a parameter descriptive of the structure of the spun hi-bi fiber. By Eq. (6.4a), $\phi = (1/2)\arctan(2\tau/\delta\beta)$.

Similarly, putting $\theta = \pi/2 - \phi$, $\delta = -\pi/2$ in Eq. (6.21) yields

$$\mathbf{A}(z) = \begin{bmatrix} \sin \phi \\ -j \cos \phi \end{bmatrix} e^{-jgz} \tag{6.40}$$

This is the local mode solution for another single-eigenmode transmission. The equation is identifiable to Eq. (6.16), except for an arbitrary constant phase factor $e^{-j\pi/2}$, which is trivial as far as the SOP behavior of the elliptical light is concerned. Nevertheless, if we wish, we can eliminate this extra phase factor from Eq. (6.40) simply by introducing an initial phase factor in the expression for the input elliptical light, such that $\mathbf{A}(0)$ in Eq. (6.21) is written as

$$\mathbf{A}(0) = \begin{bmatrix} \cos\theta \\ \sin\theta e^{j\delta} \end{bmatrix} e^{j\pi/2}$$

where, as before, $\theta = \pi/2 - \phi$, $\delta = -\pi/2$. When this equation is used with

TABLE 6.1   List of Inherent Characteristics of Fibers

Transfer Matrix in Local Coordinates $\mathbf{T}_l = \mathbf{O}\tilde{\Lambda}\mathbf{O}^{-1}$	Transfer Matrix in Fixed Coordinates $\hat{\mathbf{T}} = \mathbf{R}(-\tau z)\mathbf{T}_l$	Beat Length
Lo-bi fast-spun fiber		
$\begin{bmatrix} \cos(\tau z) & \sin(\tau z) \\ -\sin(\tau z) & \cos(\tau z) \end{bmatrix}$	$\begin{bmatrix} 1 & 0 \\ 0 & 1 \end{bmatrix}$	$\to \infty$ all SOPs
Linear hi-bi with small twists		
$\begin{bmatrix} e^{j(\delta\beta/2)z} & 0 \\ 0 & e^{-j(\delta\beta/2)z} \end{bmatrix}$	not defined for small twists	$2\pi/(\delta\beta)$
Strongly twisted fiber		
$\begin{bmatrix} \cos[(1-\varsigma)\tau z] & \sin[(1-\varsigma)\tau z] \\ -\sin[(1-\varsigma)\tau z] & \cos[(1-\varsigma)\tau z] \end{bmatrix}$	$\begin{bmatrix} \cos(\varsigma\tau z) & -\sin(\varsigma\tau z) \\ \sin(\varsigma\tau z) & \cos(\varsigma\tau z) \end{bmatrix}$	$\pi/\varsigma\tau$
Screw fiber		
$\begin{bmatrix} \cos[(1-\kappa)\tau z] & \sin[(1-\kappa)\tau z] \\ -\sin[(1-\kappa)\tau z] & \cos[(1-\kappa)\tau z] \end{bmatrix}$	$\begin{bmatrix} \cos(\kappa\tau z) & -\sin(\kappa\tau z) \\ \sin(\kappa\tau z) & \cos(\kappa\tau z) \end{bmatrix}$	$\pi/\kappa\tau$

Elliptical-bi fiber

$\mathbf{T}_l = \mathbf{O}\tilde{\Lambda}\mathbf{O}^{-1}$ (Transfer matrix in local coordinates)

$$\begin{bmatrix} \cos^2\phi e^{jgz} + \sin^2\phi e^{-jgz} & \sin(2\phi)\sin(gz) \\ -\sin(2\phi)\sin(gz) & \sin^2\phi e^{jgz} + \cos^2\phi e^{-jgz} \end{bmatrix}$$

$\phi = 0.5\arctan(2L_b/L_s)$, $g = (\pi/L_b)[1 + 4(L_b/L_s)^2]^{1/2}$

$\hat{\mathbf{T}} = \mathbf{R}(-\tau z)\mathbf{T}_l$ (Transfer matrix in fixed coordinates)

$$\begin{bmatrix} \cos(\tau z) & -\sin(\tau z) \\ \sin(\tau z) & \cos(\tau z) \end{bmatrix}\begin{bmatrix} \cos^2\phi e^{jgz} + \sin^2\phi e^{-jgz} & \sin(2\phi)\sin(gz) \\ -\sin(2\phi)\sin(gz) & \sin^2\phi e^{jgz} + \cos^2\phi e^{-jgz} \end{bmatrix}$$

ϕ, g: same as above   beat length undefined

Eq. (6.21), the result corresponding to Eq. (6.40) becomes

$$\mathbf{A}(z) = \begin{bmatrix} j \sin \phi \\ \cos \phi \end{bmatrix} e^{-jgz} \tag{6.40a}$$

which is exactly the same as Eq. (6.16).

In the preceding analysis, we demonstrated by an alternative method why single-eigenmode transmission occurs so rarely in spun hi-bi fiber. Mathematically, one can use any pair of values for (θ, δ) with which to derive the desired result. Technologically, however, it is scarcely possible to launch into the spun hi-bi fiber an SOP of light whose parameters are related to the fiber parameters exactly by $\theta = \phi$, $\delta = \pi/2$, or by $\theta = \pi/2 - \phi$, $\delta = -\pi/2$.

Table of Transfer Matrices for Different Kinds of Fiber

By now we have treated several different kinds of fibers: spun lo-bi fiber, linear hi-bi fiber, strongly twisted fiber, circular hi-bi (Screw) fiber, and spun hi-bi (elliptical-bi) fiber. Each fiber has a different transmission characteristic described by a specific transfer matrix. For ready reference, Table 6.1 is included here as a summary.

REFERENCES

[1]   L. Li, J. R. Qian, and D. N. Payne, "Current sensors using highly birefringent Bow-Tie fibres," *Electron. Lett.*, vol. 22, pp. 1142–1144 (1986).

[2]   H. C. Huang, "Elliptically birefringent optical fiber transmission characteristics," (invited Paper) *Fiber and Integrated Optics*, vol.15, pp. 71–80 (1996).

[3]   J. Sakai, S. Machida, and T. Kimura, "Existence of eigen polarization modes in anisotropic single-mode optical fibers," *Opt. Lett.*, vol. 6, pp. 496–498 (1981).

BIBLIOGRAPHY

L. M. Bassett, I. G. Clarke, and J. H. Haywood, "Polarization dependence in a Sagnac loop optical fibre current sensor employing a 3 × 3 optical coupler," *SPIE*, vol. 2360, pp. 596–599 (1994).

I. M. Bassett, I. G. Clarke, and X. Ma, "An analysis of spun linearly birefringent optical fibres," *Proc. 14th ACOFT*, Brisbane, pp. 137–140 (1989).

P. L. Chu, T. W. Whitbread, P. M. Allen, W. S. Wassef, X. Ma, and I. G. Clarke, "Manufacture and applications of spun elliptical optical fibre for current sensing," *Proc 14th ACOFT*, Brisbane, pp. 133–136 (1989).

I. G. Clark, "Temperature-stable spun elliptical-core optical fiber current transducer," *Opt. Lett.*, vol. 18, pp. 158–160 (1993).

I. G. Clarke, I. M. Bassett, D. Geake, S. B. Poole, M. G. Sceats, A. D. Stokes, and P.L. Chu, "A spun elliptical core optical fibre for electric current measurement," *Proc. 16th ACOFT*, Adelaide, pp. 222–225 (1991).

H. C. Huang and J. R. Qian, "Theory of imperfect nonconventional single-mode optical fibers," in *Optical Waveguide Sciences* (H. C. Huang and A. W. Snyder, eds.), Martinus Nijhoff Publishers, The Hague, 360 pages (1983).

R. I. Laming, D. N. Payne, and L. Li, "Current monitor using elliptical birefringent fibre and active temperature compensation," *SPIE*, vol. 798, pp. 283–287 (1987).

R. I. Laming, D. N. Payne, and L. Li, "Compact optical fibre current monitor with passive temperature stabilization," *Proc. OFS*, New Orleans, pp. 123–128 (1988).

R. I. Laming and D. N. Payne, "Electric current sensors employing spun highly birefringent optical fibers," *IEEE J. Lightwave Tech.*, vol. LT-7, pp. 2084–2094 (1989).

J. R. Qian, Q. Guo, and L. Li, "Spun linear birefringence fibres and their sensing mechanism in current sensors with temperature compensation," *IEE Proc.-Optoelectron.*, vol. 141, pp. 373–380 (1994).

J. R. Qian and L. Li, "Current sensors using highly-birefringent Bow-Tie fibers," *Proc. OFS, Tokyo*, pp. 85–88 (1986).

Ma Xing and P. L. Chu, "Design of temperature independent spun elliptical fibres," *Proc. 16th ACOFT*, Adelaide, pp. 254–257 (1991).

CHAPTER SEVEN

Fiber-Optic Analogs of Bulk-Optic Wave Plates

The preceding chapters dealt with *uniform* fibers of different birefringence characteristics. Analytically, the coupled-mode equations that describe these uniform fibers are differential equations of constant coefficients. Unlike the preceding chapters, the present chapter is devoted to a kind of *nonuniform* fiber whose transmission characteristics is described by coupled-mode equations whose coefficients are variables, or functions of z. While the mathematical theory for variable coupling presented in the following is applicable to nonuniform fibers in general, the topics addressed here will focus on a specific class of nonuniform fibers called "fiber-optic analogs of bulk-optic wave plates."

The substance of this chapter is essentially based on inventions for which I own three U.S. patents, successively, in 1990, 1992, and 1995 [1–3], and on my two papers just published, in June and September 1997 [4, 5], of which the first bears the same title as the present chapter.

A basic element of our concern is a variably spun birefringent fiber, with a spin rate varying from very fast to zero, or vice versa. This novel fiber element can be readily made by the existing fabrication technique, with fairly loose tolerances of the structural parameters. Analytic theory predicts that such a nonuniform fiber element will function like a bulk-optic quarter-wave plate, but with the favorable feature of being inherently *wide band*. Experimental evidence confirms the theoretical prediction. With this fiber-optic analog of a quarter-wave plate as a building block, the wide-band half-wave plate and full-wave plate can also be made in the form of variably spun birefringent fibers.

7.1 TECHNICAL BACKGROUND

Bulk-optic wave plates of different kinds are versatile elements widely used in any laboratory. While such bulk-optic wave plates can be incorporated into a fiber-optic system in order to perform the prescribed state-of-polarization (SOP) transform functions, the resulting hybrid (bulk optics plus fiber optics) layout is likely bulky, and oftentimes requires careful and deliberate manual adjustments for an overall optical alignment. It is therefore natural for one to search for fiber-optic counterparts of the bulk-optic wave plates that can be inserted in-line in a fiber-optic circuitry, to improve the overall system performances and greatly simplify the overall instrumentation layout.

In this investigation of a basic nature, I received much inspiration from the early research in the art of microwaves in general, and in the theory of variable coupling of modes in particular [6, 7]. The investigation led to the rediscovery that some early concepts and theories do offer perspectives that are relevant to the modern art of fiber optics.

The phenomena of SOP transforms actually occur intentionally or unintentionally in various places in fiber optics. In a linearly highly birefringent (hi-bi) fiber, for example, if the incident linear light is not accurately oriented along one or the other principal axis, the SOP of light will undergo all changes in half a beat wavelength. However, because the SOP is extremely sensitive to the length of fiber as well as to the orientation of light, a section of linear hi-bi fiber is not suitable for use in the SOP transform, though in principle it is.

Among the several inventions [8–10] relating to SOP control and transform, LeFevre's all-fiber polarization controller is particularly useful in laboratory work for its capability to transform in a general way one SOP to another SOP, by manual adjustment or tuning. Nevertheless, LeFevre's device is structured by conventional fiber whose simplicity is associated with a lack of the capability of SOP maintenance. For laboratory use, this does not pose a problem because the desired polarization, if distorted, can still be restored by a new manual adjustment of the fiber-optic setup.

For certain non-attendance applications, however, it is desirable to get rid of manual adjustment or aligning in achieving and maintaining a prescribed SOP transform. With this view in mind, I disclose a new approach to the realization of certain manual-free SOP transforms in fiber optics [1–5].

I focus the analysis and discussion on a simple fiber element that was made by spinning a birefringent fiber of any version with a *slowly varying* spin rate from fast to zero, or conversely from zero to fast. With regard to the technical aspect of such a novel fiber element, it is just that simple. A schematic diagram of one variety of such a fiber element is shown in Figure 7.1.

Curiously, as simple as the structure is, the fiber element has been discovered to behave like a bulk-optic quarter-wave plate, with the very favorable feature being inherently wide band. For brevity, I refer to such a fiber section as a practical polarization transformer (PPT). Appropriate cascading of two PPT units makes a composite fiber element that behaves either like a bulk-optic half-wave plate or a bulk-optic full-wave plate, depending on the PPT units selected and the manner in which they cascade.

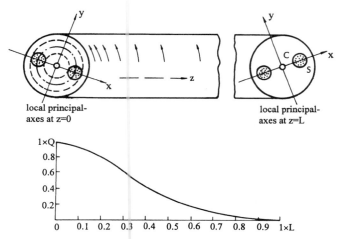

FIGURE 7.1   Fiber-optic analog of bulk-optic quarter wave plate. (Top) The structure (C, the core; S, the stress filament). (Bottom) Spin-rate variation (Q, normalized spin rate).

7.2   MATHEMATICAL REVIEW OF PREVIOUS COUPLED-MODE THEORY

We adopt the coupled-mode theoretical approach to solving the fiber-optic PPT. Since the structure concerned is not uniform, but varies in the longitudinal or transmission direction z, one must solve the coupled-mode equations with *variable* (or function) coefficients. Before dealing with this more general analytic framework, let us review the classical method of diagonalization as applied to solving coupled-mode equations with constant coefficients. This review may help us find where and how the mathematics needs to be extended in our immediate attempt to develop a generalized coupled-mode theory for variable coupling.

Coupled-Mode Equations with Constant Coefficients

While the coupled-mode theoretical approach is consistently employed throughout this book, the treatment of different fibers in the previous chapters varies in the way of mathematical simplification, and in the physical interpretation. In this preliminary section, the analytic method will be reviewed in fairly general terms, without regard to the specific characteristics of the fiber structure.

In formulating coupled-mode equations descriptive of light transmission in a particular fiber structure, we must first specify the reference coordinates.

For spun fibers, it is feasible to adopt the local coordinates, instead of the fixed (or laboratory-fixed) coordinates, because of the very great mathematical simplification made possible in the former choice. The local coordinates are defined as those that coincide with the principal axes of the spun fiber, and hence are rotating at an angular speed equal to the spin-rate of the spun fiber concerned. In matrix form, the coupled-mode equations for local modes are written as

$$\frac{d\mathbf{A}}{dz} = \mathbf{KA}$$

$$\mathbf{A} = \begin{bmatrix} A_x \\ A_y \end{bmatrix} \tag{7.1}$$

where A_x and A_y are the local modes. For a uniformly spun fiber, $\mathbf{K}$ is a square matrix of constant elements. The specific expressions that represent the elements of $\mathbf{K}$ depend not only on the choice of the reference coordinates (fixed or local), but also on the choice of the base modes (linear or circular). Following the customary use, we choose linear base modes in this chapter, while recognizing that the choice of circular reference modes is equally valid (see Section 4.6). Note that in the general case the linear base modes A_x and A_y are not simply the rectangular (x and y) components of an inclined linear light. They are such components only if A_x and A_y happen to be in phase. In the general case, however, they are both complex functions with different amplitudes and different phase factors such that, when being placed in the column matrix $\mathbf{A}$, together they represent a light of any SOP.

The Classical Method of Diagonalization

For matrices of constant elements descriptive of a uniformly spun fiber, the classical method of diagonalization is applied by the orthogonal transformation, such that

$$\mathbf{A} = \mathbf{OW} \tag{7.2}$$

$$\frac{d\mathbf{W}}{dz} = \Lambda\mathbf{W}$$

$$\Lambda = \mathbf{O}^{-1}\mathbf{KO} = \begin{bmatrix} jg & 0 \\ 0 & -jg \end{bmatrix} \tag{7.3}$$

$$\mathbf{W}(z) = \tilde{\Lambda}\mathbf{W}(0) \tag{7.4}$$

$$\tilde{\Lambda} = \begin{bmatrix} e^{jgz} & 0 \\ 0 & e^{-jgz} \end{bmatrix} \tag{7.4a}$$

where **O** is the diagonalizing matrix, and **W** is a column matrix whose elements are designated by W_x and W_y. Since the differential equations for W_x and W_y are diagonalized, the elements of Λ are the eigenvalues, and the elements of $\tilde{\Lambda}$ are the eigenfunctions. But here the subscripts x and y for W_x and W_y do not bear the same geometrical meaning as they do for the local modes A_x and A_y. In matrix theory, the elements W_x and W_y are referred to as the eigenmodes in the "normal coordinates." Generally, these normal coordinates cannot be interpreted as geometrical coordinates in real space, but are simply a mathematical artifice for the sake of diagonalization. With this artifice we are able to solve **A** from **W** via Eq. (7.2) (see Appendix B).

If the common phase factor $e^{j(\omega t - \beta z)}$ is written out in the formulation, it is identifiable that e^{jgz} and e^{-jgz} represent a fast mode and a slow mode, respectively. The solution of Eq. (7.1) is then given by

$$\mathbf{A}(z) = \mathbf{T}_l \mathbf{A}(0) \tag{7.5}$$

$$\mathbf{T}_l \equiv \mathbf{T}_l(z) = \mathbf{O}\tilde{\Lambda}\mathbf{O}^{-1} \tag{7.5a}$$

where the matrix $\mathbf{T}_l$ (called the "transfer matrix") is descriptive of the fiber structure only, wholly independent of the incident light **A**(0) that excites the fiber at its input end.

In the preceding formulation, the matrix **K** was not specified. For a uniformly spun fiber, this matrix is of the form:

$$\mathbf{K} = \begin{bmatrix} j\dfrac{\delta\beta}{2} & \tau \\ -\tau & -j\dfrac{\delta\beta}{2} \end{bmatrix} \tag{7.6}$$

where $\delta\beta$ is the phase-velocity difference between A_x and A_y, and the coupling coefficient τ is equal to the spin rate, provided the local coordinates are chosen [11]. With the above specification of **K**, we have

$$\mathbf{O} = \begin{bmatrix} \cos\phi & j\sin\phi \\ j\sin\phi & \cos\phi \end{bmatrix}$$

$$\phi = \frac{1}{2}\arctan(2Q) \tag{7.7}$$

$$Q = \frac{\tau}{\delta\beta} = \frac{L_b}{L_S}$$

$$g = \pi\left(1 + 4Q^2\right)^{1/2} \tag{7.8}$$

where L_b is the beat length of fiber in the unspun state, and L_S is the spin pitch. The parameter Q is an important figure of merit of a spun fiber, which is also called the coupling capacity, the qualification factor, the normalized spin-rate, among others. With z normalized with respect to L_b, the eigenvalues g and $-g$ are also written in this adimensional form.

7.3 MATHEMATICAL EXTENSION TO VARIABLY COUPLED-MODE THEORY

The foregoing mathematical outline concerns a uniformly spun fiber. For a variably spun fiber, the required mathematics is certainly beyond the previous scope. Nevertheless, this preliminary material is by no means redundant, inasmuch as by the simpler mathematical formulation for uniformly spun fiber it will be easier to detect where and how the mathematics needs extension in our immediate attempt to develop a generalized coupled-mode theory that is applicable to variable coupling.

I know very well that the following mathematics is not likely to be available in standard texts, having appeared only diversely in some journals. In microwave theory and technique, the initial work on variably coupled modes can be traced back to the three papers published in the same issue of the *Bell System Technical Journal* [12–14]. The third paper due to Louisell (1955) is more mathematical, contributing an approximate solution of two variably coupled modes. Afterwards, Huang (1960) [15, 16] and Keller and Keller [17] independently published their mathematical methods, by different analytic approaches, for solving a generalized variably coupled-mode system comprising N modes, with identifiable end results in the form of exponential-like functions. In that early attempt [15, 16] Huang succeeded in treating matrices of variable elements as in operational calculus, thereby enormously simplifying the mathematical manipulation.

The mathematical background initially established in microwaves is now rediscovered as being very powerful and probably unique in dealing with light transmission in a longitudinally nonuniform fiber-optic structure, such as a variably spun birefringent fiber. Again we use the coupled-mode equation, Eq. (7.1), allowing only the coupling coefficient to become a variable:

$$\frac{d\mathbf{A}}{dz} = \mathbf{K}(z)\mathbf{A} \qquad (7.9)$$

$$\mathbf{K}(z) = \begin{bmatrix} j\dfrac{\delta\beta}{2} & \tau(z) \\ -\tau(z) & -j\dfrac{\delta\beta}{2} \end{bmatrix} \qquad (7.9\text{a})$$

where $\tau(z)$ and hence $\mathbf{K}(z)$ are now functions of z.

In the previous case of constant coupling, the spirit of the classical method of diagonalization was to solve $\mathbf{A}$ from $\mathbf{W}$ by the transformation $\mathbf{A} = \mathbf{OW}$, where the diagonalizing matrix $\mathbf{O}$ is a matrix of constant elements, given by Eq. (7.7). Now for the more general case of variable coupling, if we attempt a solution again by this transformation, we will have to allow $\mathbf{O}$ to accommodate variable elements. The mathematical formulation is then written as

$$\mathbf{A} = \mathbf{O}(z)\mathbf{W} \tag{7.10}$$

$$\mathbf{O}(z) = \begin{bmatrix} \cos\phi & j\sin\phi \\ j\sin\phi & \cos\phi \end{bmatrix}$$

$$\phi \equiv \phi(z) = \frac{1}{2}\arctan[2Q(z)] \tag{7.10a}$$

$$Q(z) = \frac{\tau(z)}{\delta\beta} = \frac{L_b}{L_S(z)}$$

where $\mathbf{O}(z)$ is a matrix of variable elements, inasmuch as $\tau(z)$, $L_S(z)$, and hence $Q(z)$, $\phi(z)$ are all functions of z. Note that while the dimension of the spin rate $\tau(z)$ is L^{-1} (revolutions per unit length), the normalized spin rate $Q(z)$ is adimensional.

Up to now, one can scarcely find what distinguishes Eqs. (7.9) and (7.10) for variable coupling from the Eqs. (7.1), (7.2), (7.6), and (7.7) for constant coupling ($\tau = $ const.), except that $\tau \equiv \tau(z)$ in Eqs. (7.9) and (7.10) is now taken to be a function of z. However, a drastic analytic difference between variable coupling and constant coupling will immediately appear in the next step when the transformation (7.10) is put into the variably coupled-mode equation, Eq. (7.9).

In retrospect, I initially attempted the said step of derivation in terms of differential equations (rather than matrices), and in that way found the mathematics formidably complicated, even for three variably coupled modes. Later, I tried to treat the differential equation of matrices of variable elements in the same way as in operational calculus. The mathematics then became extremely simple, and was later found to lead to the same result as did the very tedious differential-equation formulation. Thus, in the operational-calculus approach, when the transformation $\mathbf{A} = \mathbf{OW}$ is substituted into Eq. (7.9), the left side of this equation becomes $d(\mathbf{OW})/dz = \mathbf{O}(d\mathbf{W}/dz) + (d\mathbf{O}/dz)\mathbf{W}$, and the right side is $\mathbf{K}(\mathbf{OW})$, such that [15, 16]

$$\frac{d\mathbf{W}}{dz} = \mathbf{N}(z)\mathbf{W} \tag{7.11}$$

$$\mathbf{N}(z) = \mathbf{O}^{-1}\mathbf{KO} - \mathbf{O}^{-1}\frac{d\mathbf{O}}{dz} \tag{7.11a}$$

where $\mathbf{K} \equiv \mathbf{K}(z)$, $\mathbf{O} \equiv \mathbf{O}(z)$, as understood.

On the basis of Eqs. (7.11) and (7.11a), we proceed to the task of solving $\mathbf{A}$ from $\mathbf{W}$ via Eq. (7.10). To solve the latter, we examine the matrix $\mathbf{N}(z)$ by comparing it with Λ in Eq. (7.3). The distinctive difference between the two is now apparent. While Λ for the case of constant coupling is simply represented by $(\mathbf{O}^{-1}\mathbf{KO})$, which is exactly diagonalized, the matrix $\mathbf{N}(z)$ now comprises two terms, with an extra term involving a derivative factor. Because of this extra term, $\mathbf{N}(z)$ is no longer diagonal for the case of variable-coupling. Thus, W_x and W_y do not satisfy independent differential equations separately, and hence are no longer expressible as simple exponential functions. In fact, they are coupled by the extra term $\mathbf{O}^{-1}(d\mathbf{O}/dz)$ in Eq. (7.11a). Intuitively, it is conceivable that this derivative term will be small if variation of the spin rate (i.e., $d\tau/dz$) is slow.

It was seen from Eq. (7.3) that W_x and W_y in uniformly spun fiber derived by the transformation (7.2) separately satisfy independent differential equations. They are the eigenmodes in the normal coordinates. In the present case of a variably spun fiber, W_x and W_y are no longer eigenmodes. For distinctiveness, they are referred to as *supermodes*. The word "super" implies nothing mysterious, but simply means that such modes still retain some coupling that is proportional to $d\tau/dz$. Note that supermodes simply pass into eigenmodes when the variation of τ vanishes (i.e., $d\tau/dz \to 0$). In the case of slow variation of τ, the "derivative" coupling between the supermodes can be very small, such that the supermodes are almost (or quasi-) eigenmodes.

In terms of modal description, the local modes A_x and A_y in a spun fiber refer to the independent fields in an unspun fiber whose principal axes conform with the spinning coordinates at the local point. In the actual spun fiber, A_x and A_y are therefore not independent, but are coupled to each other by the local spin rate τ. The supermodes W_x and W_y in a variably spun fiber refer to the independent fields in a uniformly spun fiber whose spin rate is equal to the spin rate of the actual fiber at the local point concerned, in addition to the conformity of its principal axes and the spinning coordinates at this local point. In the actual variably spun fiber, W_x and W_y are therefore not independent, but are coupled to each other by the rate of change of the spin rate (i.e., $d\tau/dz$) at the local point concerned. At any point where this rate of change vanishes (i.e., $d\tau/dz \to 0$), the supermodes become identical to the eigenmodes of the reference fiber (uniformly spun fiber with the local spin-rate τ). In a slow-varying nonuniform waveguide, it is perceivable that supermodes (as compared with local modes) more closely approximate the actual field structure.

Here, a fundamental concept of the supermode is worth noting. As said, the supermode is an almost independent field structure (or quasi eigenmode) under the condition of slow variation of the spin rate τ. While this quasi-eigenmode property requires $d\tau/dz$ to be small everywhere along the variably spun fiber, the overall change of τ from one point to another point is not necessarily small, but may be fairly large. Thus, while the perturbation theory will not apply to such kind of problem, the supermode approach will do.

In early publications, [15, 16], it was proved that the first and second terms for $N(z)$ on the right side of Eq. (7.11a) occupy, respectively, the diagonal and off-diagonal positions without mixing, such that

$$
\mathbf{N} = \begin{bmatrix} jg & -j\dfrac{d\phi}{dz} \\[2mm] -j\dfrac{d\phi}{dz} & -jg \end{bmatrix}
\tag{7.12}
$$

$$
g \equiv g(z) = \pi(1 + 4Q^2)^{1/2}, \qquad Q \equiv Q(z)
\tag{7.12a}
$$

$$
\begin{aligned}
\frac{d\phi}{dz} &= (1 + 4Q^2)^{-1}\left(\frac{dQ}{dz}\right) \\[2mm]
&= \frac{1}{2\pi}(1 + 4Q^2)^{-1}\left(\frac{d\tau}{dz}\right)
\end{aligned}
\tag{7.12b}
$$

where z is normalized with respect to L_b.

Apparently, the entire mathematical framework for variably coupling will reduce to the classical method of diagonalization for constant coupling, if only letting $\tau = \text{const.}$, such that $d\tau/dz = 0$.

Under the condition that the variation of τ is sufficiently slow over the entire transmission range, approximate solution of Eq. (7.11) for the supermodes can be derived by an iterative process. The zeroth-order approximation is given by

$$
W_x(z) \approx W_x(0)e^{j\int_0^z g\,dz}
\tag{7.13a}
$$

$$
W_y(z) \approx W_y(0)e^{-j\int_0^z g\,dz}
\tag{7.13b}
$$

The first-order iterative solution is given by

$$
\begin{aligned}
W_x(z) &\approx e^{j\rho}\left\{ W_x(0) - W_x(0)\int_0^z \frac{d\phi}{dz}e^{-2j\rho}\left(\int_0^z \frac{d\phi}{dz}e^{2j\rho}\,dz\right)dz \right. \\
&\qquad \left. -jW_y(0)\int_0^z \frac{d\phi}{dz}e^{-2j\rho}\,dz \right\} \\[2mm]
W_y(z) &\approx e^{-j\rho}\left\{ W_y(0) - W_y(0)\int_0^z \frac{d\phi}{dz}e^{2j\rho}\left(\int_0^z \frac{d\phi}{dz}e^{-2j\rho}\,dz\right)dz \right. \\
&\qquad \left. -jW_x(0)\int_0^z \frac{d\phi}{dz}e^{2j\rho}\,dz \right\}
\end{aligned}
\tag{7.14}
$$

where $\phi \equiv \phi(z) = 0.5 \arctan(2Q)$, $Q \equiv Q(z) = \tau(z)/(\delta\beta)$, as afore-defined, and

$$\rho \equiv \rho(z) = \int_0^z g\,dz = \int_0^z \pi(1 + 4Q^2)^{1/2}\,dz \qquad (7.14a)$$

While more iterations will yield approximate solutions of higher orders, the above-listed approximate solution will serve most practical purposes. From these supermode solutions, $W_x(z)$, $W_y(z)$, it is straightforward to derive the local-mode solutions $A_x(z)$, $A_y(z)$ by the matrix transformation, Eq. (7.10):

$$A_x(z) = \cos\phi \cdot W_x(z) + j\sin\phi \cdot W_y(z) \qquad (7.15a)$$

$$A_y(z) = j\sin\phi \cdot W_x(z) + \cos\phi \cdot W_y(z) \qquad (7.15b)$$

Via the same transformation (7.10), the initial values $W_x(0)$, $W_y(0)$ in Eqs. (7.13) and (7.14) can be derived from the given initial values $A_x(0)$, $A_y(0)$. For convenience in numerical calculation, all the terms in the mathematical formulation are defined as adimensional. We further adopt the normalization relations descriptive of the conservation of the total power in terms of either the supermodes or the local modes:

$$|W_x(z)|^2 + |W_y(z)|^2 = 1 \qquad (7.16a)$$

$$|A_x(z)|^2 + |A_y(z)|^2 = 1 \qquad (7.16b)$$

In treating a specific initial-value problem, the beat length of fiber in the unspun state (L_b, a readily measurable quantity) is given. What is to be prescribed is only the spin rate τ as a function of z. Extensive numerical computation reveals that different choices of the forms of this function, be it linear, exponential, or anything else, will make little difference in the end results, as long as the chosen function varies slowly the whole way [15, 16]. For analytic study, as well as for numerical calculation by computer, I have found it suitable to use the powered raised-cosine function to simulate the functional form in which the spin rate varies:

$$Q(z) = Q(0)\left[\frac{1}{2} + \frac{1}{2}\cos\left(\frac{\pi z}{L}\right)\right]^\gamma \qquad (7.17)$$

where, as said, the normalized spin-rate $Q = (L_b/2\pi)\tau$ is adimensional. The function is analytic, decreasing continuously over the closed interval $[0, L]$, with zero derivatives at both ends. In the equation, L (the total fiber length) is normalized with respect to L_b to become adimensional (in the same way as z). Different specifications of the value of the parameter γ result in a sooner or later occurrence of a relatively prominent sloping of the curve for $Q(z)$.

For reference, we also include here the normalized spin-rate function for $Q(z)$ that simulates a varying spin rate that increases continuously over the interval $[0, L]$:

$$Q(z) = Q(0)\left[\frac{1}{2} - \frac{1}{2}\cos\left(\frac{\pi z}{L}\right)\right]^{\gamma} \qquad (7.17a)$$

Comparing Eq. (7.17) with Eq. (7.17a), we observe that the only difference is that the $+$ sign in the former is changed to a $-$ sign in the latter. The same parameters in both equations have the same meanings. Equation (7.17a) is not used in this section, but will be useful in Section 7.7.

Simple Asymptotic Solutions

The afore-outlined mathematics provides a complete analytic framework for the study of a variably spun fiber. For a particular problem, the SOP of light that excites the fiber at the input (i.e., the initial condition) is given, such that the variably coupled-mode equations plus the initial condition constitute an initial-value problem. The particular solution of this initial-value problem describes the complete pattern of the polarization evolution of light traveling along the fiber.

In practice, it is often desirable to know only the end result of the SOP transform by a variably spun PPT, not necessarily to have detailed information about the entire course of the SOP evolution from input to output. Then we can actually adopt certain early approximations and solve the particular initial-value problem in an admirably simple way.

The simple approach is based exactly on the said two peculiar features of a PPT: the first is that the spin rate at the fast-spun end is sufficiently fast, and the second is that the variation of the spin rate from fast to zero (or vice versa) is sufficiently slow. Specifically, the spin rate at the fast-spun end requires:

$$Q_F = \frac{L_b}{L_{FS}} \gg 1 \qquad (7.18)$$

where L_{FS} is the local spin pitch at the fast-spun end of the fiber, and L_b, as before, is the unspun beat length. This is a relatively stringent condition in making a PPT, requiring a nonconventionally fast spin rate at one or the other end of the fiber. Fortunately, this condition will not pose a technological problem in fabrication practice because it is only necessary for the highest spin-rate to cover a momentarily short time interval. Provided this condition is fulfilled, then $\phi \approx \pi/4$, $\cos\phi = \sin\phi \approx 1/\sqrt{2}$ in Eq. (7.10a), such that

O_F, O_F^{-1} at the fast-spun end tend to the following very simple forms:

$$O_F \to \frac{1}{\sqrt{2}}\begin{bmatrix} 1 & j \\ j & 1 \end{bmatrix} \tag{7.19}$$

$$O_F^{-1} \to \frac{1}{\sqrt{2}}\begin{bmatrix} 1 & -j \\ -j & 1 \end{bmatrix} \tag{7.20}$$

At the unspun end ($\tau = 0$), we have $\phi = 0$ according to Eq. (7.10a), such that

$$O_0 = O_0^{-1} = \begin{bmatrix} 1 & 0 \\ 0 & 1 \end{bmatrix} \tag{7.21}$$

which, being a unit matrix, indicates that the local modes and the super-modes are identical at the unspun end of the PPT. It is easy to see that Eq. (7.21) is a mathematical description of what was described conceptually in the previous modal description of a variably spun fiber.

In regard to the second feature, it follows from Eq. (7.12) that a nonuniform PPT can be regarded as slowly varying if $|d\phi/dz| \ll 2g$ (i.e., if the magnitude of the off-diagonal term is much smaller than the difference between the diagonal terms) [15, 16]:

$$
\begin{aligned}
Q_S &= \frac{1}{2g}\left|\frac{d\phi}{dz}\right| \\
&= \frac{1}{4\pi^2(1 + 4Q^2)^{3/2}}\left|\frac{d\tau}{dz}\right| \ll 1
\end{aligned}
\tag{7.22}
$$

where Q_S is the coupling capacity of the supermodes.

Other things being equal, the longer the fiber segment, the slower the variation of the spin-rate. It is therefore fairly easy to fulfill the slow-variation condition in the case of a variably spun PPT fiber, because the length of such fiber section is practically always long enough in terms of L_b. Under the condition given by Eq. (7.22), the off-diagonal elements of the matrix N become vanishingly small. Without relying on a computer, we can estimate the upper bound of Q_S for typical specifications of a PPT fiber, say, for $L = 10^2$, $\gamma = 2$. Calculation of some relevant derivatives then yields the upper bound of Q_S to be 10^{-2}–10^{-3} by order. The average of Q_S is still orders of magnitude smaller. We are therefore tempted to ignore these small elements altogether, with the result that Eq. (7.11) for the supermodes $W_x(z)$ and $W_y(z)$ reduce to two independent differential equations. Such a vital simplifying step for Eq. (7.11) then leads to approximate solutions of the supermodes in the form of Eq. (7.13), yielding

$$\mathbf{W}(L) \approx \tilde{\Lambda}\mathbf{W}(0) \tag{7.23}$$

$$\tilde{\Lambda} \rightarrow \begin{bmatrix} e^{j\rho} & 0 \\ 0 & e^{-j\rho} \end{bmatrix} \tag{7.23a}$$

$$\rho \equiv \rho(L) = \int_0^L g\,dz$$

$$= \int_0^L \tau\left[1 + 4(L_b/L_S)^2\right]^{1/2} dz \tag{7.23b}$$

where $\rho(L_b, L_S, L)$ is a self-contained parameter descriptive of the PPT, including in a single expression all three structural figures: the unspun beat length L_b, the spin pitch L_S, and the total length L of the fiber element. For convenience, we call $\rho(L_b, L_S, L)$ the global structural parameter of a PPT.

In Eqs. (7.19), (7.20), and (7.23a), we used $\rightarrow$ (instead of =) to denote asymptotic approximations. As defined, it is by the transformation matrix $\mathbf{O}$ that the local-mode $\mathbf{A}$ is related to the supermode matrix $\mathbf{W}$. For this matrix, we observed that Eq. (7.21) referring to the unspun end of fiber (where $\tau = 0$) is an exact equality, implying no approximation. But Eqs. (7.19) and (7.20) referring to the fast-spun end are not exact equalities, but approximate asymptotic expressions. They are asymptotic in the sense that the approximation implied in the simple expressions is valid only when the spin rate (τ, or its normalized equivalent Q) tends to be indefinitely large, that is, τ or $Q \rightarrow \infty$. In practice, it requires that $Q = 10$, or higher. As regards Eq. (7.23a), it is asymptotic in the sense that the approximation is valid only in the mathematical limit of $L \rightarrow \infty$, such that variation of the spin rate is indefinitely slow. Practically, if the normalized total length of fiber L is taken to be 10^2 by order, the condition of slow variation will not be a problem.

Transfer Matrix of Nonuniform Fiber Structure

Under fairly loose conditions, the asymptotic solution of a given initial-value problem can be derived by simple matrix-algebraic manipulation. The simplification is achieved to such an extent that the mathematical task is almost effortless. To illustrate the spirit of this simple mathematical technique, take the PPT of Figure 7.1 as an example. Given the initial condition $\mathbf{A}(0)$ at the fast-spun end, the end solution can be derived straightforwardly with the aid of Eqs. (7.20), (7.23a), and (7.21) in order:

$$\mathbf{A}(L) = \mathbf{T}_l\mathbf{A}(0) \tag{7.24}$$

$$\mathbf{T}_l = \mathbf{O}(L)\tilde{\Lambda}\mathbf{O}^{-1}(0) \tag{7.24a}$$

where $\mathbf{O}^{-1}(0) = \mathbf{O}_F^{-1}$, $\mathbf{O}(L) = \mathbf{O}_0$, given by Eqs. (7.20) and (7.21), respectively.

The preceding mathematics provides a self-contained theoretical framework for treatment of the variably spun PPT fiber of different versions. Of all the approximations that have been used, neglecting the off-diagonal elements when deriving Eqs. (7.23) and (7.23a) is strikingly an unusual bold step of mathematical simplification. It is this very step that helps establish an utmostly simplified mathematical artifice (the asymptotic solution) consistently workable without relying on numerical calculation by a computer. From purely intuitive reasoning, it is difficult to judge whether or not such a simplifying mathematical step is legitimate, and whether or not the results derived therefrom are sufficiently accurate for practical uses. The mathematics becomes so unbelievably simple that at first I doubted its accuracy. Although for typical PPTs we have estimated the upper bound of Q_S as afore-said, we consider it as prudent to examine the matter more carefully by adopting this simple asymptotic approach to various initial-value problems, and compare the results thus derived with the corresponding numerical results yielded by computer calculation of the iterative solution Eq. (7.14). It is very gratifying that our simplified analytic artifice has proved to be fairly accurate. It thus becomes possible to have the main body of this paper founded on analytic theory, with most results put in closed forms. However important it is, numerical calculation by a computer is employed in only an auxiliary way.

7.4 POLARIZATION TRANSFORM BY SINGLE-SUPERMODE PROCESS

We proceed to particular SOP transforms of a variably spun PPT fiber under different initial conditions. We classify these seemingly very diverse initial-value problems in two general categories according to the involvement of only one supermode, or of both supermodes, during the light transmission process. The thought of this way of classification actually suggests itself by an analogy with the simpler case of uniform waveguide transmission, where the problems can be classified according to the involvement of a single eigenmode or of both eigenmodes (see Sections 5.6 and 5.7). In single-eigenmode transmission, the field pattern or the SOP of light remains unchanged along a uniform fiber, except for a phase shift descriptive of the wave motion. On the other hand, in dual-eigenmode transmission the occurrence of both eigenmodes makes a beating pattern whose amplitude varies periodically along the uniform fiber, with a period called the beat length. Extension of this approach to a variably spun PPT fiber appears plausible, when we recall that in a PPT fiber the supermodes are almost eigenmodes, or quasi eigenmodes.

SOP Transform in PPT by Single-Supermode Process

In view of its structure, a PPT unit is really just a spun birefringent fiber with a spin rate that varies slowly from fast to zero, or vice versa. Simple as it is, a single PPT unit has four varieties, because this fiber structure is asymmetrical such that either the fast-spun end or the unspun end can act as the input, and in each case, the sense of spin can be either clockwise or anticlockwise. Clearly, in specifying the sense of spin, or the sense of rotation of a linear light, or the handedness of a circular light, and so on, the line of sight should be defined first. This is sometimes a rather tedious work when several kinds of rotation are simultaneously involved in the process. For simplicity and neatness, therefore, the discussion in the text will be focused on one PPT variety only (say, that shown in Figure 7.1). We can even spare the specification of a rotational sense if such an omission will not cause misconceptions in the analysis. As a matter of fact, from the results derived from one PPT variety, most inferences can be drawn about other PPT varieties.

We begin with the single-supermode process of a PPT. The first question to ask is: How can we achieve such a single-supermode SOP transform process in a given variably spun PPT fiber? To answer, let the desired single supermode be $W_x(z)$, implying $W_y(z) \equiv 0$ for all z. This specifies that the initial condition for the supermodes be given by

$$\mathbf{W}(0) = \begin{bmatrix} W_x(0) \\ W_y(0) \end{bmatrix} = \begin{bmatrix} 1 \\ 0 \end{bmatrix} \tag{7.25}$$

The initial condition for the local modes can then be derived from Eq. (7.25) from Eq. (7.10):

$$\mathbf{A}(0) = \mathbf{O}(0)\mathbf{W}(0)$$
$$= \frac{1}{\sqrt{2}} \begin{bmatrix} 1 & j \\ j & 1 \end{bmatrix} \begin{bmatrix} 1 \\ 0 \end{bmatrix} = \frac{1}{\sqrt{2}} \begin{bmatrix} 1 \\ j \end{bmatrix} \tag{7.26}$$

where $\mathbf{O}(0) = \mathbf{O}_F$, given by Eq. (7.19). The resulting matrix in Eq. (7.26) is a circular light, whose handedness (whether right or left, or whether clockwise or anticlockwise) depends on from which side the observer views the light. For convenience, let Eq. (7.26) refer to a right circular light. With this specification, if we initially let the desired single supermode be $W_y(z)$, implying $W_x(z) \equiv 0$ for all z, then a similar derivation leads to $A_x(0) = 1$, $A_y(0) = -j$, which is descriptive of a left circular light.

When the appropriate input light for single supermode excitation is known, it is simple and straightforward to derive the particular solution $\mathbf{A}(z)$ from $\mathbf{W}(z)$. Let $\mathbf{A}(0)$ be the right circular light given by Eq. (7.26). Then $\mathbf{W}(0) = \mathbf{O}^{-1}(0)\mathbf{A}(0)$ takes the form of Eq. (7.25) as initially specified. By Eqs. (7.23) and (7.23a), we immediately have

$$\mathbf{W}(L) = \begin{bmatrix} e^{j\rho} & 0 \\ 0 & e^{-j\rho} \end{bmatrix} \mathbf{W}(0) \tag{7.27a}$$

$$\mathbf{A}(L) = \mathbf{O}(L)\mathbf{W}(L) = \begin{bmatrix} 1 \\ 0 \end{bmatrix} e^{j\rho} \tag{7.27b}$$

where $\mathbf{O}(L)$ is a unit matrix by Eq. (7.21) at the unspun output end $z = L$, and the exponential is a phase factor whose exponent is Eq. (7.23b):

$$\rho \equiv \rho(L) = \int_0^L \pi \left[1 + 4(L_b/L_S)^2\right]^{1/2} dz \qquad (7.27\text{c})$$

As before, $\rho \equiv \rho(L_b, L_S, L)$ is the global structural parameter of the PPT. According to Eqs. (7.26) and (7.27), a right circular light exciting the PPT at its fast-spun end will be transformed into a linear light aligned along the local principal axis x at the unspun output end. The exponential term represents the phase factor of a wave, whose common wave factor is implied. In the preceding formulation, the appearance of just one exponential term with an integral exponent (the short of a wave factor) is descriptive of the desired characteristic that a lightwave in the PPT propagates only in the fast supermode W_x throughout (see Fig. 7.2).

If the PPT fiber is excited at its fast-spun end by a left circular light, the result is

$$\mathbf{A}(0) = \frac{1}{\sqrt{2}} \begin{bmatrix} 1 \\ -j \end{bmatrix}$$

$$\mathbf{W}(0) = \mathbf{O}^{-1}(0)\mathbf{A}(0) = -j \begin{bmatrix} 0 \\ 1 \end{bmatrix}$$

$$\mathbf{W}(L) = \begin{bmatrix} e^{j\rho} & 0 \\ 0 & e^{-j\rho} \end{bmatrix} \mathbf{W}(0) \qquad (7.28)$$

$$\mathbf{A}(L) = \mathbf{O}(L)\mathbf{W}(L)$$

$$= \begin{bmatrix} 0 \\ 1 \end{bmatrix} e^{-j(\rho + \pi/2)}$$

where $\mathbf{O}(L)$ is a unit matrix, and ρ is the same structural parameter given by Eq. (7.27c). Thus, a left circular light exciting the PPT at its fast-spun end becomes at the output unspun end ($z = L$) a linear light along the local

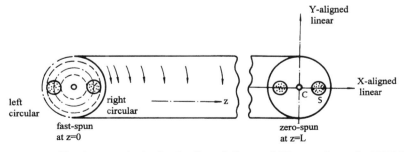

FIGURE 7.2   Circular-to-principal-axis-aligned linear SOP transform in PPT by a single supermode.

principal axis y. Light transmission along the PPT fiber then involves only the slow supermode $W_y(z)$ throughout.

Here we emphasize that the meaning of "single-supermode process" implies not only the appearance of just one exponential wave factor in the end result $A(L)$, but also the transmission behavior that, throughout the entire course of lightwave propagation from $z = 0$ to $z = L$, there is always one and only one exponential wave factor of integral exponent (or one supermode) involved in the wave-field pattern.

By intuition, the preceding analytic result of circular-to-linear SOP transform by a PPT should be predictable. It is only natural that the fast-spun end favors a circular mode, while the unspun end favors a principal-axis-aligned linear mode. In modal analysis, the *local* characteristic modes featuring the local structural properties of the PPT fiber at the fast-spun and unspun ends, respectively, are a circular light and a linear light. Thus, the SOP transform by a single-supermode process is simply a gradual evolution of the SOP, through the PPT fiber, from one local characteristic mode to the other local characteristic mode. The condition necessary for this single-supermode process to occur is that the exciting light matches the local characteristic mode at the input. From now on, we prefer using the term "local characteristic mode" in view of putting more emphasis on the role of the local characteristics at the PPT's ends, recognizing that this term is a synonym of *local* eigenmode or *local* normal mode in modal theory.

SOP Transform by Single Supermode in the Reversed Direction

In the single-supermode process, the SOP transform behavior of a PPT in the reversed direction is anticipated by intuition. Consider now a PPT variety whose spin rate varies slowly from zero to fast. The local characteristic mode at the unspun input end is a principal-axis-aligned (either x or y) linear light, while the local characteristic mode at the fast-spun output end is a circular light of either sense. The SOP transform by a single supermode in this reversed direction is naturally from a principal-axis-aligned linear light to a circular light. The mathematical derivation is simple and straightforward with the aid of asymptotic formulation. For an x-oriented linear light exciting this "reversed" PPT, we have

$$\mathbf{A}(0) = \begin{bmatrix} 1 \\ 0 \end{bmatrix}$$

$$\mathbf{W}(0) = \mathbf{O}^{-1}(0)\mathbf{A}(0) = \mathbf{A}(0)$$

$$\mathbf{W}(L) = \begin{bmatrix} e^{j\rho} & 0 \\ 0 & e^{-j\rho} \end{bmatrix}\mathbf{W}(0) \tag{7.29}$$

$$\mathbf{A}(L) = \mathbf{O}(L)\mathbf{W}(L)$$

$$= \frac{1}{\sqrt{2}}\begin{bmatrix} 1 \\ j \end{bmatrix}e^{j\rho}$$

where $\mathbf{O}^{-1}(0)$ at the unspun input end is a unit matrix, and $\mathbf{O}(L)$ at the fast-spun output end is given by Eq. (7.19). The above formulation indicates that only the fast supermode $W_x(z) = e^{j\int_0^z g\,dz}$ is involved during the whole course of the SOP transform concerned.

Similarly, for a y-oriented linear light at the input, the derivation yields

$$\mathbf{A}(0) = \begin{bmatrix} 0 \\ 1 \end{bmatrix} = \mathbf{W}(0) \qquad (7.30a)$$

$$\mathbf{W}(z) = e^{-j\rho}\mathbf{W}(0)$$

$$\mathbf{A}(L) = \frac{1}{\sqrt{2}}\begin{bmatrix} 1 \\ -j \end{bmatrix} e^{-j(\rho - \pi/2)} \qquad (7.30b)$$

which reveals that only the slow supermode $W_y(z) = e^{-j\int_0^z g\,dz}$ is involved. The results given by Eqs. (7.29) and (7.30) confirm the intuitive anticipation of the SOP transforms by a single supermode in a PPT fiber with an unspun input end.

Power Conversion in PPT by Single-Supermode Process

The foregoing analytic study of a PPT behavior by a single-supermode process was based on the asymptotic expressions, yielding the end results of the SOP transforms only. In order to view the whole pattern of the SOP evolution along the fiber, we have to recourse to numerical calculation by a computer. The formulation of the iterative solution derived in Section 7.3 is suitable for numerical calculation, with the parameters involved all being normalized (adimensional). Typical numerical examples for the SOP transforms by a single-supermode process in a PPT are shown in Figures 7.3 and 7.4.

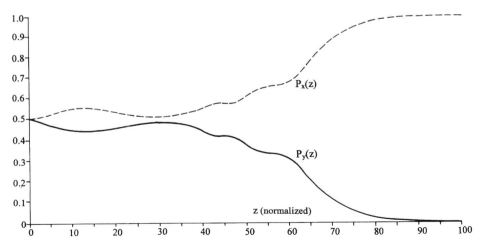

FIGURE 7.3 Evolution of power in a circular-linear SOP transform by a single supermode.

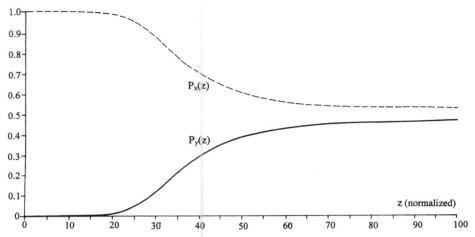

FIGURE 7.4   Evolution of power in a linear-circular SOP transform by a single supermode.

The curves in Figure 7.3 display the evolution of the local-mode powers of a PPT fiber (input fast-spun) by the single-supermode process. The numerical specifications of the relevant characteristic parameters are $Q(0) = 10$, $L = 10^2$, $\gamma = 2$. Note that what is required for numerical specification is the ratio of L_b and L_S, or the normalized spin rate Q ($= \tau/\delta\beta = L_b/L_S$). Separate numerical specifications of L_b and L_S are not necessary. Of the three specifications, the first two for $Q(0)$ and L are crucial. They are actually the necessary and sufficient conditions for a variably spun birefringent fiber to behave as a PPT, implying that the spin of the fast-spun end is sufficiently fast, and that the variation of the spin-rate is sufficiently slow. On the other hand, the numerical specification for γ, a parameter in Eq. (7.17), is relatively arbitrary. Extensive computer calculations reveal that different choices of this γ-parameter make little difference in the final numerical result.

In addition to the preceding specifications for the PPT fiber structure, we also have to specify the input light that has to match the local characteristic mode of the fast-spun end of this PPT in order to achieve the desired SOP transform by a single-supermode process. This should be a circular light. In terms of the normalized amplitudes of the local modes, let $A_x(0) = 1/\sqrt{2}$, $A_y(0) = j/\sqrt{2}$. As it is the light power that is readily measurable, the evolution of the local-mode power is of practical interest. The numerical results for $P_x(z) = |A_x(z)|^2$ and $P_y(z) = |A_y(z)|^2$ are shown as the dashed and solid curves in Figure 7.3. At the input end $z = 0$, $P_x(0) = P_y(0) = 0.5$. As z increases toward $z = L$, the evolution of power is essentially a gradual conversion from $P_y(z)$ to $P_x(z)$. The reconversion of power from $P_x(z)$ back to $P_y(z)$ is only slight. Essentially, the curve for $P_x(z)$ is continuously rising, while that for $P_y(z)$ is descending until an x-aligned linear light appears at the output end. Numerical calculation yields $A_x(L) = 0.2144 - j0.9764$ and $A_y(L) = -0.0252 + j0.0024$, giving $P_x(L) = |A_x(L)|^2 = 0.9994$ (≈ 1) and $P_y(L) = |A_y(L)|^2 = 0.0006$ (≈ 0).

The above brief description of Figure 7.3 illustrates numerically that a right circular light at the input fast-spun end of a PPT is transformed into an x-aligned linear light at the output unspun end. From the numerical data, we see that the corresponding analytic result derived previously from asymptotic approximations was correct and fairly accurate.

According to Eq. (7.27), a right circular light incident onto the concerned PPT becomes an x-aligned linear light at the output, with a phase factor $e^{j\rho}$, where $\rho = \int_0^L \pi[1 + 4(L_b/L_s)^2]^{1/2} dz$ is the global structural parameter. The numerical value of this parameter is actually implied in the preceding computer data. As seen, $A_y(L)$ is numerically negligible in comparison to $A_x(L)$, such that $A_x(L) \approx e^{j\rho}$, with $\rho = \arctan(\text{Im}/\text{Re})$, where Im and Re denote the imaginary and real parts of $A_x(L)$. According to the numerical data given in the previous paragraph, we have Re $= 0.2144$, Im $= -0.9764$, such that $\rho \approx -77.6°$ for a PPT structure specified by $Q(0) = 10$, $L = 10^2$, $\gamma = 2$. This figure seems trivial here. But in later computer calculations, this figure ($-77.6°$) proves indispensable for the numerical examples that illustrate the SOP transform behaviors by the dual-supermode process in a PPT.

Shown in Figure 7.4 are the curves descriptive of the evolution of powers in a PPT in the reversed direction (input unspun). For this PPT, we use a monotonously rising function to describe the spin rate, as given by Eq. (7.17a). As a numerical example, we again use the structural parameters $Q(L) = 10$, $L = 10^2$, $\gamma = 2$. The initial condition for this reversed PPT (by a single-supermode process) is given by $A_x(0) = 1$, $A_y(0) = 0$. Numerical computation yields $A_x(L) = 0.7246 e^{j84.6°}$, $A_y(L) = 0.6891 e^{-j5.1°}$. Thus, $P_x(L) = |A_x(L)|^2 \approx 0.525$, $P_y(L) = |A_y(L)|^2 \approx 0.475$. The phase difference between $A_x(L)$ and $A_y(L)$ is $89.7° \approx 90°$. The output light thus approximates a circular light. Figure 7.4 shows the overall pattern of the evolution of powers. The curve for $P_x(z) = |A_x(z)|^2$ falls gradually and monotonously from 1 to nearly 0.5 as $z \to L(= 10^2)$. At the same time, the curve for $P_y(z) = |A_y(z)|^2$ rises gradually and monotonously from 0 to nearly 0.5 when the output end is approached.

7.5   WIDE-BAND AND RELATED CHARACTERISTICS OF FIBER-OPTIC QUARTER-WAVE PLATE

The circular–linear or linear–circular SOP transform exhibited in a PPT reveals a striking similarity between such a fiber-optic element and a bulk-optic quarter-wave plate. It is therefore appropriate to categorize such a PPT element as the fiber-optic analog of a bulk-optic quarter-wave plate, while recognizing that the corresponding elements are not identical in all aspects.

In structure, a bulk-optic quarter-wave plate is symmetrical with respect to its two sides, so there is no need to specify which side is the input in order to realize the linear–circular and circular–linear SOP transforms. A circular light passing the plate from either side is transformed at the other side into a linear light inclined by 45° from the optic axis, and conversely, a 45° -inclined linear light passes the plate to become a circular light.

A fiber-optic PPT, on the other hand, is an asymmetrical structure with one end fast-spun and the other end unspun. In such an asymmetrical structure, it is natural that we need to specify which end is the input such that the other end is the output. In the single-supermode process, the SOP transform is either circular to the principal-axis-aligned linear, or vice versa, depending on whether the fast-spun end or the unspun end is the input. In either case, the output light will have a phase shift of $e^{j\rho}$, such that this characteristic figure ρ of the PPT plays the role that the 45° does in a bulk-optic quarter-wave plate.

An important difference in the respective performances exists between these fiber-optic and bulk-optic counterparts, and this difference favors the fiber-optic PPT element. That is, the fiber-optic PPT is inherently wide band, in distinct contrast to the inherent narrow-band property of a bulk-optic quarter-wave plate. Such an inherent difference is due to the difference in the working mechanisms of the fiber optics and the bulk optics while performing the same SOP transform. A conventional bulk-optic quarter-wave plate works on the principle that the o-wave and the e-wave acquire a phase difference of $(\pi/2)$ in passing the waveplate. This figure $(\pi/2)$ is strictly dependent on the operating frequency or wavelength such that a conventional bulk-optic quarter-wave plate is inherently narrow-band.

On the other hand, the fiber-optic PPT works on an entirely different principle. The SOP-transform process concerned takes place in a long way along the fiber, such that the input light gradually transforms itself through the slowly varying fiber structure until it reaches the output end. We recall that in the example of circular–linear SOP transform by the single-supermode process, the incident light matches the local characteristic mode (a circular light) of the PPT fiber structure at the fast-spun input $z = 0$. At the unspun output $z = L$, the local characteristic mode of this fiber structure is a principal-axis-aligned linear light. It thus transpires that the quarter-wave-plate-like behavior of the fiber-optic PPT simply represents a gradual and natural transformation from one local characteristic mode to the other local characteristic mode in a varying fiber structure that is exceedingly long in terms of unspun beat length (still orders of magnitude longer in terms of the operating wavelength).

From the previous analytic formulation it is seen that the only frequency-dependent parameter involved is L_b, the beat length of the PPT's birefringent fiber in the unspun state. Nevertheless, this frequency-dependent parameter L_b is absorbed in the phase factor of the output light, without affecting the concerned SOP transform from circular light to principal-axis-aligned linear light, or vice versa.

The necessary and sufficient condition for the design of a fiber-optic PPT, as said, only requires a sufficiently fast spin rate at one end and a sufficiently slow variation of the spin rate from fast to zero, or vice versa. That is the whole secret of the PPT that is seemingly rather strange. The bandwidth of a PPT is actually enormously wide, limited only by the restriction that light transmission through the variably spun fiber remain "single mode" in the nominal sense of this term.

More often than not, a favor is gained at a price. The wide-band feature of a PPT is gained by using an unusually long length of fiber in terms of the unspun beat-length. This may sometimes pose a problem in the microwave spectrum. Fortunately enough, in fiber optics the length requirement is no longer a problem. On the contrary, the length of a PPT in the decimeter range (not including fiber extension) is easier to draw with the aid of the existing fabrication facilities. The making of a PPT would become difficult were it desirable that such a fiber section should be short enough to be comparable with its unspun beat length.

The circumstance that none of the three structural parameters (L_b, L_S, L) requires strict specification of its exact value is advantageous both from the fabrication standpoint and from the experimental standpoint. The loose requirement on the value of L_b allows comparative freedom in the choice of the birefringent preform for fiber drawing. More importantly, the relaxed requirement on the spin-rate function $\tau(z)$ allows one to make a PPT element almost as easily as to make a common spun fiber. Were it not for this relaxed requirement, fabrication of a variably spun birefringent fiber (strictly according to a prescribed τ-function) would have been exceedingly difficult, or almost impossible. Apparently, the relaxed requirement on the fiber length L also makes laboratory or field practice more convenient, allowing more freedom in handling a fabricated PPT fiber segment, such as making fiber-end polishing, cutting, or splicing with some other fiber element. In practice, since the fabrication process necessarily yields a longer piece of fiber consisting of the PPT proper with two extension fibers on both sides, it becomes hardly possible, and scarcely necessary, to single out the PPT proper in a clear-cut way from the overall length of the fabricated fiber piece. As long as one side remains to be fast-spun, while the other side remains unspun, the variably spun fiber will work just as well, regardless of the extension fibers being longer or shorter.

Obviously, from the standpoint of system application, the behavior of principal-axis aligning of the linear light at the unspun end of a PPT is particularly attractive in maintaining the linear SOP of light in its unspun extension, which is simply a (linearly) birefringent fiber.

7.6 DUAL-SUPERMODE PROCESS IN FIBER-OPTIC QUARTER-WAVE PLATE

Previous analytic and numerical study of a PPT shows that the SOP transform by a single-supermode process (in the sense of asymptotic approximation) requires the matching of the incident light and the local characteristic mode at the input. Naturally, the next subject to take is therefore to investigate what will happen if the matching condition is not satisfied.

Analysis of PPT Behavior by Dual-Supermode Process

We again consider a PPT with a fast-spun input as the one shown in Figure 7.1. Note that, for this PPT, the SOP transform by a single-supermode requires a circular incident light. Consider the case where the incident light is not circular, but linear, for example. The initial condition is

$$A(0) = \begin{bmatrix} \cos\theta \\ \sin\theta \end{bmatrix} \tag{7.31}$$

where θ defines the orientation of the incident linear light. As before, we solve the local modes A via the supermodes W with the aid of the asymptotic expressions in Section 7.3. Thus, by Eq. (7.20), the initial condition for the super modes is given by

$$\begin{aligned} W(0) &= O^{-1}(0)A(0) \\ &= \frac{1}{\sqrt{2}} \begin{bmatrix} e^{-j\theta} \\ -je^{j\theta} \end{bmatrix} \end{aligned} \tag{7.32}$$

By Eqs. (7.23a) and (7.21),

$$W(z) = \begin{bmatrix} e^{j\rho} & 0 \\ 0 & e^{-j\rho} \end{bmatrix} W(0)$$

$$A(L) = O(L)W(L) = \frac{1}{\sqrt{2}} \begin{bmatrix} e^{j(\rho-\theta)} \\ -je^{-j(\rho-\theta)} \end{bmatrix} \tag{7.33}$$

$$\rho = \int_0^L \pi \left[1 + 4(L_b/L_S)^2 \right]^{1/2} dz \tag{7.33a}$$

where, as before, $\rho \equiv \rho(L_b, L_S, L)$ is the global structural parameter according to Eq. (7.23b).

The distinctive feature of the above equations is self-explanatory. Unlike the single-supermode process treated in Section 7.4, the present SOP transform involves both supermodes $W_x(z)$, $W_y(z)$. Both "exponential" wave factors of different integral exponents (or varying phase velocities) coexist in the course of lightwave transmission. In comparing the present dual-supermode case with the previous single-supermode case, we see that the PPT fiber structure employed in both cases is just the same, but that the initial conditions differ.

The two supermodes are quasi eigenmodes in the nonuniform fiber, such that their coexistence displays a beating-like pattern. Unlike the case of uniform fiber, however, the beating of the supermodes in a nonuniform PPT fiber has the peculiar feature that its beat length along the fiber is not constant, but varies. This is apparent by Eqs. (7.33) and (7.33a), where the integral exponents for $W_x(z)$ and $W_y(z)$ are functions of z. In approaching the fast-spun end, g and hence ρ (integral of g) increase rapidly, such that the supermode beat length rapidly becomes shorter and shorter. This peculiar beating-like characteristic of the supermodes is bound to be reflected in the coupling pattern of the local modes $A_x(z)$ and $A_y(z)$, as will be shown in Figure 7.6.

Equation (7.33) immediately reveals that, in the case of an incident linear light exciting the fast-spun end of a PPT fiber, the following amplitude relation and phase relation hold:

$$|A_x(L)|^2 = |A_y(L)|^2 = 0.5 \qquad (7.34)$$

$$\xi = 90° + 2\rho - 2\theta \qquad (7.35)$$

Equation (7.34) indicates that the two local modes have equal amplitudes, or carry equal divisions of power, at the output end. In Eq. (7.35), ξ denotes the phase difference of these two local modes, with the understanding that an integer multiple of 2π can be added on the right side of the equation for more generality.

Here it is important to note that the "equal power division" behavior of a dual-supermode PPT is reciprocal in the sense that if an incident light comprising two local modes with equal amplitudes is launched onto the unspun end, with any phase difference ξ, then the transformed SOP of light at the fast-spun end can only be linear, oriented by an angle equal to $\theta = \rho + 45° - \xi/2$.

It therefore follows from Eqs. (7.34) and (7.35) that, if the orientation of the incident linear light assumes the specific angle $\theta = \rho$, then the variably spun PPT fiber (structure the same as Figure 7.1) will transform this linear light into a circular light at the output. The global structural parameter ρ therefore remains a characteristic figure of the PPT element working by a dual-supermode process. If the incident linear light is oriented at $\theta = \rho + 45°$, then the transformed light at the output will also be linear, and inclined by 45°. For other orientation angles, the incident linear light will be transformed at the output $(z = L)$ into an elliptical light whose ellipticity is given by $b/a = \tan(45° + \rho - \theta)$.

The variety of polarization transforms due to the linear light incident onto the fast-spun end of a PPT fiber is shown in Figure 7.5. According to Eqs. (7.34) and (7.35), the emergent light at $z = L$ corresponding to different orientations of the linear light at $z = 0$ are the set of 45°-inclined elliptical SOPs all inscribed in a square whose diagonal is of unit length, with circular light and linear light as the two limiting cases.

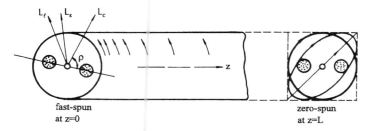

FIGURE 7.5 Linear light of different orientations incident on the fast-spun end of PPT and the consequent SOP transform by a dual-supermode process. L_c, L_ε, L_l: linear light with specific orientations at input transforming into circular, elliptical, linear light at output.

SOP Transform by Dual-Supermode Process in the Reversed Direction

Let a PPT fiber work in the reverse direction such that its input is unspun. In Section 7.4 it was shown that this PPT will work on single supermode only if the fiber is excited by the principal-axes-aligned (x or y) linear light that matches the PPT's local characteristic mode at the input end. Consider otherwise the case where the incident light is not a principal-axis-aligned linear light. It is easy to anticipate that in this case the PPT will work, not by the single-supermode process, but by the dual-supermode process. The mathematical derivations for the different initial conditions will be easy with the aid of relevant asymptotic expressions. For brevity, we shall not write out these derivations in detail, but state the end results directly, adding some descriptive explanations when necessary.

For a left circular light incident on the input unspun end of a PPT, the asymptotic solution yields a linear light oriented at an angle $(-\rho)$ at the output fast-spun end. If the incident light is right circular, then the output light is linear, but with an orientation angle of $(90° - \rho)$, multiplied by a constant phase factor $e^{j\pi/2}$. Thus, $(-\rho)$ appears as the characteristic figure of a PPT working in the reverse direction.

Consider linear light incidence in the general case, whose orientation angle is θ. The asymptotic solution for the transformed light at the output end (fast-spun) does not immediately display the "equal power division" property. We are then tempted to rotate the coordinates by an angle $-\rho$ (the characteristic angle). Interestingly enough, the end result of these rotated coordinates exactly displays the "equal power division" property plus a phase relation given by $\xi = 2\theta - 90°$. Thus, for an incident linear light with $\theta = 45°$, the output light will also be linear, which is oriented by $45°$ in the rotated coordinates, or oriented by $(45° - \rho)$ in the local coordinates.

Power Conversions–Reconversions by Dual-Supermode Process

The afore-derived asymptotic solutions are the end results only. To display the entire pattern of the power evolution along the fiber by the dual-supermode process, we need to make numerical calculation of $A_x(z)$ and $A_y(z)$, by way of $W_x(z)$ and $W_y(z)$, over the interval from $z = 0$ to $z = L$. We use the same specifications for the PPT's structure as those for Figure 7.3, that is, $Q(0) = 10$, $N = 10^2$, $\gamma = 2$, but we change the initial condition or input light so it is an x-aligned linear light given by $A_x(0) = 1$, $A_y(0) = 0$. This incident light does not match the local characteristic mode (a circular light) at the fast-spun input end, so that it excites both supermodes in the PPT. The numerical result is given in Figure 7.6. For neatness, only one curve is drawn in Figure 7.6, to display both the power evolution of $P_x(z) = |A_x(z)|^2$ and the power evolution of $P_y(z) = |A_y(z)|^2$. Here we use the power conservation relation $P_x(z) + P_y(z) = 1$, which is always accurately confirmed in the computer data to many decimal places. Thus, if the readings for $P_x(z)$ are scaled as usual along the ordinate on the left from 0 to 1 upward, then $P_y(z)$ can be read from the ordinate on the right whose scale is from 0 to 1 downward.

In Figure 7.6, the initial section (a) features a strong coupling between the two local modes $A_x(z)$ and $A_y(z)$, apparently reflecting the occurrence of the supermode beating. An almost complete power transfer takes place swiftly back and forth from one local mode to the other. The amplitudes of this almost complete power conversion–reconversion gradually decay through the intermediate section (b), which features a transition from strong coupling to weak coupling of the local modes. The end section (c) features weak coupling between the local modes, which now begin to carry nearly equal powers, with the rise and fall diminishing toward the output end. Numerical calculation shows that the higher $Q(0)$ (normalized spin rate at the fast-spun end) is, the less the eventual deviations of the two local-mode powers from 0.5 will be.

For the given initial condition $A_x(0) = 1$, $A_y(0) = 0$, the numerical data for Figure 7.6 are: $A_x(L) = 0.1533 - j0.7033$, $A_y(L) = 0.6726 - j0.1499$,

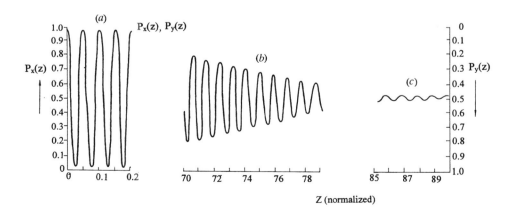

FIGURE 7.6 Evolution of power in PPT involving the dual-supermode process.

such that $|A_x(L)|^2 = 0.525$, $|A_y(L)|^2 = 0.475$. The phase difference between $A_x(L)$ and $A_y(L)$ is $\xi \approx -65.2°$, which is descriptive of an elliptical light at the output.

According to the analytical equations (7.34) and (7.35) (previously derived by the asymptotic approach), the linear–circular and linear–linear transforms occur at $\theta = \rho$ and $\theta = \rho + 45°$, respectively. To examine (by numerical calculation by a computer) if these analytic results that involve the asymptotic approximations are really accurate, we need to have the numerical value of ρ. We already obtained this from the computer data for Figure 7.3, that is, $\rho = -77.6°$. Thus, let the same PPT now work in the dual-supermode process, wherein the fast-spun end is excited by a linear light oriented at $\theta = \rho = -77.6°$. By Eq. (7.31), the relevant initial condition should be $A_x(0) = 0.2144$, $A_y(0) = -0.9764$. With this initial condition, the computer-calculated data for the output light are $A_x(L) = 0.7237e^{j89.569°}$, $A_y(L) = 0.6896e^{-j0.4585°}$. This numerical result approximates a circular light, as the analytic equations (7.34) and (7.35) predicted. It thus shows that the numerical value $-77.6°$ is really the sought-for characteristic figure ρ of the PPT fiber with the specifications: $Q(0) = 10$, $L = 10^2$, $\gamma = 2$.

Then let $\theta = \rho + 45° = -32.62°$. By Eq. (7.31), $A_x(0) = 0.8423$, $A_y(0) = -0.5390$. The computer data yield: $A_x(L) = 0.4916 - j0.5158 = 0.7126e^{-j46.38°}$, $A_y(L) = 0.4839 - j0.5080 = 0.7016e^{-j46.39°}$. The vector sum of these nearly in-phase x- and y-directed orthogonal linear modes (or rectangular components) at $z = L$ is $\mathbf{A}_x(L) = 0.9859 \cdot e^{-j46.4°} \angle(44.55°) \approx 1 \cdot e^{-j46.4°} \angle(45°)$. The asymptotic approximation predicts that the output linear light lies along a diagonal of the right square (Figure 7.5). This analytic prediction is again supported by the computer data.

For the above two cases of polarization transforms (linear–circular and linear–linear), only the end numerical results are given. To save space, the curves displaying the relevant evolutions of power are not included here. Suffice it to say that, according to the vast computer data obtained, the general trajectories descriptive of these power evolutions are actually all similar to the curve in Figure 7.6, except that each of the trajectories starts from the corresponding initial value. For a PPT working by the dual-super-mode process, there is always a violent power conversion–reconversion in the fast-spun range, with the amplitudes of the rise and fall of power gradually diminishing toward the unspun range.

"Equal Power Division" Property in Bulk-Optic Quarter-Wave Plate

In a fiber-optic PPT working by the dual-supermode process, the SOP variation of the output light follows the θ variation of the input linear light. Yet for θ varying in any manner, the total output power is always equally divided between $A_x(L)$ and $A_y(L)$, independent of the value of θ, as given by Eq. (7.34). While a change in θ results in a change of ξ, as given by Eq. (7.35), the "equal power division" property will not be affected.

In a bulk-optic quarter-wave plate, it can be shown that the "equal power division" property holds if the reference coordinates are so chosen that either x or y is inclined by 45° with respect to the optic axis. Thus, using the Jones matrix calculus, we have

$$\mathbf{A}(L) = \mathbf{J}\mathbf{A}(0) \tag{7.36}$$

in which

$$\mathbf{A}(0) = \begin{bmatrix} \cos\theta \\ \sin\theta \end{bmatrix} \tag{7.36a}$$

$$\mathbf{J} = e^{j\pi/4} \begin{bmatrix} 1 & 0 \\ 0 & j \end{bmatrix} \tag{7.36b}$$

where $\mathbf{J}$ denotes the Jones matrix for a bulk-optic quarter-wave plate with the optic axis horizontal, and θ is the angle of inclination of the incident linear light from the optic axis (tacitly assumed to be aligned with the axis x). A rotation of the coordinate axes by 45° yields

$$\mathbf{A}_{\pi/4}(L) = \frac{1}{\sqrt{2}} \begin{bmatrix} 1 & 1 \\ -1 & 1 \end{bmatrix} \mathbf{A}(L) \tag{7.37}$$

Putting Eq. (7.36) into Eq. (7.37) yields

$$\mathbf{A}_{\pi/4}(L) = \frac{e^{j\pi/4}}{\sqrt{2}} \begin{bmatrix} e^{j\theta} \\ -e^{-j\theta} \end{bmatrix} \tag{7.38}$$

and hence

$$|A_x(L)|^2 = |A_y(L)|^2 = 0.5 \tag{7.39a}$$

$$\xi = 2\theta \pm \pi \tag{7.39b}$$

referring to the 45°-rotated coordinates. Equations (7.39a) and (7.39b) for a bulk-optic quarter-wave plate correspond to Eqs. (7.34) and (7.35) for a fiber-optic PPT.

Dual-Supermode Process vs. Single-Supermode Process

A prominent difference between the two processes of a PPT is apparently revealed when one compares the respective power-evolution curves. Figures 7.3 and 7.4 show that, if the SOP transform involves a single supermode only, the transfer of power between the local modes is a gradual process. Indeed, while this process is gradual, the overall power transfer over a long length of fiber can be fairly large. On the other hand, Figure 7.6 shows that the SOP transform involving two supermodes is described by oscillatory power curves of the local modes, with violent power conversion–reconversion over the fast-spun range.

The characteristic figure ρ plays different roles in SOP transforms by the single supermode and the dual supermode. In the single-supermode case, the SOP transform is either from circular to principal-axis-aligned linear (input fast-spun), or from principal-axis-aligned linear to circular (input unspun), regardless of the value of ρ. The parameter ρ appears only in the phase factor (or wave factor) of the transformed light, and is descriptive of a phase change in the lightwave by the single-supermode process. In the dual-supermode case, a θ-oriented linear light at the fast-spun input is always transformed into an output light whose power is equally shared by the local modes $A_x(L)$ and $A_y(L)$, with a phase difference ξ related to θ and ρ, which together determine the SOP of the output light.

As far as the circular-linear or linear-circular SOP transform is concerned, both the single-supermode process and the dual-supermode process work, though in different ways. In Section 7.5, with reference to the single-supermode process, we discussed the wide-band property in favor of the fiber-optic PPT. There it was comparatively easier to interpret the wide-band property in terms of a smooth variation of the SOP over a long length of fiber, such that the input local characteristic mode gradually transforms itself into the output local characteristic mode. In the dual-supermode process, the two super (quasi-eigen) modes beat, or alternatively, the two local modes are strongly coupled, over the fast-spun part of the PPT. This would mean narrow-band, were it to refer to a uniform fiber. However, the swift oscillations of large amplitudes cover only a short range (note the different scales for the three sections of the abscissa in Figure 7.6). The PPT's overall behavior during the dual-supermode process is always governed by Eqs. (7.34) and (7.35), regardless of the operating wavelength and related factors.

Concerning in-line applications, the single-supermode process appears more attractive. In the single-supermode process, the principal-axis aligning of the linear light allows the PPT fiber to be incorporated into a fiber circuitry once and for all, without the need of later manual adjustment or tuning. In case perturbations occur, their effects will be absorbed in the integral exponent of the one and only phase factor of the single supermode, such that the existing SOP transform can still be maintained. In contrast, for a PPT working by the dual-supermode process, a perturbation may affect the two supermodes differently, resulting in an SOP of light at the output that is more liable to become unstable.

A summary of the fiber-optic PPT versus the bulk-optic quarter-wave plate in regard to SOP transforms is given in Tables 7.1 and 7.2 for ready reference.

TABLE 7.1 Fiber-Optic PPT Characteristics

Type of Optics	Input Light at $z = 0$	Output Light at $z = L$	Reference Axes
	right circular $\dfrac{1}{\sqrt{2}}\begin{bmatrix} 1 \\ j \end{bmatrix}$	x-aligned linear $\begin{bmatrix} 1 \\ 0 \end{bmatrix} e^{j\rho}$	
	left circular $\dfrac{1}{\sqrt{2}}\begin{bmatrix} 1 \\ -j \end{bmatrix}$	y-aligned linear $\begin{bmatrix} 0 \\ 1 \end{bmatrix} e^{-j(\rho + \pi/2)}$	
Fiber-optic PPT fast-to-zero spun	specific linear $\begin{bmatrix} \cos\rho \\ \sin\rho \end{bmatrix}$	left circular $\dfrac{1}{\sqrt{2}}\begin{bmatrix} 1 \\ -j \end{bmatrix}$	local principal axes for all four cases
	arbitrary linear $\begin{bmatrix} \cos\theta \\ \sin\theta \end{bmatrix}$	equal-power division $|A_x|^2 = |A_y|^2 = 1/2$ $\xi = \pi/2 + 2\rho - 2\theta$	
	right circular $\dfrac{1}{\sqrt{2}}\begin{bmatrix} 1 \\ j \end{bmatrix}$	specific linear $\begin{bmatrix} \cos(\pi/2 - \rho) \\ \sin(\pi/2 - \rho) \end{bmatrix} e^{j\pi/2}$	local principal axes
Reversed PPT	left circular $\dfrac{1}{\sqrt{2}}\begin{bmatrix} 1 \\ -j \end{bmatrix}$	specific linear $\begin{bmatrix} \cos(-\rho) \\ \sin(-\rho) \end{bmatrix}$	local principal axes
	specific linear $\dfrac{1}{\sqrt{2}}\begin{bmatrix} 1 \\ 1 \end{bmatrix}$	specific linear $\begin{bmatrix} \cos(\pi/4 - \rho) \\ \sin(\pi/4 - \rho) \end{bmatrix} e^{j\pi/4}$	local principal axes
	arbitrary linear $\begin{bmatrix} \cos\theta \\ \sin\theta \end{bmatrix}$	equal-power division $|A_x|^2 = |A_y|^2 = 1/2$ $\xi = 2\theta - \pi/2$	axes rotated by $-\rho$

Note: In the table, it is tacitly assumed that spinning of the PPT fiber is in the $+$ or clockwise sense. For $-$ or anticlockwise spinning, x-aligned linear light and y-aligned linear light listed in the table are to be interchanged, and linear light of specific orientation is to become the normal to this light. If both the spinning of PPT fiber and the circling of a circular light change their senses, then the transformed light at output will remain unchanged. As described in the text, the parameter ρ is a function of L_b, L_S, and L, characteristics of the structural features of the PPT.

TABLE 7.2 Bulk-Optic Quarter-Wave Plate Characteristics

Type of Optics	Input Light at $z = 0$	Output Light at $z = L$	Reference Axes
Bulk-optic quarter-wave plate	right circular $\frac{1}{\sqrt{2}}\begin{bmatrix} 1 \\ j \end{bmatrix}$	specific linear $\frac{1}{\sqrt{2}}\begin{bmatrix} 1 \\ -1 \end{bmatrix}e^{j\pi/4}$	optic-axis horizontal
	left circular $\frac{1}{\sqrt{2}}\begin{bmatrix} 1 \\ -j \end{bmatrix}$	specific linear $\frac{1}{\sqrt{2}}\begin{bmatrix} 1 \\ 1 \end{bmatrix}e^{j\pi/4}$	optic-axis horizontal
	specific linear $\frac{1}{\sqrt{2}}\begin{bmatrix} 1 \\ 1 \end{bmatrix}$	right circular $\frac{1}{\sqrt{2}}\begin{bmatrix} 1 \\ j \end{bmatrix}e^{j\pi/4}$	optic-axis horizontal
	arbitrary linear $\begin{bmatrix} \cos\theta \\ \sin\theta \end{bmatrix}$	equal-power division $\lvert A_x \rvert^2 = \lvert A_y \rvert^2 = 1/2$ $\xi = -2\theta$	axes rotated by $-\pi/4$

Note: For the optic-axis to be vertical, the transformed linear light from right and left circular light will interchange, while the transformed right circular light from a $\pi/4$-inclined linear light will change into left circular. For a rotation of coordinate axes by $\pi/4$, the "equal-power-division" behavior still holds, with the phase-difference relation changed to $2\theta + \pi$.

7.7 PPT ELEMENTS IN CASCADE AND WAVE-PLATE-LIKE BEHAVIORS

The preceding sections dealt with a fiber-optic PPT of different varieties, and showed how it behaves like a bulk-optic quarter-wave plate. Development of this line of thought naturally leads to the anticipation that two PPT units appropriately cascaded can form fiber-optic structures that behave like a bulk-optic half-wave plate and a full-wave plate. This section aims at an analytic study of this subject.

Two PPT units can be cascaded in many different ways, inasmuch as the sense of spin of each PPT may be either clockwise or anticlockwise, and the input may be either fast-spun or unspun. To be exact, there are $4 \times 4 = 16$ different ways. Of these many possible ways of cascading, we are interested in those composite fiber structures wherein the spinning is continuous across the joint of the two PPT units. Such variably spun fibers can be fabricated in a continuous one-step drawing process, yielding a one-piece composite fiber element without the need to join two separate fiber sections.

Fiber-Optic Analog of Bulk-Optic Half-Wave Plate

The Q-function shown in Figure 7.7 is employed to make a two-section PPT, whose behavior can be readily shown to be similar to that of a bulk-optic

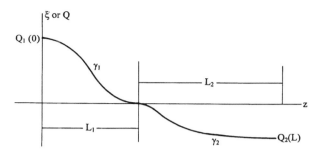

FIGURE 7.7   Spin function for fiber-optic analog of bulk-optic half-wave plate.

half-wave plate. The spin rate starts from a fast value, slows down gradually to zero, and then speeds up in an opposite sense to a fast value again. The magnitudes of the initial spin rate and the final reversed spin-rate are not required to be equal, nor are the lengths of the two fiber sections, and nor are the respective γ-values. What is required is that both the initial spin rate and the final spin rate are sufficiently fast, in either sense, and that the variation of the spin function is all-way sufficiently slow.

An analytic study of a composite PPT is just as simple as the case of a single PPT unit, if we consistently use the asymptotic approach. The mathematical steps are all the same, except that both Eq. (7.17) and Eq. (7.17a) are used for the composite Q-function, with the former to specify the decreasing part of the Q-function, and the latter to describe the negatively increasing part of this function. In regard to Figure 7.7, since its input end is fast-spun such that its local characteristic mode is circular, a circular light exciting this variably spun PPT fiber will make it work by the single-supermode process, while a linear light excitation will make the same fiber work by the dual-supermode process.

While it is simple and easy to treat the composite PPT of Figure 7.7 as an integral whole (i.e., to solve it as just one analytic initial-value problem), an even easier way is to view it in terms of the two PPT units that compose the composite structure. Then, using the previously derived results of a PPT of different varieties, the final result can be obtained, almost effortlessly, for either the single-supermode process or the dual-supermode process.

We first consider the case of the single-supermode process. Suppose the incident light is right circular. The emergent light from the first PPT section is then an x-aligned linear light in the unspun extension fiber in-between the two fiber sections, with a phase shift $e^{j\rho_1}$ descriptive of the wave motion in the single "fast supermode." This x-aligned linear light then acts as the incident light of the second PPT section. The matching condition is fulfilled at the input of the second PPT section, such that light transmission in this section remains in the same single supermode. Because of the reversal in the sense of spinning of the second PPT unit, the x-aligned incident light becomes a left circular light with an extra phase shift $e^{j\rho_2}$ at the final end of the composite fiber.

The SOP transform, which works consistently by single supermode in the respective fiber sections of the composite PPT of Figure 7.7, is really just that simple. The only point that needs additional care is that the sense of spin rate in the second section is opposite to that of the first section. The asymptotic expressions listed in Section 7.3 imply the tacit assumption that in the problem under study the spin rate τ varies only in magnitude, not in sense. For the present problem involving a reversal of the sense of spin, if we use τ to denote a clockwise spin, then an anticlockwise spin should be denoted by $-\tau$, taking it for granted that the observer's line of sight is properly specified. A reversal of the sign of τ and hence of Q results in a reversal of the sign of ϕ, such that the equivalents of Eqs. (7.19) and (7.20) for the second PPT section assume the following forms:

$$\mathbf{O}_F = \frac{1}{\sqrt{2}} \begin{bmatrix} 1 & -j \\ -j & 1 \end{bmatrix} \tag{7.19a}$$

$$\mathbf{O}_F^{-1} = \frac{1}{\sqrt{2}} \begin{bmatrix} 1 & j \\ j & 1 \end{bmatrix} \tag{7.20a}$$

Apparently, the above result is simply an interchange of Eqs. (7.19) and (7.20). As said before, at any unspun point (or over any unspun part) of a variably spun fiber where $\tau = 0$, the unit matrix expression given by Eq. (7.21) universally applies. Obviously enough, zero spin rate does not need to take care of a sign. The final result for the composite PPT of Figure 7.7 is

$$\mathbf{A}(0) = \frac{1}{\sqrt{2}} \begin{bmatrix} 1 \\ \pm j \end{bmatrix} \tag{7.40}$$

$$\mathbf{A}(L) = \frac{1}{\sqrt{2}} \begin{bmatrix} 1 \\ \mp j \end{bmatrix} \exp[\pm j(\rho_1 + \rho_2)] \tag{7.41}$$

where $L = L_1 + L_2$ is the total length, with L_1, L_2 denoting the lengths of the first and second sections, and $\rho_1 = \int_0^{L_1} g_1 \, dz$, $\rho_2 = \int_0^{L_2} g_2 \, dz$, where g_1, g_2 are related with $Q_1(z)$, $Q_2(z)$ according to Eq. (7.12a). Equations (7.40) and (7.41) show that a composite PPT with a Q-function of Figure 7.7 behaves analogously as a bulk-optic half-wave plate.

One familiar application of a bulk-optic half-wave plate is to change the orientation of linear light. Thus, if the bulk-optic half-wave plate is turned by an angle θ, the emergent light will turn by an angle 2θ. This is equivalent to the circumstance that, if the half-wave plate is fixed, while the linear light at the input turns an angle θ from the optic axis, the transformed linear light at the output will turn an angle $-\theta$ from the optic axis.

It is therefore interesting to see how this bulk-optic behavior of a half-wave plate manifests itself in the fiber-optic PPT of Figure 7.7 with an incident linear light. A linear light does not match the local characteristic mode at the fast-spun end of a PPT, and hence will excite both supermodes in this variably spun fiber. The first PPT section then works by the dual-supermode process, yielding an exit light whose power is equally divided, according to Eq. (7.34), along the two principal axes of the unspun extension fiber covering the junction. The SOP of this exit light depends on ξ (the phase difference) and hence on the orientation angle θ_0 of the incident linear light according to Eq. (7.35) with $\rho = \rho_1$. Whatever this SOP is, the light of the kind of "equal power division" will always be transformed into a linear light at the final output end of the second PPT section now working by the dual-supermode process in the reverse direction. The final orientation angle θ_L of this output linear light now depends on ξ again, in accordance with Eq. (7.35) with $\rho = \rho_2$.

The end result of the asymptotic solution of the SOP transform via the two successive dual-supermode processes is

$$\mathbf{A}(0) = \begin{bmatrix} \cos \theta_0 \\ \sin \theta_0 \end{bmatrix} \tag{7.42}$$

$$\mathbf{A}(L) = j \begin{bmatrix} \cos \theta_L \\ \sin \theta_L \end{bmatrix} \tag{7.43}$$

$$\theta_L = \rho + \pi/2 - \theta_0 \tag{7.43a}$$

Let the orientation angle of the input linear light at $z = 0$ change from θ_0 to $\theta_0 + \Delta\theta_0$. Correspondingly, the orientation angle of the output linear light at $z = L$ changes from θ_L to $\theta_L + \Delta\theta_L$. By Eq. (7.43a), we have

$$\Delta\theta_L = -\Delta\theta_0 \tag{7.44}$$

In principle, therefore, if the orientation of the input linear light does not change, but the composite fiber-optic structure is made to rotate by an angle $\Delta\theta_0$, then the output linear light will change its orientation by an angle $2(\Delta\theta_0)$, exactly analogous to a half-wave plate in bulk optics.

Fiber-Optic Analog of Bulk-Optic Full-Wave Plate

The spin-rate function shown in Figure 7.8 is fitting to make a composite PPT that behaves like a full-wave plate in bulk optics. The spin rate starts at a fast initial value, gradually slows down to zero, and then increases again in the same sense to a fast value. Like the preceding case, Q_0 and Q_L are not required to be equal, nor are L_1 and L_2, and nor are γ_1 and γ_2.

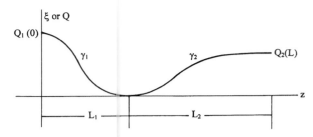

FIGURE 7.8   Spin function for fiber-optic analog of bulk-optic full-wave plate.

In the single-supermode process, circular light at the input of the PPT of Figure 7.8 will again appear circular at the output, and with the same sense (handedness). The asymptotic solution of this composite PPT is given by

$$A(0) = \frac{1}{\sqrt{2}} \begin{bmatrix} 1 \\ \pm j \end{bmatrix} \tag{7.45}$$

$$A(L) = \frac{1}{\sqrt{2}} \begin{bmatrix} 1 \\ \pm j \end{bmatrix} \exp[\pm j(\rho_1 + \rho_2)] \tag{7.46}$$

where $L = L_1 + L_2$. The resulting equation, showing an analogy between the composite PPT concerned and the bulk-optic full-wave plate, is self-explanatory.

In the dual-supermode process of a fiber-optic full-wave plate, linear light at the input also appears linear at the output, with their orientation angles related by

$$\theta_L = \theta_0 - (\rho_1 + \rho_2) \tag{7.47}$$

yielding

$$\Delta\theta_L = \Delta\theta_0 \tag{7.48}$$

Thus, in a fiber-optic full-wave plate, a rotational angle of the linear light at the input is not only accompanied by an equal rotational angle of the linear light at the output, but also with both rotations taking place in the same sense.

7.8   EXPERIMENTAL VERIFICATION OF THEORETICAL PREDICTION

This section presents the collected experimental data that I obtained ($\sim$ 1990) on my first few PPT specimens. Homemade equipment was used in fabrication as well as for the experiment. It is therefore only natural that the photos and the curves given below still retain some primitive flavor of my early attempt. I would rather keep this flavor as it was in the original exploitation than to make even a partial change, except perhaps in some of the illustrations.

Initial Prototype Specimen and Measuring Setup

Some prototype specimens of the variably spun PPT fiber were made in our laboratory. A small homemade drawing tower, incorporated with a variable-speed d.c. motor on its top, was employed for fabrication. In the design we used visible light of wavelength 0.6328 μm.

Readily available homemade preforms of the birefringent type were used as the starting material for drawing the fiber. Figure 7.9 shows the microscope photographs of the fiber ends of a typical PPT fiber section about half a meter long. Ordinarily, cross-sectional views of the fast-spun end and of the unspun end will display no difference to the microscope. Figure 7.9 shows a particular pair of photographs, in which the fast-spun slice of fiber displays some interesting whirling traces.

Relatively extensive experiments were performed on a simple traditional optical measuring bench, consisting of a 0.6328-μm He-Ne laser source emitting a linear light, a bulk-optic half-wave plate, and a bulk-optic quarter-wave plate, which together emitted a beam of light of the desired SOP to be focused onto the input end of the tested PPT fiber by a $10 \times$ microscopic objective. The SOP of the emergent light from the output end of the PPT fiber was analyzed with a Glan-Thompson analyzer, followed by a detector, whose output in the form of an electrical signal was read by a power meter and simultaneously recorded by an x-y recorder.

Experiments with Incident Light Matching the PPT's Input

We generally classify the multitude of SOP transforms of a PPT into two classes, according to whether or not the incident light matches the local characteristic mode at the input of the PPT. It is my belief that such a classification may help to clarify the otherwise seemingly very complicated and mysterious PPT characteristics. This classification agrees with our prior analytic framework, such that experimental and theoretical results can be compared with convenience. To start with, we make experiments relating to the first class.

In fiber-optic experiments on a PPT with either the fast-spun end or the unspun end as the input, it is always an embarrassing task to determine the locations of the orthogonal local principal axes acting as the reference axes for measurement. If we do not know these axes, the orientation of light cannot be adequately defined without ambiguity. For a variably spun PPT fiber, it is not easy to determine in a direct way the local principal axes that are actually rotating with a varying rate along the fiber length. Instead, the

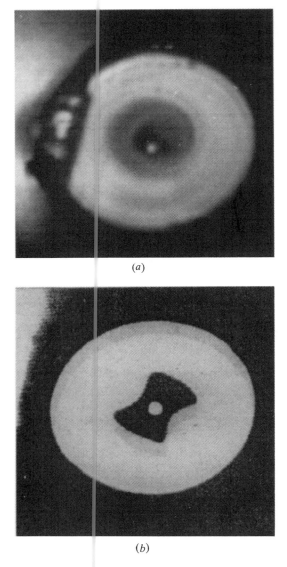

(a)

(b)

FIGURE 7.9   Microscope photographs of a prototype PPT fiber: (a) fast-spun end. (b) zero-spun end.

PPT's local principal axes are determinable indirectly by experiment on a relevant SOP transform. In this regard, it actually proves more convenient to employ a circular light, rather than the more commonly used linear light, to start with an experiment on PPT. A circular light is not associated with preferred directions that need to be determined. Also there is in general no need to know a priori the phase factor of the circular light. In fact, it is not the absolute phase, but the phase difference, that needs to be considered from the practical viewpoint. A phase-shift factor is associated with the transformed light, but does not affect the attempted SOP transform.

The afore-said SOP transform, which is suitable to start with in the experiment for a PPT, is from circular to linear, where the incident circular light matches the local characteristic mode of the PPT's fast-spun end, such that the transform process is by way of single supermode. Provided the incident light is practically circular, everything just happens that way all at once without any need of measuring adjustment or tuning. The experiment then simultaneously achieves the two tasks, that is, to determine the local principal axes of the variably spun birefringent fiber at its unspun end, and to test whether the theoretically predicted circular to (principal-axis-aligned) linear SOP transform really takes place in the actual PPT.

If instead we place the fiber with an unspun input, then the SOP transform by single supermode takes place in the form of (principal-axis-aligned) linear to circular. This therefore requires a linear light accurately matching either of the local principal axes at the input. Since these principal axes are not known a priori, the experiment will have to be performed in such a way that, while rotating the bulk-optic half-wave plate on the input side so as to vary the orientation of the incident linear light, the data of the analyzer on the output side is observed correspondingly until a circular light occurs at the output end. The specific orientation of the incident linear light yielding the circular light at the output is then one of the principal axes to be determined. In comparison, the prior-said experiment employing a circular light to excite the fast-spun end is much simpler to start with.

Shown in Figure 7.10 are the initial experimental data taken early in my simple laboratory. The curves plotted by an x-y recorder show the circular-to-linear SOP transform occurring in an early prototype PPT specimen. The input is a circular light (C in the figure), and the output is a linear light (L).

The two curves C and L in Figure 7.10 were drawn one by one by a single recorder pen. They involve some spikes and dips that do not seem to be simply the noise. One basic source of error is attributable to the mechanical imperfections in the homemade bulk-optic elements in the measuring setup. With all its faults, the initial experimental data in the form of these curves did serve to verify the theoretical prediction.

Experiments with Incident Light Not Matching the PPT's Input

An incident light not matching the local characteristic mode of the PPT fiber at its input will excite both supermodes, resulting in a variety of different SOP transforms. In this experiment, the same PPT specimen was employed as before, but with a linear light exciting its fast-spun end. According to Section 7.6, all of the possible SOP transforms via this dual-supermode process obey the amplitude relation (7.34) and the phase relation (7.35).

In order to test the validity of these fundamental equations, we performed two kinds of experiments, the first of which relates to precisely defined SOP

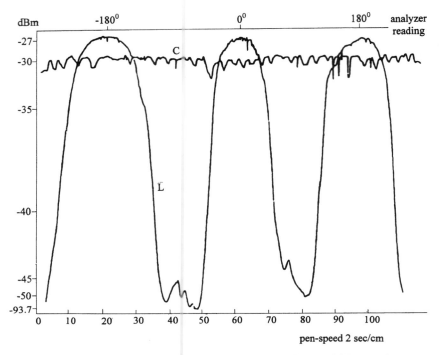

FIGURE 7.10 Experimental curve showing circular-to-linear SOP transform using a PPT.

transforms at *specific* values of θ (orientation of incident linear light), and the second relates to *random* (unpredictable) variation of θ. The recorded data of the first kind of experiment are presented here first. For clearness, the second kind of experiment will be discussed separately under a different subtitle, which deals with the very enchanting problem of random orientation of incident linear light and the "equal power division" property at the output of a PPT.

Consider the afore-said first kind of problem. If the incident orientation θ takes the specific values ρ and $\rho + 45°$, then according to Eqs. (7.34) and (7.35), we should be able to observe a circular light and a linear light, respectively, at the PPT's output. To perform this experiment, it is actually not necessary to know a priori the exact value of ρ (global structural parameter). Instead, all we need are the changes in the angular settings of certain measuring elements. Thus, we rotate the bulk-optic half-wave plate on the input side such that the orientation of the incident linear light is changed. If the rotational angle of the half-wave plate is $(\Delta\vartheta_{\lambda/2})$, light passing the plate will undergo a change of orientation equal to $\Delta\theta = 2(\Delta\vartheta_{\lambda/2})$. Since the phase-difference ξ contains the term 2θ, according to Eq. (7.35), rotating the half-wave plate by $(\Delta\vartheta_{\lambda/2})$ will cause a phase-difference change in the output light by $\Delta\xi = 4(\Delta\vartheta_{\lambda/2})$. With this simple relation in mind, it is

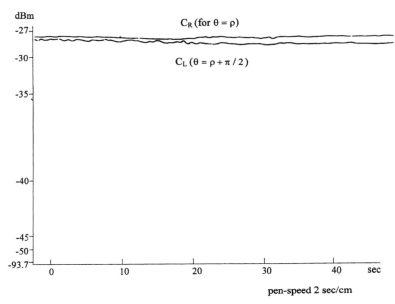

FIGURE 7.11 Experimental curve showing right and left circular light at the output due to specifically oriented linear light incident at the fast-spun end of the PPT.

actually very easy to realize the SOP transforms concerned in a few steps of successive tuning. In our early experiment, we first turned the half-wave plate in a cut-and-try manner until the analyzed output light becomes linear (distinctive "dark-and-bright"). This corresponds to an incident orientation angle of $\theta = \theta_l = \rho + 45°$. For convenience, we marked the angular reading of the half-wave plate corresponding to this linear output light as the "reference setting" in this experiment.

Then we turned the half-wave plate *precisely* by $+22.5°$ with respect to the specified reference setting (such that $\Delta\xi = 90°$), and let the x-y recorder draw a curve for the output light (power). This is shown as the curve R (i.e., C_R in Figure 7.11). And then, again with respect to the reference setting, we turned the half-wave plate *precisely* by $-22.5°$ (such that $\Delta\xi = -90°$), and let the recorder draw a second curve, shown as the curve L (i.e., C_L in Figure 7.11). Both curves are fairly flat, and almost at the same level. With the aid of a bulk-optic quarter-wave plate, curves R and L were affirmed to be circular SOPs of opposite senses. Note also that the change in the half-wave plate setting from $+22.5°$ to $-22.5°$ caused the orientation angle of the incident linear light to change by $90°$, indicating that the input linear SOPs of light, which are transformed into R and L at the output, are orthogonal in space. These experimental data provide a strong support to the reliability of the analytic relations in Eqs. (7.34) and (7.35), as well as the theoretical predictions derived therefrom.

While knowledge of the value of the ρ-parameter was not needed in the above-described SOP-transform experiments, we have investigated how to determine it experimentally. It was shown previously that linear-to-circular SOP transform implies the $\theta = \rho$ condition. But this does not mean that the value of ρ can be determined by the setting of the half-wave plate that yields a circular light at the output, inasmuch as in our theoretical framework the angular relation $\theta = \rho$ refers to the local principal axes at the *fast-spun* input whose location is yet unknown. The reference setting will not help determine this location. Under this circumstance, we attempted an experimental artifice by including a beam splitter and three plane mirrors in the measuring setup such that two beams of light are available from the same source. A bulk-optic polarizer P is placed behind the splitter. With the straight-going beam of linear SOP exciting the fast-spun end (the folding beam being screened), the setting of P that yields a circular light (say, right circular R) at the output unspun end is marked as ϑ_1 (implying $\theta = \rho$). Then the folding beam, being tuned up by the measuring bulk optics to become circular and of the same handedness as R, is led to excite the unspun end (the straight-going beam being screened) such that a linear light appears at the fast-spun end, according to Section 7.6. This linear light is analyzed by P (now acting as analyzer) at a setting marked as ϑ_2 (implying $\theta = -\rho$). The setting mid-way between ϑ_1 and ϑ_2 is the location of the principal axis x at the fast-spun end. Having found this location, it is simple and direct to determine the value of the PPT's global structural parameter ($\rho = |\vartheta_2 - \vartheta_1|/2$). We employed this experimental strategy in our laboratory, and proved its applicability and usefulness.

The "Equal Power Division" Property of PPT

This section deals with the experimental study of the PPT's behavior when a randomly oriented linear light is incident onto its fast-spun end. In our early experiment, the random orientation of the incident linear light was simply simulated by free-hand rotation of the half-wave plate, disorderly but continuously for some tens of revolutions. At the output (unspun end) of the tested PPT, the light power was detected along the two local principal axes (x, y) whose orthogonal locations had been determined in the simple preliminary experiment described in the first experiment.

While free-hand rotating the half-wave plate on the input side, the analyzer on the output side was set at a local principal axis, say x, and the x-y recorder traced a curve for P_x. The analyzer was then rotated by 90° either-wise, such that its angular setting then referred to the local principal axis y. Free-hand rotation of the input half-wave plate then yielded a second curve for P_y. In our early repetitive experiments, the x-y recorder curves for P_x and P_y never failed to nearly overlap. This is what should actually happen, according to the prediction by Eq. (7.34).

To get rid of the overlapping of curves for a clearer display of the experimental results, we changed the way of recorder-plotting, such that the two curves do not overlap, but one follows the other. Thus, after the plotting of the curve P_x is completed within the range $X \cdots X'$, the recorder's pen keeps going and the analyzer setting is turned without appreciable delay from x to y. The half-wave plate is then free-hand rotated again to produce the curve P_y within the range $Y \cdots Y'$ on the same drawing paper. As shown in Figures 7.12a and 7.12b, the curves P_x and P_y within the ranges $X \cdots X'$ and $Y \cdots Y'$, respectively, well confirm the analytically predicted equal power division property of the PPT.

In-between the two successive power curves P_x and P_y (i.e., in the interval $X'-Y$ in both Figures 7.12a and 7.12b), since the recorder's pen did continue going, a trace was left during the interval from X' and Y, which is generally valley-shaped, but seldom repeats in form. Shown in the middle of Figures 7.12a and 7.12b are the two extreme cases, of which the former is essentially flat and the latter is a deep valley. In the initial repetitive experiments of the kind ($\sim$ 1990), such an occurrence (the trace in the interval $X'-Y$) was beyond the preliminary design framework for the experiment, and hence puzzled us at first. A careful inspection of the overall measuring procedures then yielded the explanation. Interestingly enough, the valley-shaped curves

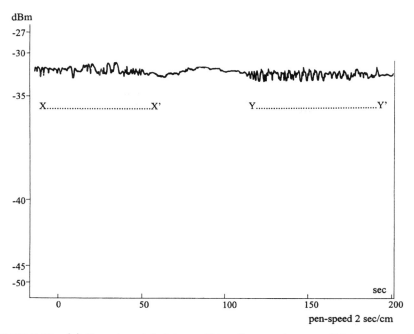

FIGURE 7.12 (a) Experimental data verifying the equal-power-division property of the PPT for random incident orientation (in $X'-Y$, $\theta \approx \rho$).

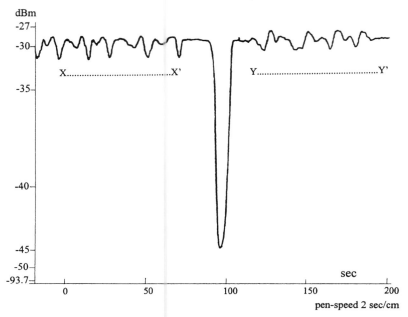

FIGURE 7.12   (*b*) Same as (*a*), except that in $X'-Y$, $\theta \approx \rho + 45°$.

are actually not ruleless, but are also governed by the fundamental equations (7.34) and (7.35), just like the curves P_x, P_y.

In examining the $X'-Y$ interval, we see that the SOP of the output light is always dependent on the orientation of the linear light incident at the fast-spun input. We then recall that, when the plotting of P_x was completed at X', the half-wave plate stopped rotating while the analyzer was quickly turned by 90° (from x to y). The temporarily stationary setting of the half-wave plate (during $X'-Y$) defined a specific input orientation angle θ. Since in this $X'-Y$ range the recorder's pen was still going, it left a trace. If during $X'-Y$, θ happens near to ρ, an essentially flat curve will be displayed, descriptive of a nearly circular light at the output as shown in Figure 7.12*a*. If during $X'-Y$, θ happens to be around $\rho + 45°$, what will be traced is a deep valley descriptive of a nearly linear light at the output as shown in Figure 7.12*b*. More frequently, θ during $X'-Y$ is intermediate between the said two limits, for which a moderately sunken curve between $X'-Y$ (not included in the text) is descriptive of an elliptical SOP. Thus, the display of different valley-shaped curves in the interval $X'-Y$ is no more mysterious, but intriguingly enough, is also governed by Eqs. (7.34) and (7.35).

It thus transpires that the three-curve plot in either Figure 7.12*a* or 7.12*b* actually involves the SOP transform of two kinds: the middle curve displays the particular SOP transform at a specific value of θ, and the outer curves for P_x and P_y display the behavior of the PPT with a randomly varying θ at its input. While the outer curves (P_x and P_y) emphatically show the "equal power division" property, this property is implied also in the middle specific SOP transform.

Experimental Confirmation of Wide-Band Property of PPT

As described in Section 7.5, the wide-band characteristic of the PPT is one unique inherence in favor of this fiber device. Because of its fundamental importance, we were prompted to confirm this property of the PPT experimentally in our laboratory. In this experiment, we use the same PPT specimen whose experimental data in the previous figures were all obtained by using the *visible* 0.6328-μm wavelength of the He-Ne laser. With this same PPT, we now measure it with a laser diode a (LD) source, which emits an *invisible* 0.83-μm light of a few MW power. This 0.83-μm wavelength, supplementing the previous 0.6328-μm wavelength, is suitable as the second wavelength for a test of the wide-band property, with its normalized frequency (the v-value) still not too small while ensuring single-mode operation in the nominal sense.

The optical bench has a comparatively small dimension, consisting simply of a 0.83-μm LD, a bulk-optic quarter-wave plate at this wavelength, a Glan-Thompson analyzer, and a detector plus an x-y recorder. At this new wavelength, we attempted the experiment of circular-to-linear SOP transform. The LD and the ($\lambda/4$) plate (both homemade, not of high quality) yield a nearly circular light whose circularity is not very high, but still serves our immediate purpose. According to the theoretical prediction, circular-to-linear SOP transform via the PPT occurs right away without any need of measuring adjustment or tuning, provided only that the incident light at the fast-spun input is admissibly circular. With the nearly circular light incident onto the PPT, the x-y recorder draws an output curve that is basically linear, as shown in Figure 7.13. We can see the invisible light of 0.83-μm with the help of a piece of "image-viewing" sheet. It is, indeed, fascinating to "see" an ever "bright" invisible light at the input transforming through the PPT to become a "bright and dark" invisible light at the output.

The photos and experimental curves, Figures 7.9–7.12, were taken in the early stage of the research ($\sim$ 1990). The experimental curves shown in Figure 7.13 were not taken until 1996 on the same PPT specimen that was used for the earlier figures. The experimental data are repetitive and fairly precise, well within the range of instrumentation and system errors. During the experiment, my associates and I were amazed to see the variety of predicted phenomena exactly manifesting themselves in the laboratory. We also felt the charm of mathematics whose intrinsic logic puts the multitude of seemingly disordered fiber-optic properties in an order all agreeing and all natural.

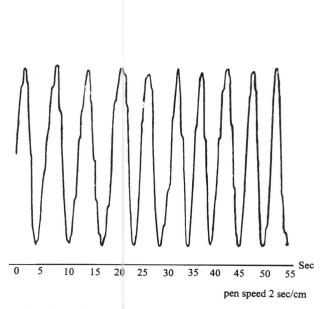

pen speed 2 sec/cm

FIGURE 7.13 Experimental confirmation of wide-band characteristics of PPT (basic properties at 0.83 μm are the same as at 0.6328 μm). Top curve: input light nearly circular (I_{max} = 2.16 MW, I_{min} = 1.92 MW). Bottom curve: output light basically linear ($I_{max} \approx$ 224 μW, $I_{min} \approx$ 12 μW). Scale of ordinate not calibrated.

7.9 TRANSFER MATRIX REPRESENTATION OF PPTs

In the preceding sections we treated the PPT transform characteristics as different initial-value problems. The asymptotic solutions descriptive of a variety of PPT problems were easy to derive with the aid of simple matrix algebra. Each solution is a *particular* solution of the relevant initial-value problem, which satisfies not only the coupled-mode equations but also the initial condition (i.e., the condition of the input light).

The solution that satisfies the coupled-mode equations alone, and not under the restriction of the initial condition, is the *general* solution. As is known, the general solution is descriptive of the waveguiding characteristics of fiber, regardless of the input light. In the previous chapters for uniform fibers, we derived the relevant general solutions by simple asymptotic approximations, and put them in matrix form called the "transfer matrix." A list of the transfer matrices was given in Table 6.1 as a summary for the varieties of *uniform* fibers treated.

Now, for *nonuniform* fiber treated in this chapter, we naturally wish to see whether the previous transfer-matrix approach for uniform fibers can be extended to apply to PPTs whose structures vary in the transmission direction z.

Transfer Matrix for Quarter-Wave Plate PPT

With the aid of asymptotic approximations, it is easy to derive the generalized transfer matrix suitable to variably spun PPT fibers. Consider a fiber-optic quarter-wave plate with a fast-spun input and a zero-spun output. By Eq. (7.24a), the transfer matrix is given by

$$\mathbf{T}_{\lambda/4} = [\mathbf{O}_0(L)][\tilde{\Lambda}][\mathbf{O}_F^{-1}(0)] \tag{7.49}$$

where

$$\mathbf{O}_F^{-1}(0) = \frac{1}{\sqrt{2}}\begin{bmatrix} 1 & -j \\ -j & 1 \end{bmatrix} \tag{7.50}$$

$$\tilde{\Lambda} = \begin{bmatrix} e^{j\rho} & 0 \\ 0 & e^{-j\rho} \end{bmatrix} \tag{7.51}$$

$$\rho = \int_0^L \pi(1 + 4Q^2)^{1/2}\, dz$$

$$\mathbf{O}_0(L) = \begin{bmatrix} 1 & 0 \\ 0 & 1 \end{bmatrix} \tag{7.52}$$

The transfer matrix of the quarter-wave plate PPT is thus given by

$$\mathbf{T}_{\lambda/4} = \frac{1}{\sqrt{2}}\begin{bmatrix} e^{j\rho} & -je^{j\rho} \\ -je^{-j\rho} & e^{-j\rho} \end{bmatrix} \tag{7.53}$$

Here we again emphasize the concept that this transfer matrix for the quarter-wave plate PPT is descriptive of the fiber structure itself, regardless of the input light.

Transfer-Matrix Approach to Quarter-Wave Plate PPT Behavior

Let a right circular light be incident onto this PPT, then the output becomes an x-aligned linear light by Eq. (7.53), that is,

$$\mathbf{A}(0) = \frac{1}{\sqrt{2}}\begin{bmatrix} 1 \\ j \end{bmatrix} \tag{7.54a}$$

$$\mathbf{A}(L) = \mathbf{T}_{\lambda/4}\mathbf{A}(0)$$

$$= \begin{bmatrix} 1 \\ 0 \end{bmatrix} e^{j\rho} \tag{7.54b}$$

Let a θ-inclined linear light be incident on the same PPT. The result is

$$\mathbf{A}(0) = \begin{bmatrix} \cos\theta \\ \sin\theta \end{bmatrix} \tag{7.55a}$$

$$\mathbf{A}(L) = \mathbf{T}_{\lambda/4}\mathbf{A}(0)$$

$$= \begin{bmatrix} e^{j(\rho-\theta)} \\ -je^{-j(\rho-\theta)} \end{bmatrix} \tag{7.55b}$$

The above examples agree as a matter of course with the previous results derived by adopting the approach of step-by-step matrix algebra. The transfer-matrix approach is straightforward and simpler in derivation, but this simplicity is gained at the price that the underlying physical meaning is concealed by the simple mathematics.

Transfer Matrix for Fiber-Optic Half-Wave Plate PPT

As shown in Figure 7.7 a regular version of the half-wave plate PPT is made by spinning the fiber first from fast to zero, and then from zero to fast in the reversed sense. Thus a half-wave plate PPT can be viewed in two different ways. One is to view the half-wave plate PPT as two cascaded quarter-wave plate PPTs. The other is to view the half-wave plate PPT as an integral whole.

In the first view, we need to have the two transfer matrices for the two quarter-wave plates with opposite senses of spinning. Let the first quarter-wave plate of length L_1 be the one treated above, whose transfer matrix is given by Eq. (7.53), that is,

$$\mathbf{T}_{\lambda/4}^{+} = \frac{1}{\sqrt{2}} \begin{bmatrix} e^{j\rho_1} & -je^{j\rho_1} \\ -je^{-j\rho_1} & e^{-j\rho_1} \end{bmatrix}$$

$$\rho_1 = \int_0^{L_1} \pi(1 + 4Q^2)^{1/2}\, dz \tag{7.56}$$

The second quarter-wave plate of length L_2 is oppositely spun, with a zero-spun input and a reversed fast-spun output, such that

$$\mathbf{O}_0(0) = \mathbf{O}_0^{-1}(0) = \begin{bmatrix} 1 & 0 \\ 0 & 1 \end{bmatrix} \tag{7.57}$$

$$\tilde{\Lambda} = \begin{bmatrix} e^{j\rho_2} & 0 \\ 0 & e^{-j\rho_2} \end{bmatrix}$$

$$\rho_2 = \int_0^{L_2} \pi(1 + 4Q^2)^{1/2}\, dz \tag{7.58}$$

$$\mathbf{O}_F(L_2) = \frac{1}{\sqrt{2}} \begin{bmatrix} 1 & -j \\ -j & 1 \end{bmatrix} \tag{7.59}$$

Substitution of the above expressions into Eq. (7.24a) yields

$$T_{\lambda/4}^- = \frac{1}{\sqrt{2}} \begin{bmatrix} e^{j\rho_2} & -je^{-j\rho_2} \\ -je^{j\rho_2} & e^{-j\rho_2} \end{bmatrix} \tag{7.60}$$

The transfer matrix of the fiber-optic half-wave plate shown in Figure 7.7 is therefore given by

$$T_{\lambda/2} = T_{\lambda/4}^{(-)} T_{\lambda/4}^{(+)}$$

$$= j \begin{bmatrix} \sin \rho & -\cos \rho \\ -\cos \rho & -\sin \rho \end{bmatrix} \tag{7.61}$$

where

$$\rho = \rho_1 + \rho_2$$

$$= \int_0^{L_1+L_2} \pi (1 + 4Q^2)^{1/2} \, dz \tag{7.61a}$$

In the second view, wherein a half-wave plate PPT is considered as an integral whole, using Eq. (7.24a) yields

$$T_{\lambda/2} = O_2 \tilde{\Lambda} O_1^{-1} \tag{7.62}$$

$$O_1^{-1} \equiv O_{+F}^{-1} = \frac{1}{\sqrt{2}} \begin{bmatrix} 1 & -j \\ -j & 1 \end{bmatrix}$$

$$\tilde{\Lambda} = \tilde{\Lambda}_2 \tilde{\Lambda}_1 = \begin{bmatrix} e^{j\rho_2} & 0 \\ 0 & e^{-j\rho_2} \end{bmatrix} \begin{bmatrix} e^{j\rho_1} & 0 \\ 0 & e^{-j\rho_1} \end{bmatrix} \tag{7.63}$$

$$O_2 \equiv O_{-F} = \frac{1}{\sqrt{2}} \begin{bmatrix} 1 & -j \\ -j & 1 \end{bmatrix}$$

where the subscripts $\pm$ refer to the senses of spin, and the subscript F means fast-spin. Substitution of Eqs. (7.63) into Eq. (7.62) yields the same results given by Eq. (7.61):

$$T_{\lambda/2} = O_{-F} \tilde{\Lambda}_2 \tilde{\Lambda}_1 O_{+F}^{-1}$$

$$= j \begin{bmatrix} \sin \rho & -\cos \rho \\ -\cos \rho & -\sin \rho \end{bmatrix} \tag{7.64}$$

where $\rho = \rho_1 + \rho_2$.

Transfer Matrix for Fiber-Optic Full-Wave Plate PPT

With reference to Figure 7.8, the transfer matrix of a fiber-optic full-wave plate PPT can be derived from Eq. (7.24a):

$$
\mathbf{T}_\lambda = \mathbf{O}_{+F} \tilde{\Lambda}_2 \tilde{\Lambda}_1 \mathbf{O}_{+F}^{-1}
$$

$$
= \begin{bmatrix} \cos\rho & \sin\rho \\ -\sin\rho & \cos\rho \end{bmatrix}
\tag{7.65}
$$

where $\rho = \rho_1 + \rho_2$. This is recognized as a rotation matrix with a rotation angle of ρ. The preceding results descriptive of the local transfer matrixes for different PPTs are listed in Table 7.3.

According the previously derived results given by Eqs. (7.53), (7.60), (7.64), and (7.65), the PPT's behavior can be readily described in terms of retardation and rotation with respect to the local coordinates. For succinctness, we shall not go further into detailed derivations. For ready reference, the relevant end results listed in the right column of Table 7.3 will serve the purpose.

TABLE 7.3   Transform Behaviors of the PPT Varieties

	Transfer Matrix in Local Coordinates (Asymptotic)	Equivalent Retarder–Rotator in Local Coordinates
$\lambda/4$ wave plate (fast $\to$ zero PPT)	$\dfrac{1}{\sqrt{2}} \begin{bmatrix} e^{j\rho} & -je^{j\rho} \\ -je^{-j\rho} & e^{-j\rho} \end{bmatrix}$	$R = \pi/2$ $\Omega = -\rho$ $\Phi = \rho - \pi/4$
$\lambda/4$ wave plate (zero $\to$ fast PPT)	$\dfrac{1}{\sqrt{2}} \begin{bmatrix} e^{j\rho} & -je^{-j\rho} \\ -je^{j\rho} & e^{-j\rho} \end{bmatrix}$	$R = \pi/2$ $\Omega = \rho$ $\Phi = -\pi/4$
half-wave plate	$j \begin{bmatrix} \sin\rho & -\cos\rho \\ -\cos\rho & -\sin\rho \end{bmatrix}$	$R = \pi$ $\Omega + 2\Phi = \rho - \pi/2$
full-wave plate	$\begin{bmatrix} \cos\rho & \sin\rho \\ -\sin\rho & \cos\rho \end{bmatrix}$	$R = 2\pi$ $\Omega = \pi - \rho$

7.10   ESTIMATION OF ERRORS

Input Fast-Spun

From the foregoing derivations, the transfer matrix of a fiber-optic quarter-wave plate PPT is given by

$$\mathbf{T}_{\lambda/4}^{+} = \mathbf{O}_0 \tilde{\Lambda} \mathbf{O}_F^{-1} = \tilde{\Lambda} \mathbf{O}_F^{-1}$$

$$\approx \begin{bmatrix} e^{j\rho} & 0 \\ 0 & e^{-j\rho} \end{bmatrix} \begin{bmatrix} \cos\phi & -j\sin\phi \\ -j\sin\phi & \cos\phi \end{bmatrix} \tag{7.66}$$

where the superscript $+$ specifies a $\lambda/4$ PPT whose spin rate is from fast to zero (said to be in the $+$ direction), and $\mathbf{O}_0$ is a unit matrix. We recall that the previous asymptotic expression, Eq. (7.53), is derived on the assumption that the ratio $\tau/(\delta\beta)$ is indefinitely large, such that $\phi \rightarrow 45°$. In the actual case, ϕ can be made close to 45°, but not equal to 45°. We then put ϕ as

$$\phi = 45° - \bar{\varepsilon} \tag{7.67}$$

where $\bar{\varepsilon}$ is a small quantity, such that

$$\mathbf{O}_F^{-1} = \begin{bmatrix} \cos\phi & -j\sin\phi \\ -j\sin\phi & \cos\phi \end{bmatrix}$$

$$\approx \frac{1}{\sqrt{2}} \begin{bmatrix} 1 & -j \\ -j & 1 \end{bmatrix} + \frac{\bar{\varepsilon}}{\sqrt{2}} \begin{bmatrix} 1 & j \\ j & 1 \end{bmatrix} \tag{7.68}$$

Substituting Eq. (7.68) into Eq. (7.66) yields

$$\bar{\mathbf{T}}_{\lambda/4}^{+}(\phi = 45° - \bar{\varepsilon}) = \mathbf{T}_{\lambda/4}^{+}(\phi = 45°) + \delta\mathbf{T}_{\lambda/4}^{+} \tag{7.69}$$

where the first term on the right side is the asymptotic expression of the transfer matrix of a fast-zero $\lambda/4$ PPT (implying the approximation $\phi = 45°$)

$$\mathbf{T}_{\lambda/4}^{+} = \frac{1}{\sqrt{2}} \begin{bmatrix} e^{j\rho} & 0 \\ 0 & e^{-j\rho} \end{bmatrix} \begin{bmatrix} 1 & -j \\ -j & 1 \end{bmatrix} \tag{7.70}$$

and the second matrix represents the error term due to the deviation of ϕ from 45°:

$$\delta\mathbf{T}_{\lambda/4}^{+} \approx \frac{\bar{\varepsilon}}{\sqrt{2}} \begin{bmatrix} e^{j\rho} & 0 \\ 0 & e^{-j\rho} \end{bmatrix} \begin{bmatrix} 1 & j \\ j & 1 \end{bmatrix} \tag{7.71}$$

A circular-to-linear SOP transformation by a $\lambda/4$ PPT therefore involves an error given by

$$\frac{P_{max}}{P_{min}} \approx \frac{1}{\bar{\varepsilon}^2} \approx (4Q)^2 \tag{7.72}$$

where $Q \equiv Q_M = \tau_M/\delta\beta$ is the maximum Q-factor that is proportional to the maximum spin rate of the PPT.

Input Zero-Spun

The transfer matrix is given by

$$\bar{T}_{\lambda/4}^-(\phi = 45° - \bar{\varepsilon}) = \mathbf{O}_F \tilde{\Lambda} \mathbf{O}_0 = \mathbf{O}_F \tilde{\Lambda}$$
$$= T_{\lambda/4}^-(\phi = 45°) + \delta T_{\lambda/4}^- \tag{7.73}$$

where the superscript $-$ specifies a $\lambda/4$ PPT whose spin rate is from zero to fast (said to be in the $-$ direction), and

$$T_{\lambda/4}^- = \frac{1}{\sqrt{2}} \begin{bmatrix} 1 & j \\ j & 1 \end{bmatrix} \begin{bmatrix} e^{j\rho} & 0 \\ 0 & e^{-j\rho} \end{bmatrix}$$
$$= \frac{1}{\sqrt{2}} \begin{bmatrix} e^{j\rho} & je^{-j\rho} \\ je^{j\rho} & e^{-j\rho} \end{bmatrix} \tag{7.74}$$

$$\delta T_{\lambda/4}^- = \frac{\bar{\varepsilon}}{\sqrt{2}} \begin{bmatrix} 1 & -j \\ -j & 1 \end{bmatrix} \begin{bmatrix} e^{j\rho} & 0 \\ 0 & e^{-j\rho} \end{bmatrix} \tag{7.75}$$

For linear-circular SOP transformation, the noncircularity of the circular light is expected to be

$$\frac{P_{max}}{P_{min}} \approx \frac{(1 + \bar{\varepsilon})^2}{(1 - \bar{\varepsilon})^2}$$
$$\approx 1 + \frac{1}{Q} \tag{7.76}$$

Comparing Eq. (7.76) with Eq. (7.72), we see that in the asymptotic approximation the error for linear-to-circular transformation is Q^{-1} by order, while the error for circular-to-linear transformation is $(4Q)^{-2}$ by order. Under the actual condition $Q \gg 1$, therefore, the transformation from linear to circular implies a larger error than the transformation from circular to linear, with Q assumed to be the same in both cases.

Fiber-Optic $\lambda/2$ PPT

As shown in Figure 7.7, the transfer matrix of the $\lambda/2$ PPT is given by

$$\overline{\mathbf{T}}_{\lambda/2} = \mathbf{O}_2 \tilde{\Lambda} \mathbf{O}_1^{-1}$$

$$= \begin{bmatrix} \cos \phi_2 & -j \sin \phi_2 \\ -j \sin \phi_2 & \cos \phi_2 \end{bmatrix} \begin{bmatrix} e^{j\rho} & 0 \\ 0 & e^{-j\rho} \end{bmatrix} \begin{bmatrix} \cos \phi_1 & -j \sin \phi_1 \\ -j \sin \phi_1 & \cos \phi_1 \end{bmatrix} \quad (7.77)$$

where $\phi_1 = 45° - \bar{\varepsilon}_1$, $\phi_2 = 45° - \bar{\varepsilon}_2$, and $\rho = \rho_1 + \rho_2$. Since $\bar{\varepsilon}_1$ and $\bar{\varepsilon}_2$ are small quantities, we expand $\mathbf{O}_1^{-1}$ and $\mathbf{O}_2$ in the neighborhood of 45° to yield

$$\begin{bmatrix} \cos \phi_1 & -j \sin \phi_1 \\ -j \sin \phi_1 & \cos \phi_1 \end{bmatrix} \approx \frac{1}{\sqrt{2}} \begin{bmatrix} 1 & -j \\ -j & 1 \end{bmatrix} + \frac{\bar{\varepsilon}_1}{\sqrt{2}} \begin{bmatrix} 1 & j \\ j & 1 \end{bmatrix} \quad (7.78a)$$

$$\begin{bmatrix} \cos \phi_2 & -j \sin \phi_2 \\ -j\sin \phi_2 & \cos \phi_2 \end{bmatrix} \approx \frac{1}{\sqrt{2}} \begin{bmatrix} 1 & -j \\ -j & 1 \end{bmatrix} + \frac{\bar{\varepsilon}_2}{\sqrt{2}} \begin{bmatrix} 1 & j \\ j & 1 \end{bmatrix} \quad (7.78b)$$

Substituting Eqs. (7.78a) and (7.78b) into Eq. (7.77) yields

$$\overline{\mathbf{T}}_{\lambda/2} = \mathbf{T}_{\lambda/2} + \delta \mathbf{T}_{\lambda/2} \quad (7.79)$$

where the first term on the right is the asymptotic expression with $\phi = 45°$:

$$\mathbf{T}_{\lambda/2} = \frac{1}{2} \begin{bmatrix} 1 & -j \\ -j & 1 \end{bmatrix} \begin{bmatrix} e^{j\rho} & 0 \\ 0 & e^{-j\rho} \end{bmatrix} \begin{bmatrix} 1 & -j \\ -j & 1 \end{bmatrix} \quad (7.80)$$

and the second term on the right of Eq. (7.79) is the error

$$\delta \mathbf{T}_{\lambda/2} \approx \bar{\varepsilon}_2 \mathbf{R}(\rho) + \bar{\varepsilon}_1 \mathbf{R}(-\rho) \quad (7.81)$$

where $\mathbf{R}$ represents a rotation matrix given by

$$\mathbf{R}(\pm \rho) \equiv \begin{bmatrix} \cos \rho & \pm \sin \rho \\ \pm \sin \rho & \cos \rho \end{bmatrix} \quad (7.82)$$

If letting $\bar{\varepsilon}_1 = \bar{\varepsilon}_2 = \bar{\varepsilon}$, Eq. (7.81) reduces to

$$\delta \mathbf{T}_{\lambda/2} \approx 2\bar{\varepsilon} \cos \rho \begin{bmatrix} 1 & 0 \\ 0 & 1 \end{bmatrix}$$

$$\approx (2Q)^{-1} \cos \rho \begin{bmatrix} 1 & 0 \\ 0 & 1 \end{bmatrix} \quad (7.83)$$

For a fiber-optic half wave plate, the upper bound of the expected error is therefore Q^{-1} by order of magnitude.

Fiber-Optic Full-Wave Plate

With reference to Figure 7.8, the transfer matrix of a full-wave plate PPT is given by

$$
\begin{aligned}
\overline{\mathbf{T}}_\lambda &= \mathbf{O}_2 \tilde{\Lambda} \mathbf{O}_1^{-1} \\
&= \mathbf{T}_\lambda + \delta\mathbf{T}_\lambda \\
&= \begin{bmatrix} \cos\phi_2 & j\sin\phi_2 \\ j\sin\phi_2 & \cos\phi_2 \end{bmatrix} \begin{bmatrix} e^{j\rho} & 0 \\ 0 & e^{-j\rho} \end{bmatrix} \begin{bmatrix} \cos\phi_1 & -j\sin\phi_1 \\ -j\sin\phi_1 & \cos\phi_1 \end{bmatrix}
\end{aligned}
\tag{7.84}
$$

Let $\phi_2 = 45° - \bar{\varepsilon}_2$ and $\phi_1 = 45° - \bar{\varepsilon}_1$, where $\bar{\varepsilon}_2$ and $\bar{\varepsilon}_1$ are small quantities. We can then expand the terms involving ϕ in the neighborhood of 45°, such that

$$
\mathbf{T}_\lambda = \frac{1}{2} \begin{bmatrix} 1 & j \\ j & 1 \end{bmatrix} \begin{bmatrix} e^{j\rho} & 0 \\ 0 & e^{-j\rho} \end{bmatrix} \begin{bmatrix} 1 & -j \\ -j & 1 \end{bmatrix}
\tag{7.85}
$$

$$
\delta\mathbf{T}_\lambda = j\left[\bar{\varepsilon}_1 \mathbf{R}(\rho - 90°) + \bar{\varepsilon}_2 \mathbf{R}(90° - \rho) \right] \begin{bmatrix} 1 & 0 \\ 0 & -1 \end{bmatrix}
\tag{7.86}
$$

Letting $\bar{\varepsilon}_1 = \bar{\varepsilon}_2 = \bar{\varepsilon}$, the above equation reduces to

$$
\delta\mathbf{T}_\lambda \approx (2Q)^{-1}\sin\rho \begin{bmatrix} 1 & 0 \\ 0 & -1 \end{bmatrix}
\tag{7.87}
$$

Thus, in the case of a fiber-optic full-wave plate, the upper bound of the error is again Q^{-1} by order of magnitude.

7.11   EXAMPLES OF COMPUTER CALCULATION OF PPT

As in Section 7.3, let the normalized spin-rate function $Q = (L_b/2\pi)\tau$ of a quarter-wave PPT of length L be described by

$$
Q(z) = Q_M[0.5 \pm 0.5\cos(\pi z/N)]^\gamma
\tag{7.88}
$$

where $N(= L/L_b)$; $z(= z/L_b)$; the normalized parameters, Q_M, are all adimensional; and γ is a numeral. For the $\pm$ signs, the upper sign $(+)$ is for a PPT fiber spun from fast to zero, while the lower sign $(-)$ is for a PPT fiber spun from zero to fast.

Computer Data of a Fast-to-Zero PPT

The $+$ sign is used in Eq. (7.88) to describe the spin rate Q of a fast-to-zero PPT. First, we give a typical example to illustrate the numerical data. More detailed data are tabulated to show the consequences of changing the value of any of the parameters.

Given the PPT's structural parameters:

$$Q_M = 10, \qquad N = 100, \qquad \gamma = 2$$

Given the input light (initial condition or excitation condition):

$$A_x(0) = 1/\sqrt{2}, \qquad A_y(0) = j/\sqrt{2}$$

Computer data for the transformed light at the output $z = N = 100$ are

$$A_x(100) = 0.2144 - j0.976 = 0.9998 e^{-j77.62°}$$

$$A_y(100) = -0.0252 + j0.0024 = 0.0254 e^{-j5.421°}$$

Since $|A_y| \ll |A_x|$ at $z = 100$, the resultant linear light is essentially given by $A_x(100) \approx 1 \times e^{j\rho}$, where $\rho \approx -77.62°$ (see Section 7.4).

For a linear light output, we are interested in the extinction ratio. For the present example, this figure is

$$\eta = \frac{|A_x(100)|^2}{|A_y(100)|^2} \approx \frac{0.9996}{0.0006} \qquad (\approx 32 \text{ dB})$$

Numerical data for other sets of parameters are tabulated in Tables 7.4–7.6.

Tables 7.4–7.6 show that the extinction ratio of the output light is somewhat improved by an increase in Q_M, but little influenced by changes of N or γ.

TABLE 7.4 $Q_M = 10$, $N = 100$; Initial Condition Unchanged; γ Varies

γ	A_x	A_y	η (dB)
0.5	$0.9825 + j0.1836$ $= 0.9995\angle 10.58°$	$0.0290 + j0.0190$ $= 0.0346\angle 33.2°$	≈ 29.2
1	$-0.0691 + j0.9974$ $= 0.9998\angle(-86°)$	$0.0249 + j0.0009$ $= 0.0249\angle 2.19°$	≈ 32
2	Data given in the text		
3	$0.9948 + j0.0991$ $= 0.9997\angle 5.69°$	$0.0014 + j0.0282$ $= 0.0283\angle 87.1°$	≈ 31

TABLE 7.5 $Q_M = 10$, $\gamma = 2$; Initial Condition Unchanged; N Varies

N	A_x	A_y	η (dB)
5	0.8337 + j0.5494 = 0.9984∠33.4°	0.0545 + j0.0203 = 0.0582∠20.4°	≈ 24.7
10	−0.5725 − j0.8198 = 0.9999∠55.1°	−0.0113 − j0.0161 = 0.0197∠54.5°	≈ 34
30	0.8205 − j0.5715 = 0.9999∠(−34.9°)	−0.0166 + j0.0155 = 0.0227∠(−43.2°)	≈ 33
100	Data given in the text		
200	0.9618 + j0.2729 = 0.9998∠15.8°	0.0058 + j0.0261 = 0.0250∠(−29.6°)	≈ 32
500	0.4436 − j0.8960 = 0.9998∠(−63.7°)	−0.0218 + j0.0124 = 0.0250∠(−29.6°)	≈ 32

TABLE 7.6   Effect of Q_M on Extinction Ratio; $N = 100$; $\gamma = 2$

Q_M	A_x	A_y	η (dB)
5	0.9311 + j0.3616 = 0.9988∠21.23°	0.0183 + j0.0471 = 0.0506∠68.74°	η ≈ 26
10	Data given in the text		
20	−0.9505 + j0.3108 ≈ 1∠(−18.1°)	0.0057 − j0.0165 = 0.0175∠(−70.9°)	≈ 35
30	−0.9489 − j0.3158 ≈ 1∠18.4°	−0.0000 − j0.0070 = 0.0070∠89.9°	≈ 43

Computer Data of a Zero-to-Fast PPT

The − sign is used in Eq. (7.88) to describe the spin rate Q of a zero-to-fast PPT. As an example, the structural parameters for this PPT are assumed to be the same as in the case of the fast-to-zero PPT, that is,

$$Q_M = 10, \qquad N = 100, \qquad \gamma = 2$$

But the initial condition is differently assumed. Let the input light be an x-oriented linear light:

$$A_x(0) = 1, \qquad A_y(0) = 0$$

Computer data for the transformed light at the output of the PPT are

$$A_x(100) = 0.3553 - j0.6315 = 0.7246\,e^{-j60.6°}$$

$$A_y(100) = 0.6006 + j0.3381 = 0.6892\,e^{j29.4°}$$

The result shows that, at the output end ($z = 100$) of the PPT, the phase of A_y leads that of A_x by

$$\angle(A_y) - \angle(A_x)$$

$$= 29.4° - (-60.6°) = 90°$$

The total power at $z = 100$ is

$$|A_x(100)|^2 + |A_y(100)|^2 = 0.5250 + 0.4750 = 1$$

It is thus seen that the output light from the fiber-optic PPT element satisfies the qualifications of a right circular light with respect to phase and power relations. The computer data for the phase difference is exactly equal to 90°, and that for the total power is exactly equal to unity, as shown above at $z = 100$. This merely signifies that the phase difference and the total power turn out to be exact within the number of the digital places taken for the numerical calculation by the computer. Should we take more digital places, ± deviations from these exact values will show up in the numerical data.

Nonetheless, the numerical data show a noticeable noncircularity of the output light in view of the appreciable difference between the amplitudes $|A_x|$ and $|A_y|$. At $z = 100$, the noncircularity is

$$\varepsilon = \frac{|A_x|^2 - |A_y|^2}{|A_x|^2 + |A_y|^2}$$

$$= 0.5250 - 0.4750 = 5\%$$

Computer data for other sets of structural parameters are tabulated in Tables 7.7–7.9.

From the preceding computer data it is clear that, if the prescribed normalized spin rate Q_M is kept a constant, a change in the form of the spin-rate function will not be effective to improve the circularity of the transformed light at the output of the PPT. As anticipated, it will be possible to improve the circularity if higher values of Q_M are achievable.

TABLE 7.7 $Q_M = 10$, $N = 100$; Initial Condition Unchanged; γ Varies

γ	A_x	A_y	$\angle(A_y - A_x)$	ε
0.25	$0.6504 - j0.1988$ $= 0.6810\angle(-17°)$	$0.6113 + j0.4047$ $= 0.7331\angle(33.51°)$	$50.51°$	$\approx 7.5\%$
1	Data given in the text			
2	$-0.4439 + j0.5726$ $= 0.7245\angle(127.8°)$	$-0.5447 - j0.4223$ $= 0.6892\angle(217.8°)$	$90°$	5%
3	$-0.6204 - j0.3742$ $= 0.7245\angle(211.1°)$	$0.3559 - j0.5902$ $= 0.6892\angle(301.1°)$	$90°$	5%

TABLE 7.8 $Q_M = 10$, $\gamma = 2$; Initial Condition Unchanged; N Varies

N	A_x	A_y	$\angle(A_y - A_x)$	ε
5	$0.7049 + j0.2249$ $= 0.7399\angle(17.70°)$	$-0.1603 + j0.6534$ $= 0.6728\angle(103.8°)$	$\approx 86.1°$	$\approx 10\%$
10	$-0.4849 + j0.5392$ $= 0.7252\angle(131.7°)$	$-0.5122 - j0.5122$ $= 0.6865\angle(221.9°)$	$\approx 90°$	$\approx 5.173\%$
30	$-0.5387 + j0.4848$ $= 0.7247\angle(222°)$	$0.4607 - j0.5124$ $= 0.6890\angle(312°)$	$90°$	5%
100	Data for A_x and A_y given in the text		$90°$	5%
200	$0.0321 - j0.7239$ $= 0.7246\angle(-87.46°)$	$0.6885 + j0.0305$ $= 0.6892\angle(2.537°)$	$90°$	5%
500	$0.1072 - j0.7166$ $= 0.7245\angle(-81.49°)$	$0.6819 + j0.1001$ $= 0.6892\angle(8.351°)$	$90°$	5%

7.12 NOTES

Poincare-Sphere Representation of SOP

In some order, everything goes in harmony through different approaches. A theory said to be more general must embody the existing theories as its special cases. Section 2.9 briefly discussed this point in connection with the retarder–rotator formulation of a uniformly spun fiber in the fixed coordinates. Meanwhile, if one has ever thought about spun fibers, one will always remember the initial work by Ulrich and Simon [18], who used the Poincare sphere to describe polarization evolutions in fiber optics. Extension of this approach to the variably spun PPT fiber should lead to the same results as we have obtained in this paper. This is, indeed, the case, as the following will show.

TABLE 7.9   Effect of Q_M on ε at Output; $N = 100$, $\gamma = 2$

Q_M	A_x	A_y	$\angle(A_y - A_x)$	ε
1	$0.3307 - j0.7837$ $= 0.8503\angle(-69.12°)$	$0.4844 + j0.2044$ $= 0.5257\angle(22.88°)$	$90°$	$\approx 45\%$
2	$0.1447 - j0.7748$ $= 0.7882\angle(-79.42°)$	$0.6050 + j0.1130$ $= 0.6155\angle(10.58°)$	$90°$	$\approx 24\%$
5	$-0.3587 + j0.6489$ $= 0.6710\angle(208.9°)$	$-0.5872 - j0.3246$ $= 0.7414\angle(118.9°)$	$90°$	10%
10	Data for A_x and A_y given in the text		$90°$	5%
20	$-0.5228 - j0.4891$ $= 0.7159\angle(223.1°)$	$0.4770 - j0.5099$ $= 0.6982\angle(313.1°)$	$90°$	2.5%
30	$-0.3066 + j0.6436$ $= 0.7129\angle(115.5°)$	$-0.6330 - j0.3018$ $= 0.7013\angle(205.5°)$	$90°$	$\approx 1.6\%$

For brevity, we take over the entire formalism, as well as the figures for the Poincare sphere, of the cited paper [18], with only some changes of the symbols that are self-clear. In the local coordinates, the rotation vector $\overline{\omega}° = (\delta\overline{\beta})° + \overline{\tau}°$, where $\overline{\tau}°$ is a function of z. At the fast-spun end, $(\delta\overline{\beta})°$ is negligibly small as compared in magnitude with $(\overline{\tau}°)$, such that $\omega°$ points to either $L°$ or $R°$ (according to the sense of spin). From the fast-spun end to the unspun end, since the vertical vector $\overline{\tau}°$ decreases continuously while the horizontal vector $(\delta\overline{\beta})°$ remains constant, the angle of elevation of $\overline{\omega}°$ changes continuously until the output unspun end is reached where $\overline{\omega}° = (\delta\overline{\beta})°$, pointing to either $H°$ or $V°$, as the case may be. The PPT structure is thus described in the Poincare sphere formulation. Specific SOP transform depends on the kind of incident light. An incident light of circular SOP (left or right) corresponds to the point $L°$ or $R°$. From input to output the trajectory of the SOP evolution will start from $L°$ or $R°$, and trace a quadrant of the large circle $L°H°R°V°$ to reach $H°$ or $V°$ at the output end. On the other hand, if the incident light is linear (azymuthal angle arbitrary), the relevant initial condition corresponds to a point always on the equator of the local Poincare sphere. As light traverses the fiber, while the angle of elevation of the rotation vector $\overline{\omega}°$ decreases continuously from $\overline{\tau}°$ to $(\delta\overline{\beta})°$, the SOP evolution will appear as a continuous turn (with $Q°-P°$ as axis) of the large circle passing $Q°$ and $P°$, with the initial large circle being the equator corresponding to the input linear light of arbitrary azimuthal angle, and with the final large circle (at the output end) to be the one that passes $L°Q°R°P°$ at the output. This is the one and only one specific large circle on the local Poincare sphere, for which the "equal power division" property holds independent of the orientation of the incident linear light. Note that this geometrical construction of the SOP trajectory implies the same two conditions of the PPT as well, i.e., the spin rate at the fast-spun end is sufficiently fast, and the variation of spin rate is sufficiently slow. The former ensures the initial $\bar{\omega}^{\circ}$ pointing to L or R, and the latter ensures that the successions of incremental rotations do approximate an integration.

The Definition of Supermode

In my analytic framework for variably coupled modes, I call the term W_x or W_y the *supermode*, laying more emphasis on the fact that it is *not* an independent field. This terminology can be traced back to my early work on microwave guide transmission. In the study of fiber optics, I initially used the term "super-local mode" in my papers (see, for example, Ref. [19], 1979; Ref. [20], 1980). Later, at two international workshops (1983, 1984), I used the shortened term "super mode" [21–23], which was then further simplified to the single word "supermode."

Here I wish to mention a terminological matter regarding this word. In the relevant literature of fiber and integrated optics, some other writers also used the same word "supermode" but with a different meaning. For example, in Ref. [24], Vasallo treated a composite waveguiding structure consisting of two uniformly coupled waveguides, and derived the independent solutions of such a composite system, which he called the supermodes. As Vasallo put it, his supermodes are "the fields with a global $e^{-j\Gamma z}$ dependence" (see p. 29 of Ref. [24]). The exponent Γ is a constant which can be obtained from the propagation constants of the local modes in the composite system by the method of diagonalization.

In the light of coupled-mode theory, it makes no difference whether the problem under study is a composite system of two coupled waveguides, or it is a single fiber in which two polarization modes coexist and are coupled. What Vasallo calls the "supermodes," or "the fields with a global $e^{-j\Gamma z}$ dependence" are thus seen to be what were previously referred to as the "normal modes" or eigenmodes by our choice of the terminology. In contrast, the supermodes that I have defined in this text are not real normal modes or real eigenmodes, but are quasi normal modes or quasi eigenmodes, that still retain some residual couplings whose strengths are proportional to the derivatives of the perturbations.

REFERENCES

[1] H. C. Huang, "Passive Fiber-Optic Polarization Control," U.S. Patent No. 4,943,132 (1990).

[2] H. C. Huang, "Passive Fiber-Optic Polarization Control Element," U.S. Patent No. 5,096,312 (1992).

[3] H. C. Huang, "Practical Circular-Polarization Maintaining Optical Fiber," U.S. Patent No. 5,452,394 (1995).

[4] H. C. Huang, "Fiber-optic analogs of bulk-optic weave plates," *Appl. Opt.*, vol. 36, pp. 4241–4258 (1997).

[5] H. C. Huang, "Practical circular-polarization maintaining optical fiber," *Appl. Opt.*, vol. 36, pp. 6968–6975 (1997)

[6] H. C. Huang, "Contributions to the theory of coupled modes and nonideal waveguides," *Inter. Symp. uber Fragen der Physik und Tech. bei Hochsten Frequenzen*, Heinrich-Hertz Institut, Akademia der Wissenschaften der DDR (1965).

[7] H. C. Huang, *Coupled Mode and Nonideal Waveguides*, Microwave Research Institute (MRI), Polytechnic of New York, 191 pages (1981).

[8] H. C. Lefevre, "Fiber Optic Polarization Controller," U.S. Patent No. 4,389,090 (1983).

[9] T. Matsumoto and H. Kano, "Fiber Optic Polarization Controller," U.S. Patent No. 4, 793,6789 (1983).

[10] H. J. Shaw, R. C. Youngquist, and J. L. Brooks, "Birefringent Fiber Narrow-Band Polarization Coupler and Method of Coupling Using Same," U.S. Patent No. 4, 801,189 (1989).

[11] P. McIntyre and A. W. Snyder, "Light propagation in twisted anisotropic media: Application to photoreceptors," *J. Opt. Soc. Am.*, vol. 68, pp. 149–157 (1978).

[12] J. S. Cook, "Tapered velocity coupler," *Bell Syst. Tech. J.*, vol. 34, pp. 807–822 (1955).

[13] A. G. Fox, "Wave coupling by warped normal modes," ibid, pp. 823–852 (1955).

[14] W. H. Louisell, "Analysis of the single tapered mode coupler," ibid, pp. 853–870 (1955).

[15] H. C. Huang, "Generalized theory of coupled local modes in multi-wave guides," *Scientia Sinica*, vol. 9, no. 1, pp. 142–154 (1960).

[16] Huang Hung-chia, "Method of slowly varying parameters," *Acta Mathematica Sinica*, vol. 11, pp. 238–247 (1961); also, *Rev. J. Math.* (in Russian), p. 39 (1961).

[17] H. B. Keller and J. B. Keller, "Exponential-like solutions of system of ordinary differential equations," *J. Soc. Ind. Appl. Math.*, vol. 10, no. 2, pp. 246–259 (1962).

[18] R. Ulrich and A. Simon, "Polarization optics of twisted single-mode fibers," *Appl. Opt.*, vol. 18, pp. 2241–2251 (1979).

[19] H. C. Huang, "On local normal modes in optical fiber and film waveguides," *Scientia Sinica*, vol. 22, pp. 1147–1155 (1979).

[20] H. C. Huang, "Weak coupling theory of optical fiber and film waveguides," *Radio Science*, vol. 16, pp. 495–499 (1981).

[21] H. C. Huang and A. W. Snyder (editors), *Optical Waveguide Sciences*, Proc. of International Symposium, Martinus Nijhoff, 360 pages, (1983).

[22] H. C. Huang, "Super mode concept and applications," contributed to the 9*th International Workshop on Optical Waveguide Theory*, Reisensburg, Germany, September 7–10 (1984).

[23] K. Petermann, "Progress in optical waveguide theory," *J. Opt. Commun.*, vol. 5, pp. 153–155 (1984).

[24] C. Vasallo, *Optical Waveguide Concepts*, Book 1 of Optical Wave Sciences and Technology (series editor: Huang Hung-Chia), ELSEVIER, Amsterdam, 322 pages (1991).

SUPPLEMENT: VARIABLY COUPLED-MODE THEORY AND APPLICATION EXAMPLES

Chapter 7 focuses on the treatment of the fiber-optic analogs of bulk-optic wave plates, a class of nonuniformly spun birefringent fibers called the practical polarization transformer (PPT). The mathematical strategy that is adopted for the treatment of the PPT is the theory of *variably* coupled modes. This theory as a mathematical–physical method is fairly general in application, applicable not only to the PPT problems, but also to other problems that involve coupled waves with variable couplings. One example to be dealt with in this supplement is the challenging problem concerning the excitation problem of the spun highly birefringent (hi-bi) fiber discussed in Chapter 6. Also described in this supplement is a new idea for the device architecture for Faraday rotation-effect measurement.

7S.1   Spun Hi-Bi Taper and Excitation of Elliptical Eigenmode

Referring to Eq. (6.3), a spun hi-bi fiber is a kind of elliptically birefringent fiber featured by the restrictive condition:

$$\tau_S \approx \delta\beta \qquad (7S.1)$$

or $L_S \approx L_b$, where τ_S and hence $L_S(= 2\pi/\tau_S)$ are constant, and $\delta\beta$ and L_b, respectively, are the unspun linear birefringence and the unspun beat length.

The existence of (right and left) elliptical eigenmodes in spun hi-bi fiber was confirmed early, but there still is no practical way to excite either of the elliptical eigenmodes because of the technical difficulty in the input light matching. A common linear light will not excite a single elliptical eigenmode (see Section 6.4).

In this supplement we disclose a simple and useful way of exciting a single elliptical eigenmode in the spun hi-bi fiber [1]. We use a single-piece two-section fiber whose structure is shown in Figure 7S.1. The main fiber section (on the right side) is a regular spun hi-bi fiber of uniform spin rate. On the left side there is a tapered fiber section. This is simply a continuous extension of the main section of spun hi-bi fiber, but with a varying spin-rate dropping from τ_S to zero from right to left, or alternatively, increasing from zero at the input end to τ_S where the taper reaches the uniformly spun hi-bi fiber section. It goes without saying that such a spun hi-bi fiber with a taper

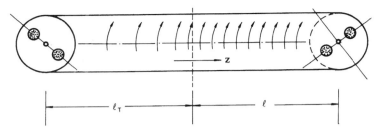

FIGURE 7S.1 A two-section spun hi-bi fiber module: l_T = the taper; l = the spun hi-bi fiber.

is, and should be, made in a one-step continuous drawing process, such that at the jointless interface ($z = z_l$) the core and stress filaments on both sides naturally and exactly match each other.

When first seen, the tapered section in Figure 7S.1 looks very much like a PPT element. This is not the case, however. The previously treated PPT element specifically refers to a nonuniform spun fiber with a spin-rate variation from *fast* to zero, or vice versa. But a spun hi-bi taper never allows its spin rate to go fast, but keeps it to a moderate rate that matches the specified moderate spin rate τ_S of the spun hi-bi fiber. This is the fundamental difference between an elliptical-bi taper and a PPT.

Nevertheless, the underlying principle of the two elements (an elliptical-bi taper and a $\lambda/4$ PPT) is that they have much in common. As a matter of fact, the latter performing the linear–circular light transform can be regarded as a special case of the former performing the linear–elliptical light transform. Thus, a principal-axis-aligned linear light exciting the zero-spun fiber end of the tapering section automatically becomes one or the other elliptical eigenmode when entering the uniform spun hi-bi fiber.

The Analytic Theory

By Eq. (7.24a), the transfer matrix in the local coordinates of a section of nonuniformly spun hi-bi fiber whose spin-rate varies from zero (at $z = 0$) to τ_ε (at z_l) is given by:

$$\mathbf{T}_l = \mathbf{O}_\varepsilon \tilde{\Lambda}_\varepsilon \mathbf{O}_0^{-1} \tag{7S.2}$$

where $\mathbf{O}_0^{-1} = \mathbf{O}_0$ is a unit matrix descriptive of the structural characteristic of the unspun end of the taper, and

$$\hat{\Lambda}_\varepsilon \approx \begin{bmatrix} e^{j\rho_\varepsilon} & 0 \\ 0 & e^{-j\rho_\varepsilon} \end{bmatrix} \tag{7S.3}$$

$$\mathbf{O}_\varepsilon = \begin{bmatrix} \cos\phi_\varepsilon & j\sin\phi_\varepsilon \\ j\sin\phi_\varepsilon & \cos\phi_\varepsilon \end{bmatrix} \tag{7S.4}$$

where we used the asymptotic expression of $\tilde{\Lambda}_\varepsilon$ (here a diagonal matrix), in which $\rho_\varepsilon = \int_0^{z_l} g\,dz$, $g = [\tau^2 + (\delta\beta/2)^2]^{1/2}$. Asymptotically, the approximation implied in $\tilde{\Lambda}_\varepsilon$ only requires slow variation of the structural parameters along the longitudinal direction, such that the off-diagonal elements in Eq. (7S.3) are sufficiently small to be neglected. However, for Eq. (7S.4) we cannot use the asymptotic approximation, inasmuch as the spin rate of the fiber-taper is not allowed to reach a sufficiently high value that satisfies the fast-spun condition $L_S \ll L_b(\tau \gg \delta\beta)$, but increases only to a moderate value as required by the uniform spun hi-bi fiber section, that is, $L_S \approx L_b$ (or $\tau = \tau_\varepsilon \approx \delta\beta$) at $z = z_l$. Thus, for Eq. (7S.4) we will have to use the exact equation in which $\phi = (1/2)\arctan(2\tau/\delta\beta)$, where $\tau \approx \delta\beta$.

Let the initial condition (input light) at the left unspun fiber end be an x-aligned linear eigenmode:

$$\mathbf{A}(0) = \begin{bmatrix} 1 \\ 0 \end{bmatrix} \tag{7S.5}$$

The particular solution of the variably spun fiber-taper can be easily derived from the transfer matrix equation (7S.2) and the initial condition, Eq. (7S.5). At the output $z = z_l$ of the fiber-taper, we have

$$\mathbf{A}(z_l) = \mathbf{O}_\varepsilon \tilde{\Lambda}_\varepsilon \mathbf{A}(0)$$

$$= \begin{bmatrix} \cos \phi_\varepsilon \\ j \sin \phi_\varepsilon \end{bmatrix} e^{j\rho_\varepsilon} \tag{7S.6}$$

This is exactly the right elliptical fast eigenmode of the spun hi-bi fiber [see Eq. (6.9)]. Our variably coupled-mode theory therefore rigorously proves that the fiber-taper performs the function of transforming an x-oriented linear light at the input to an elliptical light at its output, which naturally and exactly matches one or the other elliptical eigenmode of the spun hi-bi fiber.

Similarly, if we put a y-aligned linear light at the unspun input end of the fiber-taper, the transformed light at the output will be a left elliptical slow eigenmode of the spun hi-bi fiber:

$$\mathbf{A}(0) = \begin{bmatrix} 0 \\ 1 \end{bmatrix}$$

$$\mathbf{A}(z_l) = \mathbf{O}\tilde{\Lambda}_\varepsilon \tag{7S.7}$$

$$= \begin{bmatrix} \sin \phi_\varepsilon \\ -j \cos \phi_\varepsilon \end{bmatrix} e^{-j(\rho_\varepsilon - \pi/2)}$$

This is exactly the left elliptical slow eigenmode of the spun hi-bi fiber [see Eq. (6.10)].

The Fractional-Wave-Plate-Like Behavior of Spun Hi-Bi Fiber-Taper

The elliptical-bi taper described above, behaves *in principle*, like a fiber-optic analog of a bulk-optic fractional wave plate. Such kind of fiber section can be viewed as a "fractional" or "truncated" PPT, with "truncated" meaning that the spin rate does not go fast, but only goes to a moderate level, such that $L_S \approx L_b$. Mathematically, we can calculate the parameters of this fiber-optic fractional wave plate (in the form of a spin-rate fiber-taper), which is capable of performing a desirable linear-elliptical SOP transform. According to the variably coupled-mode theory, this fiber-optic fractional wave plate will be wide band.

In practice, however, it is by no means appropriate to take the elliptical-bi taper as a fractional wave plate in the sense of a kind of fiber-optic device. The reason is simple. Such fiber-taper cannot be fabricated separately as a single unit. As described above, in view of actual application, such fiber-taper is useful only when it is fabricated as an extension fiber of the main fiber section (the spun hi-bi fiber). That is to say, such fiber-taper and the uniformly spun hi-bi fiber can only be fabricated as an integral whole in a continuous one-step process. Should we attempt fabricating a fiber-taper of the said kind as a separate unit according to theoretical calculation, the product will not be useful as a fiber-optics device. Matching and aligning the spun end of the fabricated fiber-taper section with exterior fiber circuitry would be extremely difficult, just like that existing in the kind of spun hi-bi fiber discussed in Chapter 6.

7S.2   Magnetic-Field Tapering for Faraday-Effect Measurement

When a magnetic field is applied, the original elliptical eigenmodes of the spun hi-bi fiber will be perturbed, and the original matching condition at the output end of l_T will be spoiled. For the matching condition to be always satisfied in a magnetized fiber, we propose that the magnetic field also be tapered within the range of l_T.

Layout for Electric-Current or Magnetic-Field Sensing

For an illustration of this idea, a sketch of the overall layout is shown in Figure 7S.2. When used in Faraday-effect measurement, the two-section spun hi-bi fiber (fiber with a taper) is coiled around the metallic conductor that carries the electric current to be measured. The novelty of the circuitry structure in Figure 7S.2 is that [1], not only does the spun hi-bi fiber have a tapering section (τ from zero to τ_S) for the sake of matching, but the fiber-solenoid around the metallic conductor also has a tapering pitch, such

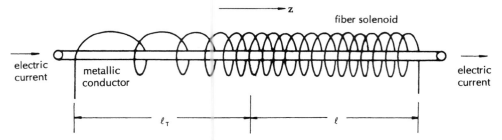

FIGURE 7S.2   Layout for electric-current or magnetic-field sensing.

that the effective longitudinal magnetic field along the coiled fiber increases from a small value in the range l_T to become a constant value in the range l.

Note that tapering of the fiber-solenoid is essentially for the tapering of the effective longitudinal magnetic field. The radius of curvature of the fiber solenoid is large enough to avoid the curvature effect for light transmission.

The reason for using a tapering pitch for the fiber-solenoid should be clear in view of the variably coupled-mode theory developed in the preceding text of this chapter. When a longitudinal magnetic field is applied to the fiber, the coupled-mode equations are modified, and the coupled-mode solution changes accordingly. For a fiber-solenoid of constant pitch, the tapering fiber section that matches a single eigenmode of the unmagnetized elliptical eigenmode of the spun hi-bi fiber will no longer match the modified eigenmode when the magnetic field is applied. The function of the magnetic-field taper is to ensure that the input light is matched to an eigenmode of the spun hi-bi fiber in the presence of any value of the applied magnetic field.

Analytic Solution of the Faraday Effect in Tapered Spun Hi-Bi Fiber

In the coupled-mode formulation, the only change caused by the magnetic field is that the coupling coefficient τ is replaced by $(\tau + \alpha_F)$, where $\alpha_F = VH$ (H is the magnetic field intensity; V is the Verdet constant).

Let an x-oriented linear light be incident at the left end of the tapering part of the fiber. By the method of quasi diagonalization for variably spun fiber, the relevant particular solution of the coupled-mode equations, before applying the electric current, is given by

$$\mathbf{A}(0) = \begin{bmatrix} 1 \\ 0 \end{bmatrix}$$

$$\mathbf{A}(l_T + l) = \begin{bmatrix} \cos \phi \\ j \sin \phi \end{bmatrix} e^{j \int_0^{l_T + l} g \, dz} \qquad (7S.8)$$

After applying the electric current, the solution becomes

$$\mathbf{A}_F(l_T + l) = \begin{bmatrix} \cos \phi_F \\ j \sin \phi_F \end{bmatrix} e^{j \int_0^{l_T+l} g_F \, dz} \tag{7.S9}$$

where the subscript F denotes a quantity related to the Faraday effect, and

$$\phi = \frac{1}{2} \arctan \left(\frac{2\tau}{\delta\beta} \right)$$

$$\phi_F = \frac{1}{2} \arctan \left[\frac{2(\tau + \alpha_F)}{\delta\beta} \right] \tag{7S.10}$$

$$g = \left[\tau^2 + \left(\frac{\delta\beta}{2} \right)^2 \right]^{1/2}$$

$$g_F = \left[(\tau + \alpha_F)^2 + \left(\frac{\delta\beta}{2} \right)^2 \right]^{1/2} \tag{7S.11}$$

where, as before, $\alpha_F = VH$. Since both the spin rate τ and the magnetic field H are tapered simultaneously, the matching condition is maintained at the output of the taper.

By Eqs. (7S.8)–(7S.11), the change of ellipticity and the change of phase are related separately to the applied magnetic field H. This facilitates the determination of H either by the change of ellipticity, or by the change of phase.

Let ε and ε_F be the ellipticity of the elliptical eigenmodes before and after application of the electric current and magnetic field, respectively. Then

$$\varepsilon = \tan \phi \tag{7S.12}$$

$$\varepsilon_F = \tan \phi_F \tag{7S.13}$$

By Eqs. (7S.12), (7S.13), and (7S.10),

$$H = \frac{\pi}{VL_b} \{ \tan[2 \arctan(\varepsilon_F)] - \tan[2 \arctan(\varepsilon)] \} \tag{7S.14}$$

By Eqs. (7S.8), (7S.9), and (7S.11), the magnetic field H can also be extracted from the phase change of the mode field due to the Faraday effect.

REFERENCE

[1]   H. C. Huang, "Elliptical-bi fiber-taper incorporated with tapering magnetic-field for current measurement," filed at the Office of the Director-General of the SPA (Shanghai Patent & Trademark Agency) (November 1996).

Postscript

Before closing the main body of the text, I would like to give two general viewpoints here as a postscript. One concerns the peculiar technological features of special fiber optics in contrast to those of microwaves. The other concerns the prospects of practical fiber versions.

In this book, I tackled special fiber-optics problems by the coupled-mode theoretical approach that was developed in microwaves fairly early. However, my analytic framework is not a simple following up of the early method constructed in microwaves. One major extension that I have made in my theoretical structure is the extensive use of asymptotic approximations (in particular, the fast-spun and the slow-varying approximations). It is easy to appreciate the fact that, in treating a variety of transmission problems in special fiber optics, analytic closed-form solutions would not have been possible if not resorting to such very unusual approximations.

At my first attempt, it was surprising for me to find that the end results based on such utmost approximations did turn out fairly accurate from the application viewpoint. The reason that this analytic accuracy is possible is due to the special feature of fiber-optics technology. Existing technology of special fiber fabrication permits a rotation of the fiber, during the linear draw, at a fast angular rate of 2000 revolutions per minute (rpm) or higher. Meanwhile, variation of this angular rate in a practically manageable length of fiber can be made as slow as desired in optical wavelength range. Thus, it is on this technology of special fiber-optics that acknowledges the legitimacy of the utmost approximations on which my analytic framework is structured.

While it is the microwave approach that consistently guides the present analytic study of special fiber optics, uniform and nonuniform, no counterpart waveguides and waveguide devices can be found in earlier well-developed microwaves. Obviously, the reason lies not in principle, but in technology.

My second viewpoint, which concerns the prospects of practical fiber versions, has been expressed here and there throughout the text, explicitly or implicitly. In summary, I have formed the conviction that the only practically useful kind of fiber that is capable of maintaining a stable state of polarization (SOP) of light in transmission is either the linear-bi fiber or the circular-bi fiber. The elliptical-bi fiber because of its attendant technological difficulties does not appear to be a practical kind of fiber from the application point of view.

Currently, when one speaks of a highly birefringent (hi-bi) fiber, it is almost always taken for granted that "hi-bi" means high *linear*-birefringence. This recalls the fact that in the exploitation of (linear and circular) hi-bi fibers over the past one or two decades, only linear hi-bi fibers have been successful, notably the Panda, the Bow-Tie, the elliptical-cladding, and the flat fibers. Despite assiduous efforts also being made in the R & D of circular hi-bi fiber, nearly in parallel with the R & D of linear hi-bi fiber, the advancement of the former is by far less fruitful. Until now, the only practical means available to produce circular birefringence in fiber is still the early-known method of postdraw twisting.

Nevertheless, currently available optical fibers, including the linearly hi-bi fiber versions, are not suitable in certain important applications such as Faraday-effect device applications, polarization-maintaining fiber-cable line with splices, and closed fiber-optic circuitry involving a Sagnac loop. The feasibility of using circular light, instead of linear light, in these and some other applications has long been recognized, in principle, by people engaged in this science. In practice, however, a variety of fiber structures have been exhaustively tested, but all in vain. This book discloses my invention of an entirely new fiber (U.S. patent issued September 1995) whose structure is simply a " spun one-eyed Panda," and amazingly, whose essential property is exactly what a practical circular hi-bi fiber should have. Also disclosed in the book is another U.S.-patented invention of mine: " fiber-optic analogs of bulk-optic quarter waveplates." These two inventions are so closely related, with one supplementing the other, that when used together they are capable of structuring completely novel fiber-optic circuitry in which only circular light power flows, and in which linear light from presently available laser diode (LD) sources is transformed into circular light to feed the fiber circuitry. Perhaps, during the future development of such *integrated fiber technology* we will be forced to make some change in the traditional idea that we used to have in dealing with fiber-optic circuitry or systems. The traditional idea is entirely based on linear light transmission in fiber.

Conversion of Maxwell's Equations into Coupled-Mode Equations

The classical method of converting Maxwell's equations into coupled-mode equations provides a complete theoretical basis for the coupled-mode theory. The foundation of this method of approach was established early in microwaves, in the initial paper by S. A. Schelkunoff entitled "Conversion of Maxwell's equations into generalized telegraphist's equations" [1]. A large number of papers on coupled modes in microwaves employed these very well-known *generalized telegraphist's equations*, or followed the line of thought underlying these equations. The so-called coupled-mode equations differ in no way from the generalized telegraphist's equations, except that the former use the traveling wave formulation, while the latter use the standing-wave formulation. In fiber optics, Snyder first derived the coupled-mode equations for optical fiber [2] based on Maxwell's equations. Among the pioneering scientists, it was Marcuse who constructed a comprehensive coupled-mode theory by conversion of Maxwell's equations [3–5]. In his book [4], Marcuse laid particular emphasis on practical application of the coupled-mode theory.

A.1 DIFFERENT SETS OF MODES

For imperfect or irregular waveguides, it is often difficult mathematically to find the independent wave fields (or "characteristic" modes, or true normal modes, or simply the eigenmodes) that individually satisfy Maxwell's equations and the imposed boundary conditions. In such cases, we can only rely on the coupled-mode formulation for a practical solution to an otherwise difficult problem.

The spirit of the method of conversion of Maxwell's equations into coupled-mode equations is to represent the wave field in the actual waveguide by the set of eigenmodes in a simpler (fictitious) waveguide called the "reference waveguide." Since the choice of the reference waveguide is in a sense arbitrary, the coupled-mode formulation is by no means unique, but involves the various reference waveguides chosen. Nevertheless, there are three kinds of modes associated with the corresponding reference waveguides (fibers) that have been found to be particularly useful to practical problems relating to imperfect and irregular fibers, namely the *ideal* modes, the *local* modes, and the *super* modes [6].

Ideal modes refer to the independent solutions of an idealized fiber model, infinitely long and straight, unspun, and without any forms of distortion. Such modes are useful for representing the wave field in an actual conventional fiber with only slight imperfections and irregularities, such as tiny variations of diameter, slight offsets, small residual ellipticity of fiber cross sections. An actual conventional fiber with only slight perturbations does not deviate too much from the said idealized fiber model, so that representation of the actual wave field in terms of the series of ideal modes is likely to converge.

Wherever appropriate, the use of ideal modes is preferable because of their mathematical simplicity. Nevertheless, the set of ideal modes is not suitable for describing the wave field of an actual fiber whose geometrical or material deviations from the idealized model are not small enough to be considered as perturbations. One of the common examples is a fiber-taper whose cross section may change by any amount from one end to the other. Also, since the ideal modes belong to an idealized model that is straight, they are not applicable to a curved or bent fiber, except when the fiber is long and laid in a natural course without sharp bends such that the fiber line can be considered as virtually straight.

Figure A.1 shows a fiber with distorted core–cladding interface. The dashed straight lines indicate the ideal core boundary, while the actual core boundary is shown by the solid slightly curved lines. If the actual core boundary deviates only slightly from the ideal core boundary for all z, the use of ideal modes for wave-field representation is adequate, as has been said. As

FIGURE A.1 Reference fibers of ideal modes and of local modes for a deformed actual fiber. Curved solid lines: actual fiber; straight dashed lines: reference fiber of ideal modes; short solid lines: reference fiber of local modes. (From Ref. [4].)

each ideal mode is z-invariant, this simple mode does not include the information of the local property of the fiber at any coordinate z.

An alternative way for representing the wave-field of the actual fiber is to use the local modes. The two short solid lines in Figure A.1 that intersect the core boundary of the actual fiber at a particular coordinate z indicate the alternative reference fiber whose eigenmodes are the local modes. Obviously, each local mode is a function of the local coordinate z, and hence more informative than the ideal mode in the wave-field representation.

Compared with the ideal modes, the local modes are often more useful because they are not restricted by the condition of slight perturbations in order to be applicable to imperfect or irregular fiber problems. The said fiber-taper (with arbitrary change in dimension from one end to the other) is one simple example for which the use of the local modes is suitable. Numerous application examples using the local modes are provided by the practical problems treated in the text of this book.

The two different mode sets (ideal modes and local modes) relating to the two reference fibers shown in Figure A.1 are generally not sufficient for practical use. Both mode sets belong to straight reference fibers. The local modes, better than the ideal modes, are informative of the transverse characteristics of the actual fiber at any particular coordinate z. But they are not able to describe the curvature characteristics at the local point z. Therefore, the local modes are inadequate for dealing with either microbending or macrobending problems of fiber.

There is still another class of modes that is of interest. Figure A.2 describes the relevant concept. Along with the local modes, Figure A.2 shows a third class of modes that are the eigenmodes in a ring-shaped fiber taken as the reference fiber. The radius of curvature of this ring-shaped fiber is identical to the local radius of curvature of the actual fiber at any particular coordinate concerned. Obviously, these modes of a new class are still not eigenmodes or independent wave fields of the actual fiber, but are neverthe-

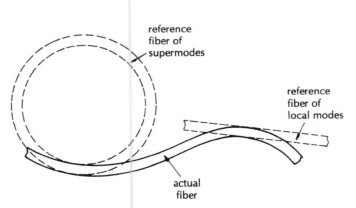

FIGURE A.2   Local modes and supermodes in a curved fiber.

less closer to the real eigenmodes than the local modes are. By intuition, we naturally expect that the couplings between the modes of the third class should be proportional to the rate of change of the curvature of the actual fiber in the transmission direction. To differentiate them from the others, it is convenient to give these new modes a name; hence, *super* modes.

In treating irregular fiber optics, using ideal modes for wave-field representation for the actual fiber is rather restrictive. The relevant field representation or field expansion is eligible only if the actual fiber irregularity is sufficiently slight in the sense of a perturbation (see Fig. A.1). This explains why, throughout the text of this book, we did not use the ideal modes in the analytic study of the variety of fiber versions whose structural specialties are far beyond the sense of perturbations. In other words, the actual fibers treated in the text differ too much from the simple idealized fiber model.

We recall that, in the earlier chapters, we always started the analysis with the local modes. To be exact, for uniformly-coupled mode problems, we used the coupled-local-mode equations and obtained the normal modes by the method of diagonalization. In the case of variably coupling in nonuniform fiber, we again started from the local modes, and obtained the supermodes by the method of quasi-diagonalization. To avoid possible confusion in terminology, we regard it as appropriate not to include the concept of ideal modes in the main body of the book (wherein this concept is not used), but to relegate it to this appendix. Really, either from the theoretical or from the practical standpoint, a complete coupled-mode theory cannot be framed if it does not include even a brief description of the basic concept of ideal modes.

In contrast to ideal modes, the coupled-mode equations in terms of local modes generally do not imply the restrictive condition in the sense of perturbation. In Figure A.1, the local modes are defined as eigenmodes in a straight reference fiber chosen to be parallel to the reference fiber for the ideal modes. This way of choosing the local reference fiber is sometimes feasible because the mathematical derivation of the coupling coefficients for such local modes is not too laborious, but only slightly more complicated than the derivation of the coupling coefficients for the ideal modes. Nevertheless, such local modes are still applicable under the restriction of slight perturbations. To enlarge the scope of applicability, the local modes shown in Figure A.2 are defined as eigenmodes in a reference fiber that is straight and tangential to the actual fiber at the coordinate concerned. Such reference fiber of local modes is therefore slanted, and the slant varies with the coordinate z. With the reference fiber so chosen for the local modes, the coupled local-mode formulation is relieved of the slight-perturbation restriction, naturally at a price of more complicated mathematics. In principle, at least, the local modes can be more freely defined, and apply to a broader class of irregular fibers in practice.

The concept of supermodes is a natural outgrowth of using the local-mode expansion in a variably coupled mode system. Figure A.2 takes as an example a fiber line with a variable curvature. As said, for the common example of a waveguide taper, except probably the rare case of very short and very slight tapering, ideal modes are not suitable for wave-field representation. The reference waveguide of the local modes is a straight and uniform waveguide whose cross section coincides with the local cross section of the actual tapered waveguide. The reference waveguide of the supermodes is thus a horn-shaped waveguide whose side-wall is tangential to the local profile of the taper. For an illustration of this example, see Refs. [6, 7].

Chapter 7 was (entirely) devoted to variably spun fibers, wherein the supermodes play an indispensable role. We did not use the ideal modes there. The local modes defined with respect to a variably spun fiber are eigenmodes in a reference fiber whose transverse structural pattern is identical to the local transverse structural pattern of the variably spun fiber, and is z-invariant (i.e., unspun). The supermodes, being quasi eigenmodes in the actual variably spun fiber structure, refer to the eigenmodes in a reference (uniformly spun) fiber whose transverse configuration coincides with the local transverse configuration of the actual fiber, and in addition, whose spin rate equals the local spin rate of the actual fiber. By intuition, it is conceivable that the defined supermode reference fiber better approximates the actual variably spun fiber.

While for different fiber structures the local modes are differently defined, the spirit of wave-field expansion in terms of local modes is the same for all cases, that is, the reference fiber of the local modes possesses all the local characteristics of the actual fiber except the rate of change of this local feature. The supermode reference fiber takes care of the rate of change at the local point, but takes it to be a uniform change. While still not being the eigenmodes in the actual fiber, the super modes represent a better approximation, and hence are termed quasi eigenmodes.

As noted in Section 7.3, the quasi-eigenmode property of the supermodes requires slow variation of the spin rate along the whole length of the variably spun fiber. Note that the condition of *slow* variation is totally different from the condition of *slight* deviation. "Slight" deviation simply means perturbation. On the other hand, "slow" variation does not imply the sense of perturbation. In the variably spun fiber studied in Chapter 7, for example, the overall change in the spin rate from one point to another point is not necessarily small, but may be fairly large. Thus, while the perturbation theory does not apply to such kind of problem, the supermode approach does.

In this book, the supermodes are obtained from the local modes by modal transformation in the method of diagonalization. A direct approach to the supermode representation of wave field is possible, and finds examples in microwaves. One example is to find the independent fields of a ring-shaped waveguide of constant curvature, and take them as supermodes in the actual waveguide of arbitrary curvature. Another example is to find the independent fields of a horn-shaped waveguide of constant slant, and take them as supermodes in the actual tapered waveguide with an arbitrarily tapering profile. While this direct approach to supermodes does work in a few examples, the mathematics involved is very complicated. The method of diagonalization that is consistently used in this book is by far simpler, and hence much more useful in practical applications. Additional mathematical treatment of the different mode sets useful to irregular nonconventional fibers will be covered in Appendix B.

A.2   METHOD OF CONVERSION OF MAXWELL'S EQUATIONS

In principle, the method of conversion of Maxwell's equations can be applied to an actual waveguide with different choices for the reference waveguide (ideal, local, or super), leading correspondingly to different coupled-mode equations for ideal, local, or supermodes. In practice, however, one does not have much freedom in these choices, inasmuch as derivation of the coupled-mode equations is more likely to succeed only with a proper choice of the reference waveguide. A formal mathematical derivation by conversion of Maxwell's equations may lead to a formal series solution that does not converge.

Generally, the mathematical derivations involved is awfully laborious. It is therefore inevitable to resort to one or another approximate condition to help simplify the relevant boundary-value or initial-value problem so that it becomes mathematically manageable. That is why in the preceding section we put due emphasis on the two distinctive types of restrictive conditions, that is, the condition of slight perturbation and the condition of slow variation.

Our immediate interest now is to gain a feeling of the mathematical flavor of the classical method of converting Maxwell's equations into coupled-mode equations. A detailed discussion of this method can be found in Ref. [4]. Here, we give only one example, based on this reference. While this example is admirably simple, the essence of the method concerned is included fairly completely in the several steps of the mathematical derivation.

Maxwell's Equations for the Actual Fiber

Consider simple harmonic function of time, such that partial differentiation with respect to t is equivalent to a complex number, that is, $\partial/\partial t \to j\omega$. In fiber optics, Maxwell's equations are written as

$$\nabla \times \mathbf{H} = j\omega n^2 \varepsilon_0 \mathbf{E}$$
$$\nabla \times \mathbf{E} = -j\omega \mu_r \mu_0 \mathbf{H}$$
$$(\text{A1})$$

where $\mathbf{E}$ and $\mathbf{H}$ are the electric and magnetic field vectors; ε_0 and μ_0 are the dielectric permittivity and magnetic permeability of free space; the refractive index is $n = \sqrt{\varepsilon_r}$, where ε_r is the relative dielectric permittivity;

and μ_r is the relative magnetic permeability. For a nonmagnetic medium, μ_r ($= 1$) is usually not written out explicitly. Note that the refractive index for the actual fiber is assumed to be a function of not only the transverse coordinates, but also the longitudinal coordinates, that is, $n = n(x, y, z)$ or (r, φ, z).

We decompose each vector field into a transverse part and a longitudinal part:

$$\mathbf{E} = \mathbf{E}_t + \mathbf{z}E_z \qquad (\text{A2a})$$

$$\mathbf{H} = \mathbf{H}_t + \mathbf{z}H_z \qquad (\text{A2b})$$

and let

$$\nabla = \nabla_t + \mathbf{z}\frac{\partial}{\partial z} \qquad (\text{A3})$$

where the subscript t denotes "transverse," and $\mathbf{z}$ denotes unit vector in the z direction. Maxwell's equations for the actual fiber are then decomposed into two sets of equations:

$$\nabla_t \times \mathbf{z}H_z + \mathbf{z} \times \frac{\partial \mathbf{H}_t}{\partial z} = j\omega\epsilon_0 n^2 \mathbf{E}_t \qquad (\text{A4a})$$

$$\nabla_t \times \mathbf{z}E_z + \mathbf{z} \times \frac{\partial \mathbf{E}_t}{\partial z} = -j\omega\mu_0 \mathbf{H}_t \qquad (\text{A4b})$$

and

$$\mathbf{z}E_z = \frac{1}{j\omega\varepsilon_0 n^2}\nabla_t \times \mathbf{H}_t \qquad (\text{A5a})$$

$$\mathbf{z}H_z = -\frac{1}{j\omega\mu_0}\nabla_t \times \mathbf{E}_t \qquad (\text{A5b})$$

Direct solutions for eigenmode fields in the actual fiber are difficult to obtain, because the refractive index n is not z-invariant. We therefore resort to the aid of some reference fiber whose structure is simpler than the actual fiber structure, and whose eigenmodes can be obtained by solving Maxwell's equations with respect to the simpler reference fiber model.

Maxwell's Equations for the Reference Fiber

We choose as reference fiber a simpler (usually idealized) fiber model whose refractive index n_0 is z-invariant, that is, $n_0 = n_0(x, y)$ or $n_0(r, \varphi)$. The reference eigenmodes in such an idealized model are available, inasmuch as the linear polarization (LP) modes derived in Section 1.2 most likely serve the purpose. Generally, such reference modes in principle are derivable with respect to the boundary-value problem relating to the reference fiber model. Each reference eigenmode is labeled by ν or μ, to keep the notation simple. In the case of the LP mode, either ν or μ stands for the subscript ml of the mode LP_{ml}. Since such eigenmodes individually satisfy Maxwell's equations, we have

$$\nabla_t \times (\mathbf{z}h_{\nu z}) - j\beta_\nu (\mathbf{z} \times \mathbf{h}_{\nu t}) = j\omega\varepsilon_0 n_0^2 \mathbf{e}_{\nu t} \qquad (A6a)$$

$$\nabla_t \times (\mathbf{z}e_{\nu z}) - j\beta_\nu (\mathbf{z} \times \mathbf{e}_{\nu t}) = -j\omega_0 \mathbf{h}_{\nu t} \qquad (A6b)$$

and

$$\mathbf{z}e_{\nu z} = \frac{1}{j\omega\varepsilon_0 n_0^2} \nabla_t \times \mathbf{h}_{\nu t} \qquad (A7a)$$

$$\mathbf{z}h_{\nu z} = -\frac{1}{j\omega\mu_0} \nabla_t \times \mathbf{e}_{\nu t} \qquad (A7b)$$

The above field equations look much like the field equations (A4a), (A4b), (A5a), and (A5b), for the actual fiber. As a matter of fact, Eqs. (A4a), (A4b), (A5a), and (A5b), as well as Eqs. (A6a), (A6b), (A7a), and (A7b), all represent reduced forms of Maxwell's equations. Nevertheless, fundamental differences exist between the above equations and the previous equations. One difference is that, the electric and magnetic components in Eqs. (A4a), (A4b), (A5a), and (A5b) refer to the *whole* field in the actual fiber, while in Eqs. (A6a), (A6b), (A7a), and (A7b) the electric- and magnetic-field components refer to individual eigenmodes of the reference fiber. The second difference is afore-said: the refractive index $n(x,y,z)$ is z-variant for the actual fiber, but the refractive index $n_0(x, y)$ is z-invariant for the reference fiber. The direct consequence is that in Eqs. (A4a) and (A4b) partial differentiation with respect to z is retained, while in Eqs. (A6a) and (A6b) partial differentiation with respect to z is replaced by a complex number, $\partial/\partial z \to -j\beta_\nu$, where β_ν is the propagation constant of any reference eigenmode labeled ν.

Field Expansion in the Actual Fiber

To derive the coupled-mode equations, the transverse electric and magnetic fields are expressed as summations of the eigenmodes in the reference fiber:

$$\mathbf{E}_t = \sum_\nu A_\nu e^{-j\beta_\nu z} \mathbf{e}_{\nu t} \qquad (A8a)$$

$$\mathbf{H}_t = \sum_\nu A_\nu e^{-j\beta_\nu z} \mathbf{h}_{\nu t} \qquad (A8b)$$

In the above summations we include the forward modes with phase factor $e^{j\beta_\nu z}$ only, while not writing out the backward modes with phase factor $e^{+j\beta_\nu z}$ for simplicity and neatness. In most practical problems of concern, the optical power that is transferred to the backward waves from an incident forward wave is negligibly small because of the large difference in propagation constants between the forward wave and backward wave. However, backward waves are sometimes not neglectful. Such is the case of a periodic waveguiding structure, wherein a forward wave and a backward wave can be strongly coupled when the beat length $2\pi/(\beta_\nu - \beta_\mu)$ of the two counter-propagating waves happens to be coincident with the period of the periodic waveguiding structure. Except for certain specific examples, the neglect of backward waves will pose no problem in field expansion.

If we put the field expansions Eqs. (A8a) and (A8b) into Maxwell's equations (A4a), (A4b), (A5a), and (A5b), the result will be a great number of equations each involving a great number of terms. Such awfully complicated equations would be hardly manageable by any processes of mathematical derivation. The usefulness of the field-expansion approach, called "orthogonal expansion" in mathematical language, rests on the existence of a certain orthogonality relation for field terms in the expansion. In essence, the orthogonal-expansion method is a generalization of the Fourier analysis, and is widely employed in many branches of applied science. We recall that in Fourier analysis of a function of time t, an orthogonality relation exists in the form of an integral from 0 to 2π of the product of two harmonic functions, which equals zero for two harmonic functions of different angular frequencies. It is with the aid of this orthogonality relation that a time function under rather loose conditions can be expanded into a summation of harmonic functions with determinable coefficients. In an analogous way, the usefulness of the field-expansion approach in optical fiber also rests on some form of orthogonality relation. According to Refs. [3, 4], the orthogonality relation for eigenmodes of the reference fiber, in the simplified case of isotropic media, is given by

$$\frac{1}{2} \int_{-\infty}^{\infty} \int_{-\infty}^{\infty} (\mathbf{e}_{\nu t} \times \mathbf{h}_{\mu t}^*) \cdot \mathbf{z} \, dx \, dy = \delta_{\nu\mu} \qquad (A9)$$

where $\delta_{\nu\mu}$ is the Kronecker delta, which equals 1 for $\nu = \mu$, and equals 0 for $\nu \neq \mu$. The superscript (*) denotes conjugate of a complex number (complex conjugate). The number $(1/2)$ before the double integral is a normalization factor.

The field expansions Eqs. (A8a) and (A8b) are then put into the transverse Maxwell's equations, Eqs. (A4a) and (A4b), wherein the longitudinal components are eliminated by substitution of Eqs. (A5a) and (A5b). The two resulting equations are then written in terms of summations of the reference eigenmodes. In view of making use of the orthogonality relation, Eq. (A9), we take, with the aid of Eqs. (A5a) and (A5b), the scalar product of the first resulting equation with $e_{\mu t}^*$, and the scalar product of the second resulting equation with $e_{\mu t}^*$, and then integrate over the entire cross section of fiber. The derivation leads to numerous mathematical equations (see Ref. [4] for details). The method, however, is simple and straightforward. The use of orthogonality relation, Eq. (A9), nullifies a large number of terms such that the eventual equations are reduced to the simplest form, as follows:

$$\frac{dA_\mu}{dz} = -j\beta_\mu A_\mu + \sum_\nu k_{k\nu} A_\nu \qquad \text{(A10)}$$

which is the desired coupled-mode equation. The coefficient $k_{\mu\nu}$ for $\nu \neq \mu$ is the coupling coefficient relating to modes labeled by ν and μ, and the coefficient $k_{\mu\mu}$ for $\nu = \mu$ is a modification term of the propagation constant as a result of mode coupling. The expressions of these coefficients are obtained directly by the preceding process of derivation, as given by

$$k_{\mu\nu} = \frac{\omega\varepsilon_0}{4j} \int_{-\infty}^{\infty} \int_{-\infty}^{\infty} (n^2 - n_0^2) \left[e_{\mu t}^* \cdot e_{\nu t} + \frac{n_0^2}{n^2} (z e_{\mu z}^*) \cdot (z e_{\nu z}) \right] dx\, dy \qquad \text{(A11)}$$

Simultaneous ordinary differential equations, Eq. (A10), can be put into matrix form:

$$\frac{d\mathbf{A}}{dz} = \mathbf{KA} \qquad \text{(A12)}$$

where $\mathbf{A}$ is a column matrix whose elements $A_1 A_2, \ldots, A_n$ denote a set of modes that are eigenmodes in the reference fiber, but are coupled modes in the actual fiber; and $\mathbf{K}$ is a square matrix whose diagonal elements are propagation coefficients of the reference modes (with modification terms due to mode couplings), and whose off-diagonal elements are coupling coefficients.

From the preceding example we see that the spirit of the coupled-mode theory is fairly simple. We have an actual fiber whose field solutions are difficult to obtain directly by solving Maxwell's equations under the given boundary conditions. So we consider a simpler fiber model, or reference fiber, whose field solutions are available. Such field solutions, called the "reference modes," do not individually represent the field in the actual fiber, but as a summation they do. The reference modes are coupled in the actual fiber according to the coupled-mode formulation. By intuition, it is conceivable that the coupling strength is in proportion to how much the actual fiber deviates from the reference fiber either in geometry or in fiber parameters. The secret of the method is to carefully search where and what the deviation is between the actual fiber and the reference fiber. It is this "deviation term" that gives rise to some nonzero off-diagonal terms that are termed the coupling coefficients. In the previous example, we see that the source of this "deviation term" comes from the use of $n(x, y, z)$ of the actual fiber, instead of $n_0(x, y)$.

In the case of anisotropic media, derivation of the coupled-mode equations follows a line similar to that described above for isotropic media [5]. The essential difference is simply that, in place of scalar permittivities, tensor permittivities are employed for the actual fiber and the reference fiber, which are now anisotropic. Conversion of Maxwell's equations to coupled-mode equations is made by tensor calculus. As expected, the resulting formulation of the coupling coefficient involves a deviation term that is now the difference between the two permittivity tensors (see Ref. [5] for details.)

In summary, by the classical method a boundary-value problem described by partial differential equations is reduced to a coupled-mode problem described by ordinary differential equations, which are easier to solve. In principle, the method is just that simple. For most problems met in practice, however, the mathematical manipulation often becomes very complicated, or scarcely tractable. In that case, we shall be forced to establish the coupled-mode formulation directly by adopting the phenomenological approach.

REFERENCES

[1] S. A. Schelkunoff, "Conversion of Maxwell's equations into generalized telegraphist's equations into generalized telegraphist's equations," *Bell Syst. Tech. J.*, vol. 34, pp. 995–1043 (1955).

[2] A. S. Snyder, "Coupled mode theory for optical fibers," *J. Opt. Soc. Am.*, vol. 62, pp. 1267–1277 (1972).

[3] D. Marcuse, "Coupled mode theory of round optical fibers," *Bell System. Tech. J.*, vol. 52, pp. 817–842 (1973).

[4] D. Marcuse, *Theory of Dielectric Optical Waveguides*, Academic Press, 257 pages (1974).

[5] D. Marcuse, "Coupled mode theory for anisotropic optical waveguides," *Bell Syst. Tech. J.*, vol. 54, pp. 985–995 (1975).

[6] H. C. Huang, "On local normal modes in optical fiber and film waveguides," *Sci. Sin.*, vol. 22, pp. 1147–1155 (1979).

[7] H. C. Huang, "Weak coupling theory of optical fiber and film waveguides," *Radio Science*, vol. 16, pp. 495–499 (1981).

BIBLIOGRAPHY

S. V. Stoyanov and V. V. Shevchenko, "Nonuniform anisotropic optical fibers with locally diagonal permittivity," English translation of Russian J. (Radiotekh Elektron), *J. Commun. Technol. and Electron.*, vol. 40, pp.34–40 (1995).

Variably Coupled-Mode Theory

The coupled-mode equations are simultaneous ordinary differential equations descriptive of the traveling waves and their interactions in microwave guides or optical fiber. Treating these *ordinary* differential equations is, in a mathematical sense, an easier task than treating partial differential equations descriptive of Maxwell's field equations.

The ordinary differential equations for coupled modes are established either by conversion of Maxwell's equations, or in a phenomenological way with or without an apparent connection to Maxwell's equations. Despite the mathematical elegance of the method of conversion of Maxwell's equations, the phenomenological way is indispensable in developing the coupled-mode theory. A point of special importance is that, for a variably coupled system whose coupling coefficients are not constants, but some functions of the coordinate z, the method of conversion of Maxwell's equations in the strict sense becomes virtually useless because of insurmountable mathematical difficulty. Under such circumstances, the only thing we can do is simply to allow the coefficients of the ordinary differential equations to be variables, in a phenomenological manner.

This appendix concerns the ordinary differential equations descriptive of the coupled modes in general terms, without regard to the way by which these equations are established. A brief account on the classical method of diagonalization for coupled-mode equations with constant coefficients is given below, prior to the more involved theory of variable couplings.

B.1   DIAGONALIZATION OF COUPLED-MODE EQUATIONS WITH *CONSTANT* COEFFICIENTS

Consider a set of coupled-mode equations involving N modes. For succinctness, these equations are put in matrix form:

$$\frac{d\mathbf{A}}{dz} = \mathbf{KA}$$

$$\mathbf{A} = \begin{bmatrix} A_1 \\ A_2 \\ \vdots \\ A_n \end{bmatrix} \tag{B1}$$

$$\mathbf{K} = \begin{bmatrix} K_{11} & K_{12} & \cdots & K_{1n} \\ K_{21} & K_{22} & \cdots & K_{2n} \\ \vdots & \vdots & & \vdots \\ K_{n1} & K_{n2} & \cdots & K_{nn} \end{bmatrix}$$

where the column matrix $\mathbf{A}(A_1, A_2, \ldots, A_n)$ can be a set of modes of any type, and the square matrix $\mathbf{K}$ is the corresponding "coupling" matrix.

In matrix theory, a column matrix is also called a "vector." For a matrix of the third rank, the elements of the matrix can be thought of as the three components of a space vector. When the rank is higher than three, a column matrix has no correspondence with a vector in real space. Nevertheless, as a mathematical abstraction, a column matrix is still called a vector in an n-dimensional space whose dimension can be indefinitely large.

The essence of the method of diagonalization is to introduce a transformation so that the matrix of the differential equations can be diagonalized. Let this transformation be given by

$$\mathbf{A} = \mathbf{OW} \tag{B2}$$

where $\mathbf{O}$ is the diagonalizing matrix. Substituting Eq. (B2) into Eq. (B1) yields

$$\frac{d\mathbf{W}}{dz} = \Lambda\mathbf{W} \tag{B3}$$

$$\Lambda = \mathbf{O}^{-1}\mathbf{KO} \tag{B3a}$$

If Λ is now a diagonal matrix of elements $\lambda_1, \lambda_2, \ldots, \lambda_n$, then Eq. (B3) becomes n independent differential equations, such that $W_1, W_2, \ldots, W_n$ (elements of $\mathbf{W}$) are readily solvable, and hence $A_1, A_2, \ldots, A_n$ are solvable by the transformation (B2). In the language of matrix algebra, Λ is said to be "similar" to $\mathbf{K}$.

The elements $W_1, W_2, \ldots, W_n$ are called "normal" modes, or eigenmodes, in the normal coordinates. The terminology employed here actually implies some mathematical abstraction. What are the "normal" modes $W_1, W_2, \ldots, W_n$? What are the "normal" coordinates? From a practical standpoint, we can view these normal modes simply as a set of independent modes of Eq. (B3) that are related *mathematically* by Eq. (B2) with the set of coupled modes $A_1, A_2, \ldots, A_n$. The normal coordinates, unlike a system of coordinates in real space (well-defined geometrically and measurable by experiment), do not have a geometrical meaning in real space. We cannot determine the geometrical location of the normal coordinates by an experimental method. In reality, the concept of the normal coordinates is an outgrowth of mathematics. In mathematical physics, the subject matter concerned is related to the theory of orthogonal expansion in n-dimensional, or so-called Hilbert space. But here we shall not go further into more abstraction. In view of applications, it suffices to understand that the normal modes $W_1, W_2, \ldots, W_n$ can be dealt with by the simple mathematical transformation (B2).

Two questions still await answers: (1) What are the elements $\lambda_1, \lambda_2, \ldots, \lambda_n$ of the diagonal matrix Λ? (2) What should the matrix $\mathbf{O}$ that makes Λ ($= \mathbf{O}^{-1}\mathbf{KO}$) diagonal be?

For answers, we write Eq. (B3a) as $\mathbf{KO} = \mathbf{O}\Lambda$, such that

$$
\begin{bmatrix} k_{11} & k_{12} & \cdots & k_{1n} \\ k_{21} & k_{22} & \cdots & k_{2n} \\ \vdots & \vdots & & \vdots \\ k_{n1} & k_{n2} & \cdots & k_{nn} \end{bmatrix} \begin{bmatrix} o_{11} & o_{12} & \cdots & o_{1n} \\ o_{21} & o_{22} & \cdots & o_{2n} \\ \vdots & \vdots & & \vdots \\ o_{n1} & o_{n2} & \cdots & o_{nn} \end{bmatrix}
$$
$$
= \begin{bmatrix} o_{11} & o_{12} & \cdots & o_{1n} \\ o_{21} & o_{22} & \cdots & o_{2n} \\ \vdots & \vdots & & \vdots \\ o_{n1} & o_{n2} & \cdots & o_{nn} \end{bmatrix} \begin{bmatrix} \lambda_1 & & & \\ & \lambda_2 & & \\ & & \ddots & \\ & & & \lambda_n \end{bmatrix}
\tag{B4}
$$

The elements in the first column on each side of the resulting matrix Eq. (B4), are

$$
k_{11}o_{11} + k_{12}o_{21} + \cdots + k_{1n}o_{n1} = \lambda_1 o_{11}
$$
$$
k_{21}o_{11} + k_{22}o_{21} + \cdots + k_{2n}o_{n1} = \lambda_1 o_{21}
$$
$$
\cdots\cdots\cdots\cdots\cdots\cdots\cdots\cdots\cdots\cdots\cdots\cdots
$$
$$
k_{n1}o_{11} + k_{n2}o_{21} + \cdots + k_{nn}o_{n1} = \lambda_1 o_{n1}
\tag{B5}
$$

The above equations can be written in the matrix form: $\mathbf{KO} = \lambda_1 \mathbf{O}_1$, where $\mathbf{O}_1$ is a column matrix of elements $o_{11}, o_{21}, \ldots, o_{n1}$. Generally, for the elements in the ith column of the resulting matrix on each side of Eq. (B4), we have

$$\mathbf{KO}_i = \lambda_i \mathbf{O}_i$$

$$(\mathbf{K} - \mathbf{I}\lambda_i)\mathbf{O}_i = 0, \qquad i = 1, 2, \ldots, n \tag{B6}$$

$$\mathbf{Q}_i = \begin{bmatrix} o_{1i} \\ o_{2i} \\ \vdots \\ o_{ni} \end{bmatrix}$$

where $\mathbf{I}$ is unit matrix.

In mathematical physics, (B6) is called the eigen-equation of the matrix $\mathbf{K}$. The nonzero solution of the column matrix $\mathbf{Q}_i$ requires that the determinant of the matrix in the bracket in Eq. (B6) is equal to zero. The problem is thus reduced to solving the following equation:

$$\det(\mathbf{K} - \mathbf{I}\lambda) = 0 \tag{B7}$$

This is an algebraic equation of degree n, from which the n values of λ (i.e., $\lambda_1, \lambda_2, \ldots, \lambda_n$) can be determined. Since the elements of the diagonal matrix Λ can take these and only these specific values, these values of λ are called the eigenvalues of the matrix $\mathbf{K}$. For $\mathbf{K}$ satisfying some usual conditions, the column matrices of $\mathbf{O}$ are orthogonal in the sense

$$\sum_{k=1}^{n} o_{kr} \cdot o_{ks}^* = 0, \qquad s \neq r$$

$$= 1 \qquad s = r \tag{B8}$$

With the eigenvalue λ_i being determined by the eigenequation, the homogeneous algebraic equations, Eqs. (B6), give the ratios of $o_{1i}, o_{2i}, \ldots, o_{ni}$. One more relation is needed in order to determine the values of $o_{1i}, o_{2i}, \ldots, o_{ni}$. This is provided by the preceding normalization relation for $s = r$.

B.2   QUASI DIAGONALIZATION OF *VARIABLY* COUPLED-MODE EQUATIONS

In the more general case, some or all coefficients in the differential equations are functions of the independent variable. Such differential equations are of fundamental importance in many branches of applied science, yet the method of solving these equations could not be found in mathematical publications until 1955 when microwave scientists at Bell Labs made public their work on variably coupled modes [1, 2, 3]. The third paper by W. Louisell (1955) gave the end result of an approximate solution for the variable coupling of two modes. Thus, the desired solution of variable coupling in the two-mode case has existed since then. Later, Huang [4, 5] and Keller and Keller [6] independently published their methods, by different analytic approaches, for solving a generalized variably coupled system comprising N modes, with identifiable end results in the form of exponential-like functions. Immediately after the publication of Huang (1960), the fundamental equation [Eq. (B11); below] was used in the work by B. Z. Kazenelenbaum [7].

In Refs. [4, 5], I succeeded in an early attempt to treat matrices of variable elements in a way like operational calculus, thereby significantly simplifying the mathematical formulation. In matrix formulation, the simultaneous linear ordinary differential equations are expressed as

$$\frac{d\mathbf{A}}{dz} = \mathbf{K}(z)\mathbf{A} \tag{B9}$$

where $\mathbf{A}$ is a column matrix of rank n, with elements $A_1, A_2, \ldots, A_n$, and $\mathbf{K}$ is a square matrix with elements k_{rs}, $r = 1, 2, \ldots, n$, $s = 1, 2, \ldots, n$. While $\mathbf{A}$ still can be any type of modes (ideal modes, local modes, supermodes), we chose to have it denote the local modes, so that the notation of this appendix conforms with that of the main body of the book.

For a solution of Eq. (B9), in which the elements of $\mathbf{K}$ are variables in general, we introduce the following transformation:

$$\mathbf{A} = \mathbf{O}(z)\mathbf{W} \tag{B10}$$

which is put into Eq. (B9) to yield

$$\frac{d\mathbf{W}}{dz} = \mathbf{N}\mathbf{W} \tag{B11}$$

$$\mathbf{N} = \mathbf{O}^{-1}\mathbf{K}\mathbf{O} - \mathbf{O}^{-1}\frac{d\mathbf{O}}{dz} \tag{B11a}$$

where $\mathbf{O} \equiv \mathbf{O}(z)$, and $\mathbf{K} \equiv \mathbf{K}(z)$, $\mathbf{N} \equiv \mathbf{N}(z)$. $\mathbf{N}(z)$ is similar to $\mathbf{K}(z)$ in the sense of Lowey [8]. Here, the meaning of "in the sense of Lowey" is simply a matter of terminology, implying no difficult mathematical abstraction. In classical matrix algebra treating matrices of constant elements, Λ and $\mathbf{K}$ in Eqs. (B1) and (B3) are termed "similar" matrices (in the ordinary sense). For variable elements, Lowey [8] first investigated the extended concept of similarity of matrices with variable elements using Eq. (B11a), and hence $\mathbf{N}(z)$ is so termed.

From Eq. (B11a) we see that $\mathbf{N}$ contains two terms. The first term $\mathbf{O}^{-1}\mathbf{K}\mathbf{O}$ is a diagonal matrix. The second term $\mathbf{O}^{-1}(d\mathbf{O}/dz)$ is proportional to the derivative of **O**. If the coefficients of the variably coupled-mode equations vary slowly, this second term can be small enough such that **N** becomes quasidiagonal, and Eq. (B11) is solvable by an iterative process. The necessary and sufficient condition for the iterative condition to converge is [4, 5]

$$Q = \frac{N_{rs}(z)}{N_{rr}(z) - N_{ss}(z)} \ll 1 \qquad (s \neq r) \qquad (B12)$$

where the quotient Q is called the coupling capacity of the mode set, because it is this quotient, not the coupling coefficient N_{rs} alone, that determines the strengths of conversions and reconversions between the modes. A reduction of the Q-value results in faster convergence of the iterative solution of the differential equations. Under the condition (B12), the first-order iterative solution of Eq. (B11) is derived as

$$W_k(z) \approx e^{\rho_k(z)} \left\{ W_k(0) + \sum_{r \neq k} W_r(0) \int_0^z N_{kr}(z) e^{\rho_r(z) - \rho_k(z)} \, dz \right.$$

$$\left. + \sum_{r \neq k} \int_0^z e^{\rho_r(z') - \rho_k(z')} N_{kr}(z') \sum_{s \neq r} W_s(0) \int_0^{z'} e^{\rho_s(z'') - \rho_r(z)} N_{rs}(z'') \, dz'' \, dz' \right\}$$

$$(B13)$$

where

$$\rho_k(z) = \int_0^z \lambda_k(z) \, dz$$

$$\lambda_k(z) = N_{kk}(z)$$

$$(B14)$$

The initial conditions **W**(0) can be derived from the initial conditions **A**(0) by the transformation (B10).

B.3   DIFFERENT SETS OF MODES AND THE "EQUIVALENCE FORMULAS"

The varieties and complexities involved in the treatment of irregular optical fibers suggest the importance of a proper choice of the reference fiber with respect to which the wave field is expanded. Generally speaking, for *slight* perturbation of fiber it is often convenient to use the ideal modes for wave-field expansion on the ground of relative simplicity in mathematical manipulation, while for *slow*-variation problems the use of supermodes by transformation of local modes is likely to prove advantageous. Here we again note that the condition of "slight" perturbation is generally more restrictive than the condition of "slow" variation.

In connection with Eq. (B9), it was noted that, for the sake of using the same notation throughout the book, we chose to keep using **A** to denote the set of local modes. As before, **W** thus denotes the normal modes for constant coupling and denotes the supermodes for variable coupling. As regards the ideal modes, we need a new symbol. Here we choose the small letter **a** for the ideal modes, and choose the lowercase letter **k** to denote the corresponding coupling matrix. In a general study of the coupled modes in an irregular fiber, we need to deal with all these types of modes: the ideal modes **a**, the local modes **A**, as well as the supermodes **W** [9–13].

The matrices of the coupling coefficients of the three sets of modes (**a**, **A**, **W**) are **k**, **K**, **N**, respectively. A question of theoretical and practical importance is therefore raised: Are there any connections between these different coupling coefficients?

In his book, Marcuse [9] derived the coupling coefficients for ideal modes, by conversion of Maxwell's equations. And then, again by conversion of Maxwell's equations, he derived the coupling coefficients for local modes. By comparing the mathematical forms of the two end results (coupling coefficients for the ideal modes and coupling coefficients for the local modes), he obtained, on the basis of perturbation approximations, a simple "equivalence formula" that relates the ideal-mode coupling coefficients to the local-mode coupling coefficients (Ref. [9, p. 121]):

$$f(z) \rightarrow -\frac{j}{\beta_\nu - \beta_\mu}\frac{df}{dz}$$

where $f \equiv f(z)$, termed the deformation function, is proportional to the ideal-mode coupling coefficient. Thus, with the aid of this equivalence formula, when the coupling coefficients of one kind are known, we immediately have the coupling coefficients of the other kind, without the need to repeat the tedious derivation from Maxwell's equations. The restrictive condition for the preceding equivalence formula to be valid is that the perturbations are sufficiently slight.

Marcuse did not use the term "super modes." However, in his treatment of microbendings of a fiber line, he introduced (without derivation) a second "equivalent formula," which relates the local-mode coupling coefficients to the coupling coefficients of the "local modes of a second kind" (see Ref. [9, pp. 165–166]):

$$f(z) \rightarrow -\frac{1}{(\beta_\nu - \beta_\mu)^2}\frac{d^2f}{dz^2}$$

Since the second equivalent formula is not derived in Ref. [9], we presume that it is the form of the first equivalence formula that suggested this form for the second equivalence formula. To apply, the restrictive condition for the above formula again requires slight perturbations of the fiber irregularity.

In summary, three types of modes are used in Marcuse's theory on dielectric optical fibers; the ideal modes, the (ordinary) local modes, and the local modes of the second kind [9]. Further, the three types of modes defined by Marcuse are identical, or at least similar, to the three types of modes (ideal, local, and super) used in my early work on microwave guides [4, 5].

Interestingly enough, we have found in our analytic framework that it is the method of diagonalization that helps establish the relations between the ideal modes and local modes, and between the local modes and the super-modes. The local-mode coupling matrix $\mathbf{K}$ and the ideal-mode coupling matrix $\mathbf{k}$ are similar in the sense of Lowey, such that

$$\mathbf{K} = \mathbf{O}_k^{-1}\mathbf{k}\mathbf{O}_k - \mathbf{O}_k^{-1}\frac{d\mathbf{O}_k}{dz} \tag{B15}$$

where $\mathbf{O}_k$ is the quasi-diagonalizing matrix for $\mathbf{k}$. Under the restrictive condition of slight perturbation, Eq. (B15) immediately yields:

$$K_{ij} \approx \frac{1}{j(\beta_j - \beta_i)}\frac{dk_{ij}}{dz} \tag{B15a}$$

where the capital letter K_{ij} and the lowercase-letter k_{ij} denote, respectively, the local-mode coupling coefficient and the ideal-mode coupling coefficient [10, 11].

By a similar mathematical process with respect to $\mathbf{K}$ and $\mathbf{N}$, we have

$$\mathbf{N} = \mathbf{O}^{-1}\mathbf{K}\mathbf{O} - \mathbf{O}^{-1}\frac{d\mathbf{O}}{dz} \tag{B16}$$

$$N_{ij} \approx \frac{-1}{j(\beta_j - \beta_i)}\frac{dK_{ij}}{dz}$$

$$\approx \frac{-1}{(\beta_j - \beta_i)^2}\frac{d^2K_{ij}}{dz^2} \tag{B16a}$$

where the first approximate identity is valid under the condition of slow variation of perturbations, while the second approximate identity is valid under the more restrictive condition of slight perturbations [10, 11]. The preceding results derived by the method of quasi diagonalization agree with the two equivalence formulas. Application examples can be found elsewhere [12, 13] besides Ref. [9].

B.4   REPEATED QUASI NORMALIZATION OF MODES

The method of diagonalization is alternatively called the method of orthogonalization, or the method of normalization, in linear algebra. In the standard theory of linear algebra, only matrices with constant elements are treated. Nevertheless, in highly irregular special fiber optics we will have to deal with matrices involving variable elements. Since an exact diagonalization of the matrix with variable elements is generally not achievable, the best we can do is a quasi diagonalization (or quasi normalization) of this matrix.

The line of thought leading to the exponential-like iterative solution Eq. (B13) suggests the possibility of again applying the mathematical process of quasi normalization, and securing a set of differential equations whose couplings are further reduced.

At the 9th International Workshop on Optical Waveguide Theory [14, 15], some questions were raised on the concept of supermodes. A question of fundamental importance was whether the set of supermodes can really be found in an actual fiber structure. The answer was: That depends on the fiber structure considered. In the case of a nonuniformly twisted fiber, I did find the supermodes using matrix–algebraic manipulation. Another question concerns the possibility of performing repeatedly the process of quasi diagonalization so that sets of supermodes of higher rank can be derived with indefinitely diminishing coupling strength. In regard of this question, it was agreed at the workshop that, in principle at least, this mathematical possibility exists. Practically speaking, it is conceivable that the three sets of modes (ideal modes, local modes, and supermodes) appear to be the best choices among all the possibilities.

B.5   NOTES

The coupled-mode theory, which was well-developed in the microwaves research, was rediscovered to be also a useful tool in the study of fiber optics. Fiber optics as an important and fast-developing modern high technology actually embraces a vast number of new problems awaiting solutions and answers. The consequence is then a further advancement of the coupled-mode theory in its areas of application. Of particular interest is application of this theory to single-mode fiber—the main kind of fiber used in modern optical fiber. In an irregular single-mode fiber, it is the two polarization modes that are coupled to each other.

For uniformly coupled-mode problems, the method of diagonalization has been well-established, and has achieved relative maturity, in the theory of matrix-algebra. A fiber optics theoretician seldom needs to take care of the restrictive condition for the matrix **K** of the coupled-mode equations before starting to solve the coupled-mode equations by the method of diagonalization. The matrix elements necessarily satisfy the required mathematical restriction, so long as the principle of conservation of power is ensured in the coupled-mode formulation. Two kinds of coupling coefficients (i.e., the off-diagonal elements) were dealt with in different chapters of the book. For one kind, which occurs extensively in problems involving the choice of linear light as the base modes, the off-diagonal elements (τ and $-\tau$) are real, but opposite in sign. For the other kind, occurring when choosing circular light as the base modes, the off-diagonal elements (jc and jc) are purely imaginary, and same in sign. In both cases, the off-diagonal elements or coupling coefficients satisfy the "negative conjugate" relation as required by the conservation of power. For the variety of problems dealt with in this book, we therefore can always find the appropriate diagonalizing matrix **O** which diagonalizes **K** into a diagonal matrix. In other words, the possibility of finding the diagonalizing matrix seems to be something that is taken for granted. It therefore becomes scarcely necessary to question whether or not the matrix **K** that defines the coupled-mode equations is actually diagonalizable.

Nevertheless, an applied mathematician is often not satisfied with a final determination of a solution. It is sometimes of more theoretical interest to ask: what sort of **K** can be diagonalized, and what cannot? For coupled-mode equations with constant coefficients, here we simply give the answer without proof. The answer is: in mathematical language, the restrictive condition that a matrix **K** can be diagonalized is that it is either Hermitian or anti-Hermitian. This is a more general condition that includes the negative conjugate relation. The matrix that is able to diagonalize **K** is a unitary matrix (see p. 41 of Ref. [16] and Section 6.6 of Ref. [17]).

Solutions of variably coupled-mode problems are generally more difficult to obtain. Intuitive reasoning, as well as judicious use of mathematical approximations, are therefore inevitable. A careful reading of Section B.3 may have detected that, while we did derive the second equivalence formula by the mathematical technique of quasi diagonalization, our derivation still implies intuition to some extent. For example, it is by intuition that the type of modes produced by quasi diagonalization of **k** (for ideal modes) are identifiable to the local modes. Indeed, a search of the mathematical techniques for solutions of variably coupled-mode problems presents a challenge to applied mathematicians and theoretical engineers.

In principle, either the uniform coupled-mode theory or the variably coupled-mode theory applies equally well to a single waveguiding structure which supports two or more wave-fields, and to a composite system of coupled transmission lines or waveguides each transmitting a single wave-field.

Some authors used "supermodes" to refer to the *independent* fields of a *composite* waveguiding structure comprising two uniformly coupled waveguides. Such "supermodes" are different from the supermodes that we have defined for the quasi independent wave-fields in a nonuniform fiber structure (see "Definition of Supermodes" in Section 7.12).

REFERENCES

[1] J. S. Cook, "Tapered velocity coupler," *Bell Cyst. Tech. J.*, vol. 34, pp. 807–822 (1955).

[2] A. G. Fox, "Wave coupling by warped normal modes," ibid, pp. 823–852 (1955).

[3] W. H. Louisell, "Analysis of the single tapered mode coupler," ibid, pp. 823–852 (1955).

[4] H. C. Huang, "Generalized theory of coupled local modes in multi-wave guides," *Sci. Sin.*, vol. 9, no. 1, pp. 142–154 (1960).

[5] H. C. Huang, "Method of slowly varying parameters," *Acta Math. Sin.*, vol. 11, pp. 238–247 (1961); also, *Rev. J. Math.* (in Russian), p. 39 (1961).

[6] H. B. Keller and J. B. Keller, "Exponential-like solutions of system of ordinary differential equations," *J. Soc. Ind. Appl. Math.*, vol. 10, no. 21, pp. 246–259 (1962).

[7] B. Z. Kazenelenbaum, *Theory of Irregular Waveguides with Slowly Varying Parameters*, Soviet Academy of Sciences Printer, Moscow, p. 54 (1961).

[8] A. Lowey, "Uber Matrizen- und Differentialkomplexe," *Math. Ann.*, vol. 78, pp. 1–15, 343–358, 359–368 (1918).

[9] D. Marcuse, *Theory of Dielectric Optical Waveguides*, Academic Press, 257 pages (1974). Refer in particular to Eq. (3.5-22) on p. 121, and Eq. (4.6-13) on p. 166.

[10] H. C. Huang, "On local normal modes in optical fiber and film waveguides," *Sci. Sin.*, vol. 22, pp. 1147–1155 (1979).

[11] H. C. Huang, "Weak coupling theory of optical fiber and film waveguides," *Radio Sci.*, *vol.* 16, *pp.* 495–499 (1981).

[12] Y. Chen and H. C. Huang, "Microbending losses of bow-tie and similar anisotropic fibres," *Electron. Lett.*, vol. 23, pp. 157–159 (1987).

[13] Y. Chen and H. C. Huang, "Microbending and macrobending behavior of bow-tie fibers and similar anisotropic optical fibers," *J. Opt. Soc. Am.*, vol. 5, pp. 380–386 (1988).

[14] K. Petermann, "Progress in optical waveguide theory," *J. Opt. Commun.*, vol. 5, pp. 153–155 (1984).

[15] H. C. Huang, "Super mode concept and applications," contributed to the *9th International Workshop on Optical Waveguide Theory*, Reisensburg, Germany, September 7–10 (1984).

[16] H. C. Huang, *Coupled Mode and Nonideal Waveguides*, Microwave Institute (MRI), Polytechnic of New York, 191 pages (1981).

[17] J. Mathews and R. L. Walker, *Mathematical Methods of Physics*, W. A. Benjamin, Inc., 475 pages (1964).

Notes on Making Special Optical Fibers

Generally speaking, fiber fabrication consists of two parts: the making of preform, and fiber drawing. Fabrication facilities vary enormously in size and complexity, depending on whether they are to be used for research, development, or industrial production. For research purposes, simple and small-sized equipment is often convenient to perform the varied tasks. In research, a desirable feature is flexibility in operation in the facilities.

C.1 THE DRAWING TOWER

The drawing tower is to pull the tip of the preform at a high temperature, thereby yielding a continual length of fiber drawn and to be taken up. The fiber-drawing setup is put in a vertical structure that looks like a tower; hence, the name "fiber-drawing tower." In rare circumstances the setup may also be arranged horizontally. The size and complexity of a fiber-drawing tower varies from laboratory to laboratory and from factory to factory. In one extreme, a fiber-drawing tower may be as small as a book shelf, while at the other extreme, it may be some stories tall.

A standard conventional drawing tower generally works for the study of special fibers as well. The incorporation of a spinner (i.e., a speed-controlled d.c. motor) on-top of the fiber-drawing tower is a major modification of the conventional setup. The spinner may be fastened to the drawing tower as an integral part of the fiber-drawing setup, or it may be specially structured as a separate unit. In the latter case, the spinner is encased in a upright cylindrical container whose upper plane cover is machined with an affixed axial rod which is inserted like a preform into the originally available chuck on top of the tower, and whose lower plane cover is holed to allow the spinner's axis to emerge from inside the container. Attached to this spinner's axis is a second chuck that is used to hoop the preform. In any workable construction, it is a prerequisite to ensure an accurate aligning of the entire fiber-drawing equipment from the top chuck to the revolving drum that winds up the line of fiber.

C.2   MAKING CIRCULAR Hi-Bi SCREW-FIBER PREFORM

The cross-sectional form of the circular-polarization-maintaining fiber (called the Screw fiber) [1] is as simple as is shown in Figure 5.1. It looks like a one-eyed panda. This single eye is the stress filament (doped most commonly with boron) that spins around the core to produce the desired circular-polarization-maintaining property. By intuitive reasoning, we expect that the design of the circular highly birefringent (hi-bi) screw fiber should be simpler than the linear-polarization-maintaining Panda fiber is. The latter has two "eyes" placed symmetrically on opposite sides of the central core. Slight asymmetry of the two eyes with respect to the central core will cause substantial degradation of the linear-polarization-maintaining behavior of the Panda fiber. The one-eyed Panda is an asymmetrical cross-sectional configuration, in which the placement of the one eye should be much less restrictive. For a Screw fiber, the distance between the single stress-applying filament and the core is optimally chosen from technological considerations, subject to the requirements of low loss and certain other factors. To simplify the preform design, it should be possible to use the established design framework for the Panda fiber with one of its two eyes taken out.

Fabrication of the Screw-fiber preform can be easily accomplished with the aid of any of the existing methods, for example, the method of burying a stress rod through a drilled off-axis hollow cylinder parallel to the axis, or the method of fitting the stress rod and the core rod, together with a number of dummy glass rods, all in a bigger glass tube (Fig. C.1), in exactly the same way

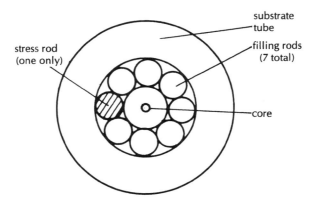

FIGURE C.1   Screw-fiber preform by the "rod-in-tube" method.

as is done in making a Panda fiber or a Panda-like fiber similar to the linear polarization-preserving Corguide. Substantial simplification in fabricating a one-eyed Panda preform is a unique advantage existing only in a preform whose geometrical configuration is asymmetrical.

The versatile modified chemical vapor deposition (MCVD) method, supplemented by the gas-etching technique, which successfully produces the Bow-Tie fiber, should also work in making a Screw-fiber preform. The making of a Bow-Tie preform consists essentially of four steps: e.g., depositing B_2O_3–SiO_2; gas-etching with fluorine, depositing GeO_2–SiO_2; and collapsing. One possible way in making a preform for the Screw fiber is to introduce a second gas-etching step to remove one of the remaining two stress sectors after the first gas-etching step and before depositing the GeO_2. At any cross section of the fiber, the structural configuration is no longer a bow-tie, but roughly a fan-shaped stress filament off-axis from the core.

Other special techniques to make the peculiar, asymmetrically shaped configuration, which comprises an on-axis core and an off-axis stress filament, can be conceived with more or less complexity. After collapse the resulting preform is likewise suitable to be spun in the next step to yield the Screw fiber.

The one stress-filament structure is the simplest of the several possible structures of the Screw fiber. Figure C.2a shows a second possible structure that contains two stress filaments. Unlike the Panda fiber, the two stress filaments are not collinear with the core, but are so placed that a line joining one stress filament and the core is perpendicular to the line joining the other stress-filament and the core. The reason for using this peculiar placement of the two stress filaments in Figure C.2a is to eliminate, or to greatly reduce, the residual intrinsic linear birefringence of the two-filament version, thus more effectively preserving the transmission of light in the circular state of polarization (SOP).

A four-filament version of the Screw fiber is shown in Figure C.2b. These four filaments are placed this way in an effort to further reduce the residual intrinsic linear birefringence in order to enhance the circular-polarization

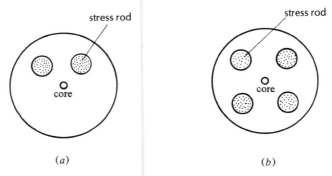

(a)   (b)

FIGURE C.2 Screw-fiber preform with (a) two stress filaments and (b) four stress filaments.

behavior of the fiber. This configuration can be extended, in principle, to include more stress filaments in a fiber, as long as the placement of the multifilaments is to get rid of the residual linear birefringence to the greatest extent. In practice, however, such sophisticated versions of fiber technologically become increasingly difficult to fabricate.

C.3   MAKING FIBER-OPTIC WAVE PLATES

In structure the fiber-optic wave plates (PPTs) are simply spun birefringent fibers with a varying spin rate [2]. Although the general class of nonuniformly spun fibers should be useful beyond the scope of the topic area of fiber-optic wave plates, for convenience we still use the short name "PPT" for the general class of this kind. Such variably spun fibers can be fabricated with the aid of the existing facility (a drawing tower complete with spinner). Nevertheless, being variably spun, such PPT fibers are designed differently from uniformly spun fibers. In particular, the structural parameters of a PPT fiber are not independent of the transmission direction, but become a function of that direction.

One very distinctive feature in the making of nonuniform spun fiber as compared with uniform spun fiber concerns the variable rotational speed of the spinner (motor). In Section 7.3, for analytic purposes we chose the powered raised-cosine function to simulate this speed:

$$\tau = \tau_0 \left[\frac{1}{2} \pm \frac{1}{2} \cos\left(\frac{\pi}{L} z \right) \right]^\gamma \tag{C1}$$

where τ is the spin rate as a function of z; τ_0 is the initial value of τ, L is the fiber length; and γ is a parameter that governs the pattern in which the spin rate varies. The upper sign ($+$) is for the case where τ descends from fast to zero, while the lower sign ($-$) is for the opposite case, that is, rises from zero to fast. A family of curves for the former case of this spin-rate function is shown in Figure C.3.

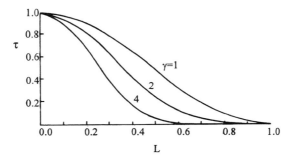

FIGURE C.3   Simulation of the speed function of the spinner (motor).

While specification of the parameters in Eq. (C1) is required in an analytic–numeric study of the variably spun fiber, in actual fabrication the tolerances of these parameters are fairly loose. The exact operational form of the spin function is not strictly required, as long as the function is continually and smoothly varying. The requirement for the length of fiber L is also fairly loose, as long as the spin rate varies slowly enough.

Technologically, the only stringent condition for making the variably spun PPT fiber is to achieve the required high value of τ_0 (maximum spin-rate). In machinery, the motor speed ϖ is customarily specified as revolutions per minute (rpm). In fiber fabrication, the unit of spin rate τ is revolutions per unit length of fiber, and is usually measured as "revolutions per meter," with the initial letters of the three words also reading rpm. But this is not to be confused with the common unit *rpm* for motor speed. These two units are simply related by

$$\tau \ (\text{revolutions per meter}) \ = \ \frac{\varpi(\text{rpm})}{v_l} \tag{C2}$$

where rpm means "revolutions per minute" for the rotational speed of motor (spinner), and v_l is the linear drawing speed in "meter per minute."

According to the afore-developed analytic theory, the transmission quality of a PPT fiber depends on the value of the qualification factor:

$$Q = \frac{\tau}{\delta\beta} = \frac{L_b}{L_s} \tag{C3}$$

that is, the ratio of unspun beat length to spin pitch. It is thus seen that we can achieve a high value of Q_0 (maximum value of Q) by three means:

1. Increase the rpm of the motor (spinner), which is in the numerator of Eq. (C2).
2. Lower the linear-drawing speed v_l in the denominator of Eq. (C2).
3. Choose the unspun beat length L_b in the numerator of Eq. (C3) not too short, but only short enough to be able to hold the desired polarization light along the output extension fiber.

In summary, the maximum value of Q_0 plays a dominant role in achieving the desired polarization performance. Provided the parameter L_b is given, then what can be done is simply to increase the angular speed of the motor ϖ while simultaneously lowering the linear-drawing speed v_l. The fiber making therefore requires an unconventionally fast spinning speed associated with an unconventionally slow linear drawing speed. But this stringent technological requirement is not insurmountable in practice, because the high spinning or rotational speed is not required to continue constantly over the entire course of fiber drawing, but only needs to run momentarily for some short period.

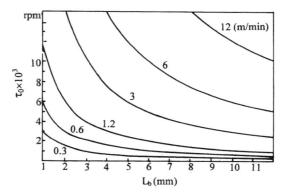

FIGURE C.4   Design diagrams for PPT. Numerals on curves: linear drawing speed.

Appropriate combinations of the values of the pertinent parameters for practical device making are shown in Figure C.4. The figure is drawn with Q_0 taken to be 10. For larger Q_0, the value of the corresponding spinning speed is scaled up proportionally.

C.4   THE MOVING MICROHEATER SETUP

The special fiber optics studied in Chapter 7 concerns a linear birefringent fiber that is spun at a varying rate: from fast to zero, or vice versa. Such variably spun birefringent fiber can be readily fabricated by the conventional approach, as described earlier. A standard fiber-drawing tower with a speed-controlled spinner (d.c. motor) will serve the purpose. The spin rate is varied by varying the d.c. voltage that is applied to the motor.

Alternatively, the variably spun birefringent fiber section can be made by a simple fabrication method: the "moving microheater technique." This was a continuation-in-part (CIP) invention of mine [3]. The distinctive feature of this method is that it uses a linear-bi fiber, not a linear-bi preform, as the starting substance.

In this novel approach, the facility that is required is not a drawing tower, but the specially devised apparatus shown in Figure C.5. Interestingly enough, the apparatus looks just like a miniature drawing tower, with everything scaled down to submeter dimensions. The symbol F in the diagram marks a length of the linear-bi fiber used as the starting substance for fabrication. The numerals 1 and 2 are two fiber-fasteners on the top and at the bottom, respectively. These two fasteners are similarly structured, except that the bottom one allows the lower part of the fiber to pass through its inside coaxial hole to form a fiber tail, T. A series of holes (numeral 3) drilled on the left side of the frame (four in number in the figure, for illustration only)

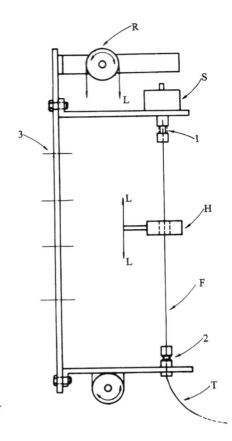

FIGURE C.5 The moving microheater apparatus.

allow different lengths of fiber to be fastened on top and at the bottom, simply by fixing the arm of the spinner at a higher or lower level. In the figure, S is a speed-controlled spinner (a small d.c. motor), and R represents a roller driven by a second motor (d.c. or a.c.), which transforms the rotational motion of this motor to the linear motion (up and down) of the microheater, H.

The moving microheater, H, heats locally the fiber that is being spun at a variable spin-rate. The entire operation takes a very short time (no more than 10 s, say). The range of fiber along which the microheater has traveled (say, 20 cm or less) will become the variably spun fiber element PPT.

We made such an apparatus in our laboratory according to the design principle given earlier [3]. We succeeded in making some short PPT fiber sections whose polarization transform behaviors were found to be exactly the same as those of PPT fiber specimens fabricated by a spinning-drawing tower.

The special advantage of this fabrication method is that the starting substance is not a preform rod, but simply a section of birefringent fiber. The drawing tower is not needed. A small, simple setup like that shown in Figure C.5 serves the purpose. Two of the shortcomings of this method at its present stage of development are (1) low product yield, because the spinning fiber exposed to local heating is apt to break during the heating-spinning process, and (2) the necessity of recoating the locally heated part of the fiber device after the process.

REFERENCES

[1] H. C. Huang, U.S. Patent No. 5,452,394 (1995).
[2] H. C. Huang, U.S. Patent No. 4,943,132 (1990).
[3] H. C. Huang, U.S. Patent No. 5,096,312 (1992).

List of Symbols

PRELIMINARY NOTES

The symbols included in this list have been arranged essentially in order of their appearance in the text. But this way of ordering is not always followed. For symbols looking alike in form, but actually denoting different meanings, we prefer to list them in groups, so that the distinctiveness of their meanings can be more impressive.

Wherever feasible, the same symbol is used to denote different meanings in different places. A prominent example is the bold capital letter **W**, which denotes the *normal modes* for constant couplings, as well as the *supermodes* for variable couplings. The is because the supermodes approach asymptotically to the normal modes when the rate of variation becomes vanishingly small. Thus, in this case, use of different symbols seems more likely to cause confusion. See Sections 2.4 and 7.3.

ABBREVIATIONS

bi:	birefringent; birefringence
hi-bi:	highly birefringent; high birefringence
lo-bi:	low birefringent; low birefringence
AT:	ampere-turns of an electric current solenoid
PPT:	passive polarization transform(er)
SOP:	state of polarization
SPSM:	single polarization single mode

SYMBOLS AND NOTATION

n:	refractive index
n_e:	(modal) effective index

296

N:	numerical aperture of fiber
ε_r:	relative dielectric permittivity
ε_0:	free-space permittivity
μ_r:	relative permeability
μ_0:	free-space permeability
ω:	angular frequency of wave
β:	propagation constant; propagation coefficient (also for average of β_1 and β_2)
H:	magnetic-field vector
E:	electric-field vector
x, y:	transverse Cartesian coordinates
r, φ:	transverse cylindrical polar coordinates
z:	longitudinal coordinate
Z_0:	plane-wave impedance in free space
Δ:	relative index difference
λ:	plane wavelength
λ_g:	modal wavelength
λ_1, λ_2:	eigenvalues
c:	coupling coefficient for linear base modes (also velocity of light in free space)
jk:	coupling coefficient for circular base modes ($k = 2\pi/\lambda$, also the wave number)
k_{11}, k_{22}:	propagation coefficients of modes 1 and 2
k_{12}, k_{21}:	coupling coefficients between modes 1 and 2, or 2 and 1
*	(superscript on right): complex conjugate
K:	modified Hankel function
J:	Bessel function
v:	normalized frequency
a:	radius of fiber core
D:	diameter of fiber core
α:	attenuation constant; attenuation coefficient
$\delta\beta$:	difference of propagation constants; phase-velocity difference
$Q\ (= \tau/\delta\beta)$:	coupling capacity; quality factor; normalized spin rate
L:	length of fiber
$L_b\ (= 2\pi/\delta\beta)$:	beat length
L_b^l:	beat length in local coordinates
$L_S\ (= 2\pi/\tau)$:	pitch of spin or twist

$\tau \ (= \varpi/\nu)$:	spin rate in revolutions per meter
ϖ:	rotational speed of motor (spinner) in revolutions per minute
ν:	fiber-drawing linear speed in meters/minute
B:	birefringence
B_c:	geometrical birefringence
B_S:	stress birefringence
A_x, A_y:	linear modes in local coordinates
$\hat{A}_x, \hat{A}_y$:	linear modes in fixed coordinates
$\wedge$ (circumflex):	indicating reference to fixed coordinates
$\hat{\mathbf{A}}$:	matrix of linear modes in fixed coordinates

SYMBOLS FOR UNIFORMLY COUPLED MODES

$\mathbf{A}$:	matrix of local modes
$\mathbf{K}$:	coupling matrix of local modes
$\mathbf{O}$:	diagonalizing matrix for $\mathbf{K}$
ϕ	$= 0.5 \arctan(2\tau/\delta\beta)$: characteristic figure of fiber (ϕ, also diameter of circling of a fiber)
g	$= [\tau^2 + (\delta\beta/2)^2]^{1/2}$: global propagation coefficient
$\mathbf{W}$:	matrix of normal modes
$\Lambda \ (= \mathbf{O}^{-1}\mathbf{K}\mathbf{O})$:	diagonalized matrix for coupled modes with constant coupling
$\tilde{\Lambda}$:	integration of Λ with respect to z
$\mathbf{T}_l \ (= \mathbf{O}\tilde{\Lambda}\mathbf{O}^{-1})$:	transfer matrix in local coordinates
$\mathbf{R}(\tau z)$:	rotation matrix for coordinates spinning at rate τ
$\mathbf{R}(-\tau z)$:	rotation matrix for coordinates spinning at rate $-\tau$
ϑ:	angular rotation of the coordinate axes
$\delta\vartheta$:	incremental angular rotation of the coordinate axes
$\hat{\mathbf{T}}$:	transfer matrix in fixed coordinates (identical to Jones matrix)
$\mathbf{J}$:	Jones matrix
$R(L)$:	retardation of fiber of length L
$\Omega(L)$:	optical rotation of fiber of length L
$\Phi(L)$:	fast-axis angle of retarder–rotator formulation for fiber of length L
θ, δ:	parameters in (θ, δ)-representation of elliptical light

ε, ψ: parameters in (ε, ψ)-representation of elliptical light

ε: ellipticity of elliptical light

ψ: orientation of elliptical light

η: extinction ratio

$\langle \eta \rangle$:   ensemble average of η

P_x, P_y: power for mode x and mode y

$\langle P_x \rangle, \langle P_y \rangle$:   ensemble averages of P_x, P_y

h: polarization-holding parameter; h-parameter

τ_ε: coupling coefficient due to small twists

Q_ε: coupling capacity due to small twists

$n_e (\lambda/\lambda_g)$:   modal-effective index

λ_g: modal wavelength of signal

$\sigma_x - \sigma_y$:   core-clad differential stress

σ: offset angle at a joint of two fiber sections

α: twist-induced optical rotation rate

$\Omega (= \alpha z)$:   twist-induced optical rotation angle

$\varsigma (= \alpha/\tau)$:   elastooptic coefficient for externally applied stress (in twisted fiber)

κ: elastooptic coefficient for built-in stress (in Screw fiber)

E_R, E_L: *complex amplitudes* of right and left circular modes

$\mathbf{E}_R, \mathbf{E}_L$:   column matrices of right and left circular modes

$\mathbf{E}$:   light of arbitrary SOP composed of right and left circular base modes

$\mathbf{O}_c$:   diagonalizing matrix for coupled-mode equations with circular base modes

$\Lambda_c (\mathbf{O}_c^{-1}\mathbf{K}\mathbf{O}_c)$:   diagonalized matrix for coupled-mode equations with circular base modes

$\tilde{\Lambda}_c$:   integration of Λ_c with respect to z

$\mathbf{T}_{lc}(= \mathbf{O}_c\tilde{\Lambda}_c\mathbf{O}_c^{-1})$:   transfer matrix with circular base modes

α_F: optical rotation rate due to Faraday effect

H_z: applied longitudinal magnetic field

V: Verdet constant

SYMBOLS FOR VARIABLY COUPLED MODES

$\mathbf{A}$:   matrix of local modes

$\mathbf{K}$:   coupling matrix of local modes

$\mathbf{Q}$:   quasi-diagonalizing matrix for $\mathbf{K}$

$\mathbf{N}$:   quasi-diagonalized matrix

W: matrix of supermodes
Q_S: coupling capacity of supermodes
ρ $\equiv \rho(L_b, L_s, L) = \int_0^L \pi [1 + 4(L_s/L_b)^2]^{1/2} \, dz$:
                                global structural parameter of PPT
ξ: phase between A_x and A_y for equal-power-division of PPT
a: matrix of ideal modes
k: coupling matrix of ideal modes
$\mathbf{O}_k$:   quasi-diagonalizing matrix for **k**

SYMBOLS FOR APPENDIX A

$\mathbf{E}_t, \mathbf{T}_t$:   transverse electric- and magnetic-field vectors
E_z, H_z: longitudinal electric- and magnetic-field vectors
z: unit vector in z-direction
$\mathbf{e}_{\nu z}, \mathbf{h}_{\partial z}$:   transverse vector fields of eigenmodes of subscript ν in reference fiber
$e_{\nu z}, h_{\nu z}$:   longitudinal vector fields of eigenmodes of subscript ν in reference fiber
$\delta_{\nu\mu}$:   Kronecker delta

MISCELLANEOUS

$\equiv$:   defined as
$\approx$:   approximately equal to
$\rightarrow$:   asymptotically tends to
                                ($\rightarrow$, also approaches to 0, tends to ∞, etc.)
                                ($\rightarrow$, also equivalent to, e.g., $\partial/\partial t \rightarrow j\omega$, $\partial/\partial z \rightarrow j\beta$, and also equivalence relations in Appendix B)
$\gg$:   much larger than
$\ll$:   much smaller than
$|A|$:   magnitude of complex quantity A
$\angle$:   argument (corresponding to exponential phase-factor) of complex quantity

Index

WILEY SERIES IN MICROWAVE AND OPTICAL ENGINEERING

KAI CHANG, Editor
Texas A&M University

FIBER-OPTIC COMMUNICATION SYSTEMS, Second Edition • *Govind P. Agrawal*

COHERENT OPTICAL COMMUNICATIONS SYSTEMS • *Silvello Betti, Giancarlo De Marchis and Eugenio Iannone*

HIGH-FREQUENCY ELECTROMAGNETIC TECHNIQUES: RECENT ADVANCES AND APPLICATIONS • *Asoke K. Bhattacharyya*

COMPUTATIONAL METHODS FOR ELECTROMAGNETICS AND MICROWAVES • *Richard C. Booton, Jr.*

MICROWAVE RING CIRCUITS AND ANTENNAS • *Kai Chang*

MICROWAVE SOLID-STATE CIRCUITS AND APPLICATIONS • *Kai Chang*

DIODE LASERS AND PHOTONIC INTEGRATED CIRCUITS • *Larry Coldren and Scott Corzine*

MULTICONDUCTOR TRANSMISSION-LINE STRUCTURES: MODAL ANALYSIS TECHNIQUES • *J. A. Brandão Faria*

PHASED ARRAY-BASED SYSTEMS AND APPLICATIONS • *Nick Fourikis*

FUNDAMENTALS OF MICROWAVE TRANSMISSION LINES • *Jon C. Freeman*

MICROSTRIP CIRCUITS • *Fred Gardiol*

HIGH-SPEED VLSI INTERCONNECTIONS: MODELING, ANALYSIS, AND SIMULATION • *A. K. Goel*

PHASED ARRAY ANTENNAS • *R. C. Hansen*

HIGH-FREQUENCY ANALOG INTEGRATED CIRCUIT DESIGN • *Ravender Goyal (ed.)*

MICROWAVE APPROACH TO HIGHLY IRREGULAR FIBER OPTICS • *Huang Hung-Chia*

NONLINEAR OPTICAL COMMUNICATION NETWORKS • *Eugenio Iannone, Francesco Matera, Antonio Mecozzi, and Marina Settembre*

FINITE ELEMENT SOFTWARE FOR MICROWAVE ENGINEERING • *Tatsuo Itoh, Giuseppe Pelosi and Peter P. Silvester (eds.)*

SUPERCONDUCTOR TECHNOLOGY: APPLICATIONS TO MICROWAVE, ELECTRO-OPTICS, ELECTRICAL MACHINES, AND PROPULSION SYSTEMS • *A. R. Jha*

OPTICAL COMPUTING: AN INTRODUCTION • *M. A. Karim and A. S. S. Awwal*

INTRODUCTION TO ELECTROMAGNETIC AND MICROWAVE ENGINEERING • *Paul R. Karmel, Gabriel D. Colef, and Raymond L. Camisa*

MILLIMETER WAVE OPTICAL DIELECTRIC INTEGRATED GUIDES AND CIRCUITS • *Shiban K. Koul*

MICROWAVE DEVICES, CIRCUITS AND THEIR INTERACTION • *Charles A. Lee and G. Conrad Dalman*

ADVANCES IN MICROSTRIP AND PRINTED ANTENNAS • *Kai-Fong Lee and Wei Chen (eds.)*

OPTOELECTRONIC PACKAGING • *A. R. Mickelson, N. R. Basavanhally, and Y. C. Lee (eds.)*

ANTENNAS FOR RADAR AND COMMUNICATIONS: A POLARIMETRIC APPROACH • *Harold Mott*

INTEGRATED ACTIVE ANTENNAS AND SPATIAL POWER COMBINING • *Julio A. Navarro and Kai Chang*

FREQUENCY CONTROL OF SEMICONDUCTOR LASERS • *Motoichi Ohtsu (ed.)*

SOLAR CELLS AND THEIR APPLICATIONS • *Larry D. Partain (ed.)*